A PRIMER OF ALGEBRAIC GEOMETRY

MONOGRAPHS AND TEXTBOOKS IN PURE AND APPLIED MATHEMATICS

1. *K. Yano*, Integral Formulas in Riemannian Geometry (1970)
2. *S. Kobayashi*, Hyperbolic Manifolds and Holomorphic Mappings (1970)
3. *V. S. Vladimirov*, Equations of Mathematical Physics (A. Jeffrey, ed.; A. Littlewood, trans.) (1970)
4. *B. N. Pshenichnyi*, Necessary Conditions for an Extremum (L. Neustadt, translation ed.; K. Makowski, trans.) (1971)
5. *L. Narici et al.*, Functional Analysis and Valuation Theory (1971)
6. *S. S. Passman*, Infinite Group Rings (1971)
7. *L. Dornhoff*, Group Representation Theory. Part A: Ordinary Representation Theory. Part B: Modular Representation Theory (1971, 1972)
8. *W. Boothby and G. L. Weiss, eds.*, Symmetric Spaces (1972)
9. *Y. Matsushima*, Differentiable Manifolds (E. T. Kobayashi, trans.) (1972)
10. *L. E. Ward, Jr.*, Topology (1972)
11. *A. Babakhanian*, Cohomological Methods in Group Theory (1972)
12. *R. Gilmer*, Multiplicative Ideal Theory (1972)
13. *J. Yeh*, Stochastic Processes and the Wiener Integral (1973)
14. *J. Barros-Neto*, Introduction to the Theory of Distributions (1973)
15. *R. Larsen*, Functional Analysis (1973)
16. *K. Yano and S. Ishihara*, Tangent and Cotangent Bundles (1973)
17. *C. Procesi*, Rings with Polynomial Identities (1973)
18. *R. Hermann*, Geometry, Physics, and Systems (1973)
19. *N. R. Wallach*, Harmonic Analysis on Homogeneous Spaces (1973)
20. *J. Dieudonné*, Introduction to the Theory of Formal Groups (1973)
21. *I. Vaisman*, Cohomology and Differential Forms (1973)
22. *B.-Y. Chen*, Geometry of Submanifolds (1973)
23. *M. Marcus*, Finite Dimensional Multilinear Algebra (in two parts) (1973, 1975)
24. *R. Larsen*, Banach Algebras (1973)
25. *R. O. Kujala and A. L. Vitter, eds.*, Value Distribution Theory: Part A; Part B: Deficit and Bezout Estimates by Wilhelm Stoll (1973)
26. *K. B. Stolarsky*, Algebraic Numbers and Diophantine Approximation (1974)
27. *A. R. Magid*, The Separable Galois Theory of Commutative Rings (1974)
28. *B. R. McDonald*, Finite Rings with Identity (1974)
29. *J. Satake*, Linear Algebra (S. Koh et al., trans.) (1975)
30. *J. S. Golan*, Localization of Noncommutative Rings (1975)
31. *G. Klambauer*, Mathematical Analysis (1975)
32. *M. K. Agoston*, Algebraic Topology (1976)
33. *K. R. Goodearl*, Ring Theory (1976)
34. *L. E. Mansfield*, Linear Algebra with Geometric Applications (1976)
35. *N. J. Pullman*, Matrix Theory and Its Applications (1976)
36. *B. R. McDonald*, Geometric Algebra Over Local Rings (1976)
37. *C. W. Groetsch*, Generalized Inverses of Linear Operators (1977)
38. *J. E. Kuczkowski and J. L. Gersting*, Abstract Algebra (1977)
39. *C. O. Christenson and W. L. Voxman*, Aspects of Topology (1977)
40. *M. Nagata*, Field Theory (1977)
41. *R. L. Long*, Algebraic Number Theory (1977)
42. *W. F. Pfeffer*, Integrals and Measures (1977)
43. *R. L. Wheeden and A. Zygmund*, Measure and Integral (1977)
44. *J. H. Curtiss*, Introduction to Functions of a Complex Variable (1978)
45. *K. Hrbacek and T. Jech*, Introduction to Set Theory (1978)
46. *W. S. Massey*, Homology and Cohomology Theory (1978)
47. *M. Marcus*, Introduction to Modern Algebra (1978)
48. *E. C. Young*, Vector and Tensor Analysis (1978)
49. *S. B. Nadler, Jr.*, Hyperspaces of Sets (1978)
50. *S. K. Segal*, Topics in Group Kings (1978)
51. *A. C. M. van Rooij*, Non-Archimedean Functional Analysis (1978)
52. *L. Corwin and R. Szczarba*, Calculus in Vector Spaces (1979)
53. *C. Sadosky*, Interpolation of Operators and Singular Integrals (1979)
54. *J. Cronin*, Differential Equations (1980)
55. *C. W. Groetsch*, Elements of Applicable Functional Analysis (1980)

56. *I. Vaisman*, Foundations of Three-Dimensional Euclidean Geometry (1980)
57. *H. I. Freedan*, Deterministic Mathematical Models in Population Ecology (1980)
58. *S. B. Chae*, Lebesgue Integration (1980)
59. *C. S. Rees et al.*, Theory and Applications of Fourier Analysis (1981)
60. *L. Nachbin*, Introduction to Functional Analysis (R. M. Aron, trans.) (1981)
61. *G. Orzech and M. Orzech*, Plane Algebraic Curves (1981)
62. *R. Johnsonbaugh and W. E. Pfaffenberger*, Foundations of Mathematical Analysis (1981)
63. *W. L. Voxman and R. H. Goetschel*, Advanced Calculus (1981)
64. *L. J. Corwin and R. H. Szczarba*, Multivariable Calculus (1982)
65. *V. I. Istrățescu*, Introduction to Linear Operator Theory (1981)
66. *R. D. Järvinen*, Finite and Infinite Dimensional Linear Spaces (1981)
67. *J. K. Beem and P. E. Ehrlich*, Global Lorentzian Geometry (1981)
68. *D. L. Armacost*, The Structure of Locally Compact Abelian Groups (1981)
69. *J. W. Brewer and M. K. Smith, eds.*, Emmy Noether: A Tribute (1981)
70. *K. H. Kim*, Boolean Matrix Theory and Applications (1982)
71. *T. W. Wieting*, The Mathematical Theory of Chromatic Plane Ornaments (1982)
72. *D. B. Gauld*, Differential Topology (1982)
73. *R. L. Faber*, Foundations of Euclidean and Non-Euclidean Geometry (1983)
74. *M. Carmeli*, Statistical Theory and Random Matrices (1983)
75. *J. H. Carruth et al.*, The Theory of Topological Semigroups (1983)
76. *R. L. Faber*, Differential Geometry and Relativity Theory (1983)
77. *S. Barnett*, Polynomials and Linear Control Systems (1983)
78. *G. Karpilovsky*, Commutative Group Algebras (1983)
79. *F. Van Oystaeyen and A. Verschoren*, Relative Invariants of Rings (1983)
80. *I. Vaisman*, A First Course in Differential Geometry (1984)
81. *G. W. Swan*, Applications of Optimal Control Theory in Biomedicine (1984)
82. *T. Petrie and J. D. Randall*, Transformation Groups on Manifolds (1984)
83. *K. Goebel and S. Reich*, Uniform Convexity, Hyperbolic Geometry, and Nonexpansive Mappings (1984)
84. *T. Albu and C. Năstăsescu*, Relative Finiteness in Module Theory (1984)
85. *K. Hrbacek and T. Jech*, Introduction to Set Theory: Second Edition (1984)
86. *F. Van Oystaeyen and A. Verschoren*, Relative Invariants of Rings (1984)
87. *B. R. McDonald*, Linear Algebra Over Commutative Rings (1984)
88. *M. Namba*, Geometry of Projective Algebraic Curves (1984)
89. *G. F. Webb*, Theory of Nonlinear Age-Dependent Population Dynamics (1985)
90. *M. R. Bremner et al.*, Tables of Dominant Weight Multiplicities for Representations of Simple Lie Algebras (1985)
91. *A. E. Fekete*, Real Linear Algebra (1985)
92. *S. B. Chae*, Holomorphy and Calculus in Normed Spaces (1985)
93. *A. J. Jerri*, Introduction to Integral Equations with Applications (1985)
94. *G. Karpilovsky*, Projective Representations of Finite Groups (1985)
95. *L. Narici and E. Beckenstein*, Topological Vector Spaces (1985)
96. *J. Weeks*, The Shape of Space (1985)
97. *P. R. Gribik and K. O. Kortanek*, Extremal Methods of Operations Research (1985)
98. *J.-A. Chao and W. A. Woyczynski, eds.*, Probability Theory and Harmonic Analysis (1986)
99. *G. D. Crown et al.*, Abstract Algebra (1986)
100. *J. H. Carruth et al.*, The Theory of Topological Semigroups, Volume 2 (1986)
101. *R. S. Doran and V. A. Belfi*, Characterizations of C*-Algebras (1986)
102. *M. W. Jeter*, Mathematical Programming (1986)
103. *M. Altman*, A Unified Theory of Nonlinear Operator and Evolution Equations with Applications (1986)
104. *A. Verschoren*, Relative Invariants of Sheaves (1987)
105. *R. A. Usmani*, Applied Linear Algebra (1987)
106. *P. Blass and J. Lang*, Zariski Surfaces and Differential Equations in Characteristic $p > 0$ (1987)
107. *J. A. Reneke et al.*, Structured Hereditary Systems (1987)
108. *H. Busemann and B. B. Phadke*, Spaces with Distinguished Geodesics (1987)
109. *R. Harte*, Invertibility and Singularity for Bounded Linear Operators (1988)
110. *G. S. Ladde et al.*, Oscillation Theory of Differential Equations with Deviating Arguments (1987)
111. *L. Dudkin et al.*, Iterative Aggregation Theory (1987)
112. *T. Okubo*, Differential Geometry (1987)

113. *D. L. Stancl and M. L. Stancl*, Real Analysis with Point-Set Topology (1987)
114. *T. C. Gard*, Introduction to Stochastic Differential Equations (1988)
115. *S. S. Abhyankar*, Enumerative Combinatorics of Young Tableaux (1988)
116. *H. Strade and R. Farnsteiner*, Modular Lie Algebras and Their Representations (1988)
117. *J. A. Huckaba*, Commutative Rings with Zero Divisors (1988)
118. *W. D. Wallis*, Combinatorial Designs (1988)
119. *W. Więsław*, Topological Fields (1988)
120. *G. Karpilovsky*, Field Theory (1988)
121. *S. Caenepeel and F. Van Oystaeyen*, Brauer Groups and the Cohomology of Graded Rings (1989)
122. *W. Kozlowski*, Modular Function Spaces (1988)
123. *E. Lowen-Colebunders*, Function Classes of Cauchy Continuous Maps (1989)
124. *M. Pavel*, Fundamentals of Pattern Recognition (1989)
125. *V. Lakshmikantham et al.*, Stability Analysis of Nonlinear Systems (1989)
126. *R. Sivaramakrishnan*, The Classical Theory of Arithmetic Functions (1989)
127. *N. A. Watson*, Parabolic Equations on an Infinite Strip (1989)
128. *K. J. Hastings*, Introduction to the Mathematics of Operations Research (1989)
129. *B. Fine*, Algebraic Theory of the Bianchi Groups (1989)
130. *D. N. Dikranjan et al.*, Topological Groups (1989)
131. *J. C. Morgan II*, Point Set Theory (1990)
132. *P. Biler and A. Witkowski*, Problems in Mathematical Analysis (1990)
133. *H. J. Sussmann*, Nonlinear Controllability and Optimal Control (1990)
134. *J.-P. Florens et al.*, Elements of Bayesian Statistics (1990)
135. *N. Shell*, Topological Fields and Near Valuations (1990)
136. *B. F. Doolin and C. F. Martin*, Introduction to Differential Geometry for Engineers (1990)
137. *S. S. Holland, Jr.*, Applied Analysis by the Hilbert Space Method (1990)
138. *J. Okninski*, Semigroup Algebras (1990)
139. *K. Zhu*, Operator Theory in Function Spaces (1990)
140. *G. B. Price*, An Introduction to Multicomplex Spaces and Functions (1991)
141. *R. B. Darst*, Introduction to Linear Programming (1991)
142. *P. L. Sachdev*, Nonlinear Ordinary Differential Equations and Their Applications (1991)
143. *T. Husain*, Orthogonal Schauder Bases (1991)
144. *J. Foran*, Fundamentals of Real Analysis (1991)
145. *W. C. Brown*, Matrices and Vector Spaces (1991)
146. *M. M. Rao and Z. D. Ren*, Theory of Orlicz Spaces (1991)
147. *J. S. Golan and T. Head*, Modules and the Structures of Rings (1991)
148. *C. Small*, Arithmetic of Finite Fields (1991)
149. *K. Yang*, Complex Algebraic Geometry (1991)
150. *D. G. Hoffman et al.*, Coding Theory (1991)
151. *M. O. González*, Classical Complex Analysis (1992)
152. *M. O. González*, Complex Analysis (1992)
153. *L. W. Baggett*, Functional Analysis (1992)
154. *M. Sniedovich*, Dynamic Programming (1992)
155. *R. P. Agarwal*, Difference Equations and Inequalities (1992)
156. *C. Brezinski*, Biorthogonality and Its Applications to Numerical Analysis (1992)
157. *C. Swartz*, An Introduction to Functional Analysis (1992)
158. *S. B. Nadler, Jr.*, Continuum Theory (1992)
159. *M. A. Al-Gwaiz*, Theory of Distributions (1992)
160. *E. Perry*, Geometry: Axiomatic Developments with Problem Solving (1992)
161. *E. Castillo and M. R. Ruiz-Cobo*, Functional Equations and Modelling in Science and Engineering (1992)
162. *A. J. Jerri*, Integral and Discrete Transforms with Applications and Error Analysis (1992)
163. *A. Charlier et al.*, Tensors and the Clifford Algebra (1992)
164. *P. Biler and T. Nadzieja*, Problems and Examples in Differential Equations (1992)
165. *E. Hansen*, Global Optimization Using Interval Analysis (1992)
166. *S. Guerre-Delabrière*, Classical Sequences in Banach Spaces (1992)
167. *Y. C. Wong*, Introductory Theory of Topological Vector Spaces (1992)
168. *S. H. Kulkarni and B. V. Limaye*, Real Function Algebras (1992)
169. *W. C. Brown*, Matrices Over Commutative Rings (1993)
170. *J. Loustau and M. Dillon*, Linear Geometry with Computer Graphics (1993)
171. *W. V. Petryshyn*, Approximation-Solvability of Nonlinear Functional and Differential Equations (1993)

172. *E. C. Young*, Vector and Tensor Analysis: Second Edition (1993)
173. *T. A. Bick*, Elementary Boundary Value Problems (1993)
174. *M. Pavel*, Fundamentals of Pattern Recognition: Second Edition (1993)
175. *S. A. Albeverio et al.*, Noncommutative Distributions (1993)
176. *W. Fulks*, Complex Variables (1993)
177. *M. M. Rao*, Conditional Measures and Applications (1993)
178. *A. Janicki and A. Weron*, Simulation and Chaotic Behavior of α-Stable Stochastic Processes (1994)
179. *P. Neittaanmäki and D. Tiba*, Optimal Control of Nonlinear Parabolic Systems (1994)
180. *J. Cronin*, Differential Equations: Introduction and Qualitative Theory, Second Edition (1994)
181. *S. Heikkilä and V. Lakshmikantham*, Monotone Iterative Techniques for Discontinuous Nonlinear Differential Equations (1994)
182. *X. Mao*, Exponential Stability of Stochastic Differential Equations (1994)
183. *B. S. Thomson*, Symmetric Properties of Real Functions (1994)
184. *J. E. Rubio*, Optimization and Nonstandard Analysis (1994)
185. *J. L. Bueso et al.*, Compatibility, Stability, and Sheaves (1995)
186. *A. N. Michel and K. Wang*, Qualitative Theory of Dynamical Systems (1995)
187. *M. R. Darnel*, Theory of Lattice-Ordered Groups (1995)
188. *Z. Naniewicz and P. D. Panagiotopoulos*, Mathematical Theory of Hemivariational Inequalities and Applications (1995)
189. *L. J. Corwin and R. H. Szczarba*, Calculus in Vector Spaces: Second Edition (1995)
190. *L. H. Erbe et al.*, Oscillation Theory for Functional Differential Equations (1995)
191. *S. Agaian et al.*, Binary Polynomial Transforms and Nonlinear Digital Filters (1995)
192. *M. I. Gil'*, Norm Estimations for Operation-Valued Functions and Applications (1995)
193. *P. A. Grillet*, Semigroups: An Introduction to the Structure Theory (1995)
194. *S. Kichenassamy*, Nonlinear Wave Equations (1996)
195. *V. F. Krotov*, Global Methods in Optimal Control Theory (1996)
196. *K. I. Beidar et al.*, Rings with Generalized Identities (1996)
197. *V. I. Arnautov et al.*, Introduction to the Theory of Topological Rings and Modules (1996)
198. *G. Sierksma*, Linear and Integer Programming (1996)
199. *R. Lasser*, Introduction to Fourier Series (1996)
200. *V. Sima*, Algorithms for Linear-Quadratic Optimization (1996)
201. *D. Redmond*, Number Theory (1996)
202. *J. K. Beem et al.*, Global Lorentzian Geometry: Second Edition (1996)
203. *M. Fontana et al.*, Prüfer Domains (1997)
204. *H. Tanabe*, Functional Analytic Methods for Partial Differential Equations (1997)
205. *C. Q. Zhang*, Integer Flows and Cycle Covers of Graphs (1997)
206. *E. Spiegel and C. J. O'Donnell*, Incidence Algebras (1997)
207. *B. Jakubczyk and W. Respondek*, Geometry of Feedback and Optimal Control (1998)
208. *T. W. Haynes et al.*, Fundamentals of Domination in Graphs (1998)
209. *T. W. Haynes et al.*, Domination in Graphs: Advanced Topics (1998)
210. *L. A. D'Alotto et al.*, A Unified Signal Algebra Approach to Two-Dimensional Parallel Digital Signal Processing (1998)
211. *F. Halter-Koch*, Ideal Systems (1998)
212. *N. K. Govil et al.*, Approximation Theory (1998)
213. *R. Cross*, Multivalued Linear Operators (1998)
214. *A. A. Martynyuk*, Stability by Liapunov's Matrix Function Method with Applications (1998)
215. *A. Favini and A. Yagi*, Degenerate Differential Equations in Banach Spaces (1999)
216. *A. Illanes and S. Nadler, Jr.*, Hyperspaces: Fundamentals and Recent Advances (1999)
217. *G. Kato and D. Struppa*, Fundamentals of Algebraic Microlocal Analysis (1999)
218. *G. X.-Z. Yuan*, KKM Theory and Applications in Nonlinear Analysis (1999)
219. *D. Motreanu and N. H. Pavel*, Tangency, Flow Invariance for Differential Equations, and Optimization Problems (1999)
220. *K. Hrbacek and T. Jech*, Introduction to Set Theory, Third Edition (1999)
221. *G. E. Kolosov*, Optimal Design of Control Systems (1999)
222. *N. L. Johnson*, Subplane Covered Nets (2000)
223. *B. Fine and G. Rosenberger*, Algebraic Generalizations of Discrete Groups (1999)
224. *M. Väth*, Volterra and Integral Equations of Vector Functions (2000)
225. *S. S. Miller and P. T. Mocanu*, Differential Subordinations (2000)

226. *R. Li et al.*, Generalized Difference Methods for Differential Equations: Numerical Analysis of Finite Volume Methods (2000)
227. *H. Li and F. Van Oystaeyen*, A Primer of Algebraic Geometry (2000)
228. *R. P. Agarwal*, Difference Equations and Inequalities, Second Edition (2000)

Additional Volumes in Preparation

A PRIMER OF ALGEBRAIC GEOMETRY

Constructive Computational Methods

Huishi Li
Bilkent University
Ankara, Turkey

Freddy Van Oystaeyen
University of Antwerp (UIA)
Wilrijk, Belgium

CRC Press
Taylor & Francis Group
Boca Raton London New York

CRC Press is an imprint of the
Taylor & Francis Group, an **informa** business

CRC Press
Taylor & Francis Group
6000 Broken Sound Parkway NW, Suite 300
Boca Raton, FL 33487-2742

First issued in paperback 2019

ISBN-13: 978-0-8247-0374-5 (hbk)
ISBN-13: 978-0-367-39896-5 (pbk)

Visit the Taylor & Francis Web site at
http://www.taylorandfrancis.com

and the CRC Press Web site at
http://www.crcpress.com

Preface

A popular saying is: *to see is to believe*! Perhaps in this saying seeing should be understood as understanding more than merely observing, and applied to objects encountered in daily life this means first measuring. At this point, the description of seemingly complex geometric shapes is linked to numbers, counting and, indeed, algebra. Thinking of this further, one may appreciate that algebraic geometry is born of the confrontation between very concrete problems—one may call them daily life problems—and far-reaching abstraction. It may be a cause for debate whether classical Greek geometry is already a version of algebraic geometry; nevertheless, there was from the beginning a close relation with the definition of numbers and with the problem of finding solutions of equations, be they linear, quadratic, or Diophantine. This relation had at times an almost mystical quality, for example, the symbol of the Pythagorean pentagon and its relation to irrational numbers.

It seems one has to understand algebra before one can get a grip on geometry. Indeed, in the hands of Descartes, algebra definitely entered geometry at a very basic level. The coordinatization of purely geometric problems, sometimes by some kind of numbers much harder to understand than rational numbers, in order to solve the problems by the kind of symbol manipulation that elementary algebra is, may nowadays look so natural and common that it is hard to convey how revolutionary the idea was. On the other hand, questions of a topological nature were dealt with in an intuitive way, e.g., is there always another point between two different points on a line? But, when formulated in coordinates, that topological intuition may be lost, or worse, it may prove to be false and misleading. Again, a very simple fact such as the existence of an ordering on certain fields of numbers and its relation to certain topological properties such as completeness, is at the basis of a new interaction between algebraic and analytical methods in geometry.

Few topics in mathematics have undergone as many groundshaking revolutions as algebraic geometry. Let us just mention here the transition from the intuitive approach of the Italian school to the abstraction of scheme theory, at the hands of A. Grothendieck and followers in the 20th century. At this point a student may have difficulty in discerning the difference between geometry and abstract category theory!

Yet, algebraic geometry did not develop apart from real applications; on the contrary, the computer age provided new types of problems. Some of these are related to effective calculations and algorithms or to a "discretization" of otherwise continuous phenomena, but there are also completely new applications of the abstract-looking theory. These range from the singularity theory in "Robot Motion Planning" to codes associated to specific algebraic varieties like, e.g., elliptic curves, and to the creation of an algorithmic algebraic geometry with applications in computational geometry, to name just a few.

Introducing modern applications of an old subject in mathematics inevitably raises a recurring dilemma: to sacrifice mathematical complexity and highlight the more down-to-earth practical aspects or just to use the applications as a motivation for the dull (to the non-specialist) theory? The first attitude shifts interest away from the beautiful (to the specialist) fundamental issues that were and will always be at the heart of new developments. The second attitude, usually showcased by the presence of many fancy pictures, is didactically unsatisfactory because, like false propaganda creating a lot of unfulfilled promises, it results in general disillusionment for those attracted to the subject solely by an interest in these applications. In our opinion, the second attitude could endanger the development of future mathematics if it were to become the dominating teaching strategy. Hence the aforementioned dilemma was also ours! So we carefully set our goals for writing this book, as follows. First, we want to convince the reader of the fact that algebraic geometry is, at a basic but (therefore) fundamental level, not only comprehensible but also "realizable" and "visualizable," so that many applications can really be understood from the theoretical foundations. Second, we hope to make it clear to the student—in particular, the student mainly interested in concrete applications such as motion planning and vision for robots or morphogenesis of plants—that without well-developed mathematical principles and theoretical depth such realizations would not be possible. You cannot have one without the other!

We try to realize the goals set out above by presenting an approach that is as constructive as possible, focusing on effective calculations where it is possible. However, we did not want to give priority to developing algorithms over proving theorems. Therefore we do not provide algorithms written in pseudo-code, but we do make the theory as "effective" as possible by a consistent use of Groebner basis techniques. We believe that, at places, this effective approach has led us to new insights of theoretical nature, e.g., some facts about dimension and some relations with the notion of transcendence degree, all from an effective calculation point of view.

To conclude, we return to folk-wisdom and quote another popular saying: *to know is to love*! With this in mind, we have added a special chapter on elliptic curves; here all techniques developed in the book find applications. We believe this topic, in combination with its links to number theory and complex analysis, will forever be one of the most beautiful subjects in mathematics. We hope it may stimulate the student to explore algebraic geometry further. Mathematics does not have to be made fun—it *is* fun!

Scope: One should view this book as a first course in algebraic geometry aimed at senior undergraduate students and first-year graduate students. We tried to keep in mind that students and researchers not specializing in pure mathematics (but also not too afraid of it) should find this book useful if they have an interest in computer algebra, robotics and computational geometry, certain areas of theoretical computer science, or mathematical methods of technology. We do assume that the reader has some knowledge of linear algebra, is familiar with algebraic structures like groups, rings, fields, ideals, homomorphisms of rings, principal ideal domains and unique factorization domains, and polynomial rings over a field, and has basic knowledge of elementary topology. Apart from these prerequisites we have strived for a completely self-contained text.

Context: Intuitively, we view a variety as a set of points the coordinates of which are solutions for a finite set of polynomials in some polynomial ring $k[x_1,\ldots, x_n]$ over a field k. Such a set will be called an *algebraic set* in k^n, being the affine space of dimension n over k, that is, the set of all possible coordinate n-tuples $(\lambda_1,\ldots, \lambda_n)$ with $\lambda_i \in k$, $i = 1,\ldots, n$. After even a superficial investigation of this idea of algebraic set one comes up with several very natural, but not necessarily trivial, questions.

1. If V is an algebraic set as above, given by $f_1,\ldots, f_s$ in $k[x_1,\ldots, x_n]$, can we decide whether $V = \emptyset$ or not? In fact, is there an algorithm to decide this?

2. If we already know that $V \neq \emptyset$, can we decide when V is finite? Again, can we do this in an algebraic (theoretical) as well as an algorithmic (effective) way?

3. Is there a method for describing or recognizing points of V? Can we draw the shape of V or approximate it?

4. With regard to "shape," can we introduce a suitable topology on V, and what class of function should one study on V in order to control its properties? Is this an algebraic question (or does it become analysis)?

5. What about natural operations with algebraic sets, e.g., union, product, intersections? Can one describe points common to several algebraic sets by algebraic methods? Why do we have to introduce projective space?

6. Is the transition from affine to projective theory controllable by

algebraic or algorithmic techniques?

7. How are solution spaces for different sets of defining polynomials different or related? Can we define obvious geometric-algebraic invariants depending on V and not on the set of polynomials describing it? Can we calculate these invariants from knowledge concerning the defining set of polynomials?

8. Is the intuitive notion of "dimension" or "size" of V well defined mathematically? Can we calculate it?

9. Do the geometric properties of V, or "around" a point of V, in an affine or projective space depend on the ambient space containing that point?

10. Do the affine and projective theories fit into a unified theory? If so, what are the important functions to study? What is then the underlying topology?

11. What kind of maps between two algebraic sets V and W preserve certain geometric properties? Can we describe these maps in an algebraic way? How closely are V and W related if particularly nice maps exist between them; e.g., are notions of "geometric isomorphisms" expressable in algebraic terms?

12. What classes of algebraic sets can we handle best? Can they be used practically? If we have some preferred classes can we then select these out by algebraic or algorithmic methods?

By developing the basic theory of algebraic geometry, we answer most of the foregoing questions in a theoretical way; and by dealing with the basic theory using effective and constructive methods we make the algorithmic solution clear where possible.

A more theoretical mathematical philosophy is present in the final chapters on divisors and elliptic curves. It should be noted that divisors are in fact at the basis of some calculation techniques in the arithmetical theory of function fields, so they fit unexpectedly well in an "effective" oriented theory. Moreover, the linear space $\mathcal{L}(D)$ associated to some divisor D, in particular on an elliptic curve, is exactly used for some Reed-Solomon code with excellent error-correcting capacity. This goes to prove that many a theorem is nothing but an application in disguise!

Since we believe in *learning by doing*, we greatly encourage the reader to do his own thing with the exercises in each section, in particular, we suggest using any available computer algebra system, e.g., MATHEMATICA, MAPLE, MACAULAY, or GNUPLOT, in order to perform calculations and plotting of curves and surfaces. Most of all, choose, construct and work on your own examples.

Huishi Li
Freddy Van Oystaeyen

Contents

Notation

$\mathbb{Z}$	the set of integers
$\mathbb{N}$	the set of integers ≥ 0
$\mathbb{Q}$	the field of rational numbers
$\mathbb{R}$	the field of real numbers
$\mathbb{C}$	the field of complex numbers
id_A	the identity map on a set A
$\lvert A \rvert$	the cardinality of a set A
$m \gg 0$	for m sufficiently large
$\mathrm{char} k$	the characteristic of a field k

Let f: $A \to B$ be a map from a set A to another set B. If $V \subset A$, we write $f|_V$ for the restriction map $V \to B$; if $U \subset B$, then we write $f^{-1}(U)$ for the set of preimages of U in A. If f is a linear map (homomorphism) of vector spaces (rings), then we write $\mathrm{Im} f$ for the set of images of f in B and write $\mathrm{Ker} f$ for the kernel of f in A.

CHAPTER I
Affine Algebraic Sets and the Nullstellensatz

As argued in the introduction, the set of solutions of a set of polynomials over a certain ground-field is at the heart of algebraic geometry. So that is our starting point here.

§1. Polynomials and Affine Space

Let k be a *commutative field*, and let

$$k[\mathbf{x}] = k[x_1, ..., x_n]$$

be the polynomial ring over k in variables $x_1, ..., x_n$. The quotient field of $k[\mathbf{x}]$ is denoted by $k(\mathbf{x}) = k(x_1, ..., x_n)$.

1.1. Definition A *monomial* in $k[\mathbf{x}]$ is a product of the form

$$x_1^{\alpha_1} x_2^{\alpha_2} \cdots x_n^{\alpha_n},$$

where all of the exponents $\alpha_1, ..., \alpha_n$ are non-negative integers. The *total degree* of this monomial is the sum

$$\alpha_1 + \cdots + \alpha_n.$$

1.2. Notation (i) $\mathbb{Z}^n_{\geq 0}$ denotes the set of n-tuples $\alpha = (\alpha_1, ..., \alpha_n)$ with non-negative $\alpha_i \in \mathbb{Z}$. (Note that, by extending the addition of $\mathbb{Z}$ to $\mathbb{Z}^n_{\geq 0}$, $\mathbb{Z}^n_{\geq 0}$ forms an additive semigroup: $(\alpha_1, ..., \alpha_n) + (\beta_1, ..., \beta_n) = (\alpha_1 + \beta_1, ..., \alpha_n + \beta_n)$.)
(ii) $\mathrm{M}(k[\mathbf{x}]) = \{x^\alpha = x_1^{\alpha_1} x_2^{\alpha_2} \cdots x_n^{\alpha_n} \mid \alpha \in \mathbb{Z}^n_{\geq 0}\}$, this is the set of all monomials in $k[\mathbf{x}]$.
(iii) $|\alpha| = \alpha_1 + \cdots + \alpha_n$ for $\alpha \in \mathbb{Z}^n_{\geq 0}$.
(iv) If $T \subset k[\mathbf{x}]$ is a subset, we write $\langle T \rangle$ for the ideal of $k[\mathbf{x}]$ generated by T.

The following is a well known fact.

1.3. Theorem $\mathrm{M}(k[\mathbf{x}])$ is a k-basis of $k[\mathbf{x}]$ as a k-vector space. Consequently, every element $f \in k[\mathbf{x}]$ has a unique expression

$$f = \sum_\alpha c_\alpha x^\alpha, \qquad c_\alpha \in k, \; x^\alpha \in \mathrm{M}(k[\mathbf{x}]).$$

□

1.4. Definition Let $f = \sum_\alpha c_\alpha x^\alpha$ be a polynomial in $k[\mathbf{x}]$.
(i) c_α is called the *coefficient* of the monomial x^α.
(ii) If $c_\alpha \neq 0$, then we call $c_\alpha x^\alpha$ a *term* of f.
(iii) The *total degree* of f, denoted $\deg(f)$, is $\max\{|\alpha| \mid c_\alpha \neq 0\}$.
(iv) If $f = 0$, then we put $\deg(0) = -\infty$.

The total degree of polynomials behaves well with respect to sum and product of polynomials, in the following sense.

1.5. Theorem Let $f, g \in k[\mathbf{x}]$ be nonzero polynomials. Then:
(i) $\deg(fg) = \deg(f) + \deg(g)$.
(ii) If $f + g \neq 0$, then $\deg(f + g) \leq \max(\deg(f), \deg(g))$. If, in addition, $\deg(f) \neq \deg(g)$, then equality holds.

□

Theorem 1.3 states that $k[\mathbf{x}]$ is an infinite dimensional k-vector space with k-basis $\mathrm{M}(k[\mathbf{x}])$. However, since every polynomial is a linear sum of monomials of finite total degree, $k[\mathbf{x}]$ is "growing" step by step by finite dimensional subspaces. This can be seen as follows.

For each integer $p \geq 0$, if we put

$$k[\mathbf{x}]_{\leq p} = \left\{ f \in k[\mathbf{x}] \;\middle|\; \deg(f) \leq p \right\},$$

then it is clear that $k[\mathbf{x}]_{\leq p}$ is a finite dimensional subspace of $k[\mathbf{x}]$, and

$$k = k[\mathbf{x}]_{\leq 0} \subset k[\mathbf{x}]_{\leq 1} \subset \cdots \subset k[\mathbf{x}]_{\leq p} \subset k[\mathbf{x}]_{\leq p+1} \subset \cdots,$$

$$k[\mathbf{x}]_{\leq p} \cdot k[\mathbf{x}]_{\leq q} \subset k[\mathbf{x}]_{\leq p+q} \text{ (indeed the equality holds)}, \; p, q \geq 0,$$

$$\bigcup_{p \geq 0} k[\mathbf{x}]_{\leq p} = k[\mathbf{x}].$$

In addition, it is also clear that

$$\mathrm{M}(k[\mathbf{x}])_{\leq p} = \left\{ t \in \mathrm{M}(k[\mathbf{x}]) \;\middle|\; \deg(t) \leq p \right\}$$

is a k-basis of $k[\mathbf{x}]_{\leq p}$. More precisely, we have

1.6. Lemma With notation as above:

$$\left| \mathrm{M}(k[\mathbf{x}])_{\leq p} \right| = \binom{p+n}{n}$$

for all $p \in I\!N$.

Proof Let $B = \{0,1\}^{p+n}$, i.e., the set of all $(p+n)$-tuples with entries from $\{0,1\}$. We define a map φ: $\mathrm{M}(k[\mathbf{x}])_{\leq p} \to B$ as follows. If $t = x_1^{\alpha_1} \cdots x_n^{\alpha_n} \in \mathrm{M}(k[\mathbf{x}])_{\leq p}$, then we set $\varphi(t) = (a_1, ..., a_{n+p})$ with

$$a_k = \begin{cases} 0 & \text{if } k = \alpha_1 + \cdots + \alpha_i + i \text{ for some } 1 \leq i \leq n \\ 1 & \text{otherwise} \end{cases}$$

The definition of $\varphi(t)$ can be visualized as follows: Write down α_1 times 1, then a zero to mark the end of the first exponent, then α_2 times 1, and so on ..., till we reach α_n times 1, then another zero, and finally $p - \deg(t)$ times 1, i.e.,

$$\varphi(t) = (\overbrace{\underbrace{1, ..., 1}_{\alpha_1 \text{ times}}, 0, \underbrace{1, ..., 1}_{\alpha_2 \text{ times}}, 0, ..., \underbrace{1, ..., 1}_{\alpha_n \text{ times}}, 0}^{\deg(t)+n}, \underbrace{1, ..., 1}_{\beta \text{ times}})$$

where $\beta = p-\deg(t)$.
It is now an easy exercise to prove that φ is injective, and that the image of φ consists of all those $(a_1, ..., a_{p+n}) \in B$ with $a_k = 0$ for exactly n different indices k. It is clear that there are exactly $\begin{pmatrix} p+n \\ n \end{pmatrix}$ such tuples. □

Next, for each integer $p \geq 0$, if we put

$$k[\mathbf{x}]_p = \left\{ f \in K[\mathbf{x}] \;\middle|\; f = \sum_{|\alpha|=p} c_\alpha x^\alpha \right\}$$

then it is clear that $k[\mathbf{x}]_p$ is a finite dimensional k-vector space with k-basis

$$\mathrm{M}(k[\mathbf{x}])_p = \left\{ t \in \mathrm{M}(k[\mathbf{x}]) \;\middle|\; \deg(t) = p \right\},$$

and it is easy to check that

$k[\mathbf{x}]_p k[\mathbf{x}]_q \subseteq k[\mathbf{x}]_{p+q}$ (indeed the equality holds) for all $p, q \geq 0$;

$k[\mathbf{x}] = \bigoplus_{p\geq 0} k[\mathbf{x}]_p$.

A ring R is called a ***ℤ*-graded ring** if $R = \oplus_{n\in\mathbb{Z}} R_n$, where each R_n is an additive subgroup of R, and if $R_n R_m \subset R_{n+m}$ for all $n, m \in \mathbb{Z}$. If $R_n = 0$ for all $n < 0$, then R is called a ***positively graded*** ring.
For a $\mathbb{Z}$-graded ring $R = \oplus_{n\in\mathbb{Z}} R_n$, each R_n is called the n-th ***homogeneous component*** of R; and an element in R_n is called a ***homogeneous element*** of degree n.

From the foregoing discussion it is clear that $k[\mathbf{x}]$ is a positively graded ring (we also say that $k[\mathbf{x}]$ is a graded ring with the ***natural gradation*** given by the total degree of polynomials).

1.7. Lemma With notation as above:

$$\left|\mathrm{M}(k[\mathbf{x}])_p\right| = \begin{pmatrix} p+n-1 \\ n-1 \end{pmatrix}$$

for all $p \in \mathbb{N}$.

Proof This is an easy exercise using Lemma 1.6. □

Keeping the *graded* structure of $k[\mathbf{x}] = \oplus_{p \geq 0} k[\mathbf{x}]_p$ in mind, we call a polynomial in $k[\mathbf{x}]_p$ a *homogeneous element of degree p*, or a *form of degree p*. Hence, every $f \in k[\mathbf{x}]$ has a unique decomposition into homogeneous elements:

$$(*) \qquad f = F_0 + F_1 + \cdots + F_p + \cdots + F_d$$

with $F_p \in k[\mathbf{x}]_p$.
If $F_d \neq 0$ in the homogeneous decomposition $(*)$ of f, then clearly $\deg(f) = d$; moreover, the terms F_0, F_1, F_2, ... are called the *constant, linear, quadratic*, ... terms of f.

1.8. Theorem (Hilbert Basis Theorem) Every ideal I of $k[\mathbf{x}]$ is finitely generated, i.e., there are finitely many $f_1, ..., f_s \in I$ such that $I = \langle f_1, ..., f_s \rangle$. Hence $k[\mathbf{x}]$ is a Noetherian ring in the sense of Appendix §1.

Proof This can be found in any ring theory textbook, but we refer the reader to CH.IV §4 for a constructive proof. □

For use in later chapters, we mention the following theorem (leaving the proof as an exercise).

1.9. Theorem Let I be an ideal of $k[\mathbf{x}]$. Then the following are equivalent:
(i) $I = \oplus_{p \geq 0} (k[\mathbf{x}]_p \cap I)$;
(ii) I is generated by homogeneous elements;
(iii) If $f = F_0 + F_1 + \cdots + F_d \in k[\mathbf{x}]$ with $F_p \in k[\mathbf{x}]_p$, then $f \in I$ if and only if $F_p \in I$ for $p = 0, 1, ..., d$.
(iv) The quotient ring $k[\mathbf{x}]/I$ is a graded ring with the gradation $(k[\mathbf{x}]/I)_p = (k[\mathbf{x}]_p + I)/I$ for $p \geq 0$, i.e. $k[\mathbf{x}]/I = \oplus_{p \geq 0} (k[\mathbf{x}]/I)_p$ with $(k[\mathbf{x}]/I)_p = (k[\mathbf{x}]_p + I)/I$.

□

1.10. Definition An ideal I of $k[\mathbf{x}]$ is called a *graded ideal* (or a *homogeneous ideal*) if I satisfies one of the equivalent conditions of Theorem 1.9.

The following important fact concerning graded ideals will be used freely in these notes.

1.11. Proposition A graded ideal P of $K[\mathbf{x}]$ is a prime ideal if and only if P is a graded prime ideal, i.e., if and only if any homogeneous $F, G \in K[\mathbf{x}]$, $FG \in P$ implies $F \in P$ or $G \in P$.

Proof Exercise. □

Remark (i) In CH.IV some properties of polynomials and monomials, leading to algorithmic solutions of several basic problems, will be explored further.

(ii) Let $I \subset k[\mathbf{x}]$ be an ideal. Then, with notation as before, the k-vector space $k[\mathbf{x}]/I$ is also "growing " step by step by finite dimensional subspaces, i.e., we have an increasing chain of finite dimensional subspaces

$$k = (k[\mathbf{x}]/I)_{\le 0} \subset (k[\mathbf{x}]/I)_{\le 1} \subset \cdots \subset (k[\mathbf{x}]/I)_{\le p} \subset (k[\mathbf{x}]/I)_{\le p+1} \subset \cdots$$

with $(k[\mathbf{x}]/I)_{\le p} = (k[\mathbf{x}]_{\le p} + I)/I$ for $p \ge 0$. In CH.V we will see that the "growth" of the k-vector space $k[\mathbf{x}]/I$ can be measured by a polynomial $h(x) \in \mathbb{Q}[x]$, i.e., the Hilbert polynomial, or more precisely, for $p \gg 0$, $\dim_k((k[\mathbf{x}]/I)_{\le p})$ may be given by $h(p)$. The degree of $h(x)$ is one of the most important algebraic-geometric invariants for the algebraic set $\mathbf{V}(I)$ (see the definition of $\mathbf{V}(I)$ in the next section), and moreover, this polynomial $h(x)$ can be produced from any generating set of I (whenever the ground field is computable, i.e., the operations of the field can be performed on a computer).

(iii) Let I be a homogeneous ideal of $k[\mathbf{x}]$, then by Theorem 1.8 we have $k[\mathbf{x}]/I = \oplus_{p\ge 0}(k[\mathbf{x}]/I)_p$ with $(k[\mathbf{x}]/I)_p = (k[\mathbf{x}]_p + I)/I$. In CH.V we will see that, as a k-vector space, the "size" (k-dimension) of $(k[\mathbf{x}]/I)_p$ may also be measured by a polynomial $h(x)$, or more precisely, for $p \gg 0$ $\dim_k((k[\mathbf{x}]/I)_p)$ may be given by $h(p)$, and the degree of this polynomial $h(x)$ is also an important geometric invariant for the projective algebraic set $\mathbf{V}(I)$ (see the definition of $\mathbf{V}(I)$ in CH.III).

Now let us introduce affine space formally.

1.12. Definition Given a field k and a positive integer n, we define n-dimensional *affine space* (or affine n-space) over k, denoted $\mathbf{A}_k^n$ or simply $\mathbf{A}^n$, to be the set of all n-tuples of elements of k, i.e.

$$\mathbf{A}_k^n = \left\{ P = (a_1, ..., a_n) \;\middle|\; a_1, ..., a_n \in k \right\}.$$

An element $P \in \mathbf{A}_k^n$ will be called a *point*, and if $P = (a_1, ..., a_n)$ with $a_i \in k$, then the a_i will be called the *coordinates* of P.

For an example of affine space, consider the case $k = \mathbb{R}$. Here we get the familiar space $\mathbb{R}^n$ from calculus and linear algebra. In general, we call $\mathbf{A}^1 = k$ the *affine line* and $\mathbf{A}^2$ the *affine plane.*

How do polynomials relate to affine space? The *key* idea is that a polynomial $f = \sum_\alpha c_\alpha x^\alpha \in k[\mathbf{x}]$, where $\alpha = (\alpha_1, ..., \alpha_n) \in \mathbb{Z}_{\geq 0}^n$ and $x^\alpha = x_1^{\alpha_1} \cdots x_n^{\alpha_n}$, determines a function:

$$\begin{array}{cccc} f: & \mathbf{A}^n & \longrightarrow & k \\ & P = (a_1, ..., a_n) & \mapsto & \sum\limits_\alpha c_\alpha a_1^{\alpha_1} \cdots a_n^{\alpha_n} \end{array}$$

The possibility to view a polynomial as a function is at the origin of the link between algebra and geometry.
This dual nature of polynomials has some unexpected consequences. For example, the question "is $f = 0$?" now has two meanings: is f the zero polynomial?, which means that all of its coefficients c_α are zero, or is f the zero function?, which means that $f(a_1, ..., a_n) = 0$ for all $(a_1, ..., a_n) \in k^n$. The surprising fact is that these two statements are not equivalent in general. For example consider the finite field $\mathbf{F}_2$ consisting of the two elements 0 and 1. Now consider the polynomial $x^2 - x = x(x-1) \in \mathbf{F}_2[x]$. Since this polynomial vanishes at 0 and 1, we have found a nonzero polynomial which gives the zero function on the affine space $\mathbf{F}_2^1$. One may also find examples in the case of higher dimension.
However, as long as k is infinite, there is no problem.

1.13. Proposition Let k be an infinite field, and let $f \in k[\mathbf{x}]$. Then $f = 0$ in $k[\mathbf{x}]$ if and only if $f\colon k^n \to k$ is the zero function.

Proof First write $f = \sum_i f_i x_n^i$, $f_i \in k[x_1, ..., x_{n-1}]$, then use induction on n, and the fact that $f(a_1, ..., a_{n-1}, x_n)$ has only a finite number of roots if some $f_i(a_1, ..., a_{n-1}) \neq 0$. □

1.14. Corollary Let k be an infinite field, and let $f, g \in k[\mathbf{x}]$. Then $f = g$ in $k[\mathbf{x}]$ if and only if f and g determine the same function on k^n.

□

Finally, we need to recall a special property of polynomials over the field of complex numbers $\mathbb{C}$.

1.15. Theorem (Fundamental Theorem of Algebra) Every nonconstant polynomial $f \in \mathbb{C}[x]$ has a root in $\mathbb{C}$.

□

In §5 we will prove a powerful generalization of Theorem 1.14 called the Hilbert Nullstellensatz.

Exercises for §1

In what follows, $k[\mathbf{x}] = k[x_1, ..., x_n]$.

1. $f \in k[\mathbf{x}]$ is an invertible element if and only if $f \in k$.
2. Let I be an ideal of $k[\mathbf{x}]$ and $n = 1$.
   (a) Prove that $k[\mathbf{x}]/I$ is a finite dimensional k-vector space;
   (b) Find a k-basis of $k[\mathbf{x}]/I$.
3. Let I be an ideal of $k[\mathbf{x}]$.
   (a) Show that if $I \neq 0$ then I is also an infinite dimensional k-vector space.
   (b) Let $I = \langle x, y \rangle \subset k[x, y]$. Show that $\dim_k(k[x, y]/I^n) = 1 + 2 + \cdots + n = \dfrac{n(n+1)}{2}$, $n \geq 1$.
   (c) If $n > 1$, is $k[\mathbf{x}]/I$ a finite dimensional k-vector space?
4. Let k be any field.
   (a) Show that there is an infinite number of irreducible monic polynomials in $k[x]$. (Hint: Suppose $f_1, ..., f_n$ were all of them, and factor $f_1 \cdots f_n + 1$ into irreducible factors.)
   (b) Show that any algebraically closed field is infinite. (Hint: The irreducible monic polynomials in $k[x]$ are $x - a$, $a \in k$.)
5. Let k be a field, $f \in k[\mathbf{x}]$, $a_1, ..., a_n \in k$.

(a) Show that $f = \sum \lambda_{(i)}(x_1 - a_1)^{i_1} \cdots (x_n - a_n)^{i_n}$, $\lambda_{(i)} \in k$.

(b) If $f(a_1, \dots, a_n) = 0$, show that $f = \sum_{i=1}^{n}(x_i - a_i)g_i$ for some (not unique) $g_i \in k[\mathbf{x}]$.

(c) let φ: $k[\mathbf{x}] \to k$ be the map defined by putting $\varphi(f) = f(a_1, \dots, a_n)$ with $f \in k[\mathbf{x}]$. Show that φ is a ring epimorphism with $\text{Ker}\varphi = \langle x_1 - a_1, \dots, x_n - a_n \rangle$ (hence $\langle x_1 - a_1, \dots, x_n - a_n \rangle$ is a maximal ideal of $k[\mathbf{x}]$).

6. (Euler's Theorem) Show that if $F \in k[\mathbf{x}]$ is a homogeneous polynomial of degree d, then

$$\sum_{i=1}^{n} \frac{\partial F}{\partial x_i} x_i = dF,$$

where $\partial/\partial x_i$ are the usual partial derivatives of $k[\mathbf{x}]$.

7. Complete the proof of Lemma 1.7, Theorem 1.9, and Proposition 1.11.

8. Let $R = \oplus_{n \in \mathbb{Z}} R_n$ be a $\mathbb{Z}$-graded ring. Show that R_0 is a subring of R containing the multiplicative identity 1_R.

9. Prove that the following two properties of $k[\mathbf{x}]$ are equivalent to Hilbert's basis theorem, respectively.

(a) Every set of ideals of $k[\mathbf{x}]$ has a maximal element with respect to the usual inclusion relation.

(b) For every increasing chain of ideals of $k[\mathbf{x}]$

$$I_0 \subset I_1 \subset \cdots \subset I_k \subset$$

there exists an n such that $I_n = I_{n+k}$ for all $k \geq 0$.

§2. Affine Algebraic Sets

Let k be a field, and let $\mathbf{A}^n = \mathbf{A}_k^n$ denote the affine n-space over k.

From §1 we know that every $f = f(x_1, \dots, x_n) \in k[\mathbf{x}] = k[x_1, \dots, x_n]$ may be viewed as a function from the $\mathbf{A}^n$ to k. Thus, we may talk about the set of *zeros* of f, i.e., $\mathbf{V}(f) = \{P \in \mathbf{A}^n \mid f(P) = 0\}$ where we write $f(P) = f(a_1, \dots, a_n)$ if $P = (a_1, \dots, a_n)$. More generally, if T

is any subset of $k[\mathbf{x}]$, we define the *zero set* of T to be the common zeros of all the elements of T, namely,

$$\mathbf{V}(T) = \left\{ P \in \mathbf{A}^n \;\middle|\; f(P) = 0, \text{ for all } f \in T \right\}.$$

2.1. Proposition Let T be a nonempty subset of $k[\mathbf{x}]$ and $I = \langle T \rangle$ the ideal generated by T in $k[\mathbf{x}]$. Then
(i) $\mathbf{V}(T) = \mathbf{V}(I)$;
(ii) There exist a finite number of $f_1, \ldots, f_s \in T$ such that $\mathbf{V}(T) = \mathbf{V}(f_1, \ldots, f_s)$.
(iii) If $T_1 \subseteq T_2$ are subsets of $k[\mathbf{x}]$, then $\mathbf{V}(T_1) \supseteq \mathbf{V}(T_2)$.

Proof (i) and (iii) can be checked directly. (ii) follows from the Hilbert basis theorem (Theorem 1.8). □

2.2. Definition A subset Y of $\mathbf{A}^n$ is an *affine algebraic set* if there exists a subset $T \subseteq k[\mathbf{x}]$ such that $Y = \mathbf{V}(T)$.
If $f \in k[\mathbf{x}]$ is not a constant, then the algebraic set $\mathbf{V}(f)$ is called the *hypersurface* defined by f. A hypersurface in $\mathbf{A}^2$ is called an *affine plane curve*. If $\deg(f) = 1$, $\mathbf{V}(f)$ is called a *hyperplane* in $\mathbf{A}^n$; if $n = 2$, it is a *line*.

2.3. Proposition (i) If Y_1, Y_2 are algebraic sets, then so is $Y_1 \cup Y_2$.
(ii) If $\{Y_i\}_{i \in J}$ is a family of algebraic sets, then $\cap_{i \in J} Y_i$ is an algebraic set.

Proof Exercise. □

Example (i) $\mathbf{A}^n$ and $\emptyset$ are algebraic sets (indeed $\mathbf{A}^n = \mathbf{V}(0)$, $\emptyset = \mathbf{V}(1)$).

(ii) We will frequently use the algebraic set $V = \mathbf{V}(y - x^2, z - x^3) \subset \mathbf{A}^3_{\mathbb{R}}$, the *twisted cubic* in $\mathbb{R}^3$, as one of the main examples.

(iii) Since for $a_i \in k$, $\mathbf{V}(x_1 - a_1, \ldots, x_n - a_n) = \{P = (a_1, \ldots, a_n)\} \subset \mathbf{A}^n$, it follows from Proposition 2.3(i) that any finite subset of $\mathbf{A}^n$ is an algebraic set.

Proposition 2.1 above establishes the connection between ideals and algebraic sets. Since for ideals we have algebraic operations like:

sum, product, and intersection, it is natural to look at the geometric meaning of these algebraic operations.

2.4. Theorem Let I and J be ideals in $k[\mathbf{x}]$. Then
(i) $\mathbf{V}(I+J) = \mathbf{V}(I) \cap \mathbf{V}(J)$.
(ii) $\mathbf{V}(I \cdot J) = \mathbf{V}(I) \cup \mathbf{V}(J)$.
(iii) $\mathbf{V}(I \cap J) = \mathbf{V}(I) \cup \mathbf{V}(J)$.

Proof Exercise. □

Let $I = \langle f_1, ..., f_s \rangle$, $J = \langle g_1, ..., g_m \rangle$ be ideals of $k[\mathbf{x}]$. For sum, resp. product, of I and J, the generators of the new ideals, or equivalently, defining equations of $\mathbf{V}(I+J)$ resp. of $\mathbf{V}(IJ)$ are easy to find. However, it is by no means trivial to find a generating set of $I \cap J$ from one for I and J. From the above theorem it is clear that $I \cdot J$ and $I \cap J$ define the same algebraic set. Thanks to the Groebner basis theory, in CH.IV §5 we will introduce (from [LCO'] p.181) an algorithm for computing generators of $I \cap J$ from $\{f_1, ..., f_s\}$ and $\{g_1, ..., g_m\}$. This result will be useful in CH.III §6.

Exercises for §2

1. Prove Proposition 2.3 and Theorem 2.4.
2. Let f be a nonconstant polynomial in $k[x_1, ..., x_n]$, k being algebraically closed. Show that $\mathbf{A}^n - \mathbf{V}(f)$ is infinite if $n \geq 1$, and $\mathbf{V}(f)$ is infinite if $n \geq 2$. Conclude that the complement of any algebraic set is infinite. (Hint: Use Proposition 1.13.)
3. Consider the plane curve $C = \mathbf{V}(f) \subset \mathbf{A}_k^2$ where f is a polynomial of degree n in $k[x, y]$. If L is a line in $\mathbf{A}_k^2$ and $L \not\subset C$, show that $L \cap C$ is a finite set of no more than n points. (Hint: Suppose $L = \mathbf{V}(y - (ax + b))$, and consider $f(x, ax + b) \in k[x]$.)

§3. Parametrizations of Algebraic Sets

Many elementary applications of algebraic geometry to other mathematical fields or other scientific subjects are related to the parametrization of algebraic sets. For example, the parametrization of curves leads in a natural way to certain aspects connecting algebraic geometry and number theory. This may be illustrated by considering Diophantine equations. Suppose $f(x_1, ..., x_n)$ is a polynomial with rational integer coefficients. The problem to be solved is to find all solutions of the equation $f(x_1, ..., x_n) = 0$ where the x_i are rational integers. One of the most practical applications of parametrization of algebraic sets stems from Computer Aided Geometric Design (CAGD). When creating complex shapes like automobiles hoods or airplane wings, design engineers need curves and surfaces that are varied in shape, easy to describe, and quick to draw on a computer. Parametric equations involving polynomial and rational functions satisfy these requirements.

Parametrization of an algebraic set V comes down to describing points of V by means of certain polynomial or rational functions. These topics are studied in more detail later in CH.II and CH.VI. Therefore, we introduce the notion of parametrization here by just giving some clarifying examples. First let us recall some familiar examples from classical "analytic" geometry.

Example (i) Suppose $ax + by + c = 0$ is the equation of a straight line in $\mathbf{A}^2_{\mathbf{R}}$. Then a and b are not both zero. We wish to obtain a parametrization of this line. Assume that $a \neq 0$ and that x_0, y_0 is a point on the line. (It is trivial to see that there must be at least one such point.) Then $ax_0 + by_0 + c = 0$ and so a point (x, y) is on the line if and only if

$$a \cdot (x - x_0) + b \cdot (y - y_0) = 0. \tag{1}$$

Set

$$t = \frac{y - y_0}{a}.$$

Then

$$\begin{cases} x = x_0 + bt, \\ y = y_0 + at. \end{cases} \tag{2}$$

These equations give parametric equations for the line since, if t is specialized arbitrarily in k, the point (x, y) given by (2) will lie on the line. The same result is obtained if we start from the assumption

$b \neq 0$. It is easy to verify that, if the line can also be parametrized in the form

$$\begin{cases} x = x_0 + b's, \\ y = y_0 + a's, \end{cases}$$

then there is a nonzero element $c \in k$ such that $a' = ca$ and $b' = cb$.

Remark In elementary analytic geometry of the Euclidean plane, a straight line is ultimately defined in terms of the intuitively acceptable characterization that for any two points (x_1, y_1) and (x_2, y_2) on the line, the ratio of $x_1 - x_2$ and $y_1 - y_2$ is a constant.

A parametrization similar to (2) can be obtained in the higher-dimensional cases of planes and hyperplanes. This can be done best by using techniques of linear algebra (see below example (ii)). Here, however, we prefer to adopt a more naive procedure which, although less elegant, is quite sufficient for our purposes. Suppose a hyperplane is given by the equation

$$a_1x_1 + \cdots + a_nx_n + b = 0.$$

Suppose $a_1 \neq 0$ and that $P = (c_1, ..., c_n)$ is a point on the hyperplane. Then

$$a_1 \cdot (x_1 - c_1) + \cdots + a_n \cdot (x_n - c_n) = 0.$$

For $i = 2, ..., n$ set

$$t_i = \frac{c_i - x_i}{a_1}.$$

Then for any point $Q = (x_1, ..., x_n)$ on the hyperplane

$$(*) \qquad \begin{cases} x_1 = c_1 + a_2t_2 + \cdots + a_nt_n \\ x_2 = c_2 - a_1t_2 \\ \vdots \\ x_n = c_n - a_1t_n \end{cases}$$

Conversely, if $t_2, ..., t_n$ are specialized arbitrarily in k, then the point $(x_1, ..., x_n)$ given by $(*)$ will lie on the hyperplane and so equations $(*)$ constitute a set of parametric equations for the hyperplane. If $a_1 = 0$, the procedure above must be modified by using instead of a_1 some other coefficient which is not zero.

(ii) Consider the system of equations

$$(1) \qquad \begin{cases} x+y+z=1, \\ x+2y-z=3. \end{cases}$$

Geometrically, this represents the line in $\mathbb{R}^3$ that is the intersection of the planes $x+y+z=1$ and $x+2y-z=3$. It follows that there are infinitely many solutions. To describe the solutions, we use row operations on equations (1) to obtain the equivalent equations

$$\begin{cases} x+3z=-1, \\ y-2z=2. \end{cases}$$

Letting $z=t$, where t is arbitrary, this implies that all solutions of (1) are given by

$$(2) \qquad \begin{cases} x=-1-3t, \\ y=2+2t, \\ z=t \end{cases}$$

as t varies over $\mathbb{R}$. We call t a *parameter*, and (2) is, thus, a *parametrization* of the solutions of (1).

To see whether the idea of "parametrizing" solutions can be applied to other algebraic sets, let us look at some other well known examples.

(iii) The *cusp curve* $\mathbf{V}(y^2-x^3) \subset \mathbb{R}^2$ may be parametrized as

$$\begin{cases} x=t^2 \\ y=t^3 \end{cases}$$

(iv) The *node curve* $\mathbf{V}(y^2-x^3-x^2) \subset \mathbb{R}^2$ may be parametrized as

$$\begin{cases} x=t^2-1 \\ y=t^3-t \end{cases}$$

(v) The *twisted cubic* $V = \mathbf{V}(y-x^2, z-x^3) \subset \mathbb{R}^3$ may be parametrized as:

$$\begin{cases} x=t \\ y=t^2 \\ z=t^3 \end{cases}$$

(vi) The unit circle

$$x^2 + y^2 = 1. \tag{1}$$

may be parametrized by using trigonometric functions:

$$\begin{cases} x = cos(t), \\ y = sin(t). \end{cases}$$

However, there is also a more algebraic way to parametrize the circle, i.e, intersecting (1) by the line $y = tx$ of variable slope t, we obtain:

$$\begin{cases} x = \dfrac{2t}{1+t^2}, \\ \\ y = \dfrac{1-t^2}{1+t^2}. \end{cases} \tag{2}$$

The reader should check that the points defined by these equations lie on the circle (1).

It is also interesting to note that the parametrizations in (i)–(v) cover the whole algebraic sets, but the parametrization of the circle given by (2) does not describe the whole circle: since $y = \frac{1-t^2}{1+t^2}$ can never equal -1, the point $(0, -1)$ is not covered.

Notice that equations (2) above involve quotients of polynomials. These are examples of rational functions, and before we can say what it means to prametrize an algebraic set, we need to recall the notion of rational function.

Let k be a field. A *rational function* in $t_1, .., t_m$ with coefficients in k is an element of the quotient field $k(t_1, ..., t_m)$, i.e., a quotient $\dfrac{f}{g}$ of two polynomials $f, g \in k[t_1, ..., t_m]$, where g is not the zero polynomial. Furthermore, two rational functions $\dfrac{f}{g}$ and $\dfrac{f_1}{g_1}$ are equal, provided $g_1 f = g f_1$ in $k[t_1, ..., t_m]$.

3.1. Definition Suppose that we are given an algebraic set $V = \mathbf{V}(f_1, ..., f_s) \subset \mathbf{A}^n$. Then a *rational parametric representation* of V consists of rational functions $r_1, ..., r_n \in k(t_1, ..., t_m)$ such that the points given by

$$\begin{cases} x_1 = r_1(t_1, ..., t_m), \\ x_2 = r_2(t_1, ..., t_m), \\ \vdots \\ x_n = r_n(t_1, ..., t_m) \end{cases}$$

lie in V.

In many situations, we have a parametrization of an algebraic set V, where $r_1, ..., r_n$ are polynomials rather than rational functions. This is what we call a *polynomial parametric representation* of V. For instance, the foregoing examples (i)–(v).

From the definition and the above examples we immediately infer the following:

- If an algebraic set V has a rational parametric representation as defined above, then the rational functions

$$r_1 = \frac{g_1}{h_1}, ..., r_n = \frac{g_n}{h_n}$$

are generally only defined on a subset of $\mathbf{A}^m$, i.e., the set

$$\bigcap_{i=1}^{n} V_{r_i}, \text{ where } V_{r_i} = \left\{ P \in \mathbf{A}^m \;\middle|\; h_i(P) \neq 0 \right\};$$

- As the example of the circle made clear, a parametrization may not cover all points of V.

The most important geometric property of an algebraic set V with a rational parametric representation is that V *cannot be expressed as the union of two proper algebraic subsets* (i.e., V is *irreducible* in the sense of CH.II §1, see CH.II Proposition 1.7).

Now let us look at several applications of parametrization of algebraic sets.

Example (vii) This example illustrates an aspect of the connection between algebraic geometry and number theory.

The curve $x^6 - x^2y^3 - y^5 = 0$ may be parametrized by intersecting it with the line $y = tx$ of variable slope t:

$$(*) \qquad \begin{cases} x = t^3 + t^5 \\ y = t^4 + t^6 \end{cases}$$

The parametrization $(*)$ enables us to find all the integral points on the curve. Certainly, if we take integral values for t in $(*)$ we get integral points on the curve. We claim that these are all integral points on the curve. Suppose (x, y) is an integral point. Then x and $y = tx$ are integers. Since $x = t^3(1 + t^2)$ the only way x can be zero is for t to be zero. Thus it may be assumed that x is not zero and so $t = \dfrac{y}{x}$ is rational. We write t in the form $t = \dfrac{a}{b}$ where a and b are relatively prime integers. Now $x = t^3 + t^5$ and so $x = \dfrac{a^3b^2 + a^5}{b^5}$. This must be an integer and so $a^3b^2 + a^5 \equiv 0(\text{mod} b)$. Therefore $a^5 \equiv 0(\text{mod} b)$. Since a and b are relatively prime, it follows that $b = \pm 1$. Thus t must be an integer, as we claimed. Therefore, we obtain all solutions of the Diophantine equation $x^6 - x^2y^3 - y^5 = 0$ by letting t range over all integer values in the equation $(*)$.

As we have pointed out in the begining of this section, another main virtue of the parametric representation of a curve or surface is that it is easy to draw the shape of the curve or surface on a computer. Given the formulas for the parametrization, the computer evaluates them for various values of the parameters and then plots the resulting points.

(viii) The surface $\mathbf{V}(x^2 - y^2z^2 + z^3)$ cannot be drawn by using the defining polynomial of V directly, instead, using the parametric representation:

$$(*) \qquad \begin{cases} x = t(u^2 - t^2), \\ y = u, \\ z = u^2 - t^2. \end{cases}$$

V can be drawn by *Gnuplot* on a computer

Figure 1

(Here two parameters t and u are used since we are describing a surface, and the above picture was drawn using t, u in the range $-1 \leq t, u \leq 1$). As an exercise, one may check that this parametrization also covers the entire surface $\mathbf{V}(x^2 - y^2z^2 + z^2)$.

(ix) Similarly, by using the parametrization of $V = \mathbf{V}(y - x^2, z - x^3)$, the twisted cubic, we can also draw V on a computer. But we prefer seeing another interesting example connected to V, i.e., we want to draw the *tangent surface* of V. This can be done as follows.
Writing $r(t) = (t, t^2, t^3)$, a particular value of t will give us a point on the curve. From calculus, we know that the tangent vector to the curve at this point is given by $r(t)$ is $r'(t) = (1, 2t, 3t^2)$. It follows that the tangent line is parametrized by

$$r(t) + u \cdot r'(t) = (t, t^2, t^3) + u \cdot (1, 2t, 3t^2) = (t + u, t^2 + 2tu, t^3 + 3t^2u),$$

where u is a parameter that moves alone the tangent line. If we now allow t to vary, then we can parametrize the entire tangent surface by

$$\begin{cases} x = t + u \\ y = t^2 + 2tu \\ z = t^3 + 3t^2u. \end{cases}$$

The parameters t and u have the following interpretations: t tells where we are on the curve, and u tells where we are on the tangent line. Here is the picture of the tangent surface drawn by *Gnuplot* on computer

Figure 2

An example of parametrization of algebraic sets applied to geometric modeling is given in CH.II §4.

On the opposite side of the parametrization of algebraic set, it is often equally useful to have an implicit representation of an algebraic set. For example, suppose we want to know whether or not the point $(1, 2, -1)$ is on the surface given above in (ii). If all we had was the parametrization $(*)$, then, to decide this question, we would need to solve the equations

$$(**) \qquad \begin{cases} 1 = t(u^2 - t^2), \\ 2 = u, \\ -1 = u^2 - t^2 \end{cases}$$

for t and u. On the other hand, if we have the implicit representation $x^2 - y^2z^2 + z^3 = 0$, then it is simply a matter of plugging coordinates into this equation. Since

$$t^2 - 2^2(-1)^2 + (-1)^3 = 1 - 4 - 1 = -4 \neq 0,$$

it follows that $(1, 2, -1)$ is not on the surface (and, consequently, equations $(**)$ have no solution).

The desirability of having both types of representations leads to the following two questions:

- (Parametrization) Does every affine algebraic set have a rational parametric representation?
- (Implicitization) Given a parametric representation of an affine algebraic set, can we find the defining equations (i.e. can we find an implicit representation)?

The answer to the first question is negative. In fact, all reduced affine algebraic sets (see the definition in CH.II) cannot be parametrized in the sense described here (see CH.II Proposition 1.7). However, not all irreducible algebraic sets have rational parametric representations. A typical example is the irreducible curve $y^2 = x(x-1)(x-\lambda)$ with $\lambda \in \mathbb{C}$ (see CH.II §1 Exercise 5). The proof of this fact derives from the following nontrivial result:

- Let k be a field of characteristic $\neq 2$, $\lambda \in k$, $\lambda \neq 0,1$. If $f, g \in k(t)$ are rational functions satisfying

$$f^2 = g(g-1)(g-\lambda)$$

then $f, g \in k$.

Those algebraic sets that have rational parametric representations are called *unirational*. In general, it is difficult to tell whether a given algebraic set is unirational or not. We refer to CH.II §3 and CH.VI §3 for further discussion concerning this question. We also refer the reader to a special issue of *Journal of Symbolic Computation* (2,3)23(1997) for *Parametric Algebraic Curves and Applications*. The situation for the second question is much nicer. We refer the reader to ([CLO'] CH.3) for a definite answer to this question. In particular, there are the "*implicitization algorithm for polynomial parametrizations*" and the "*implicitization algorithm for rational parametrizations*".

Exercises for §3

1. Find the parametrizations of the following algebraic sets.
   (a) $\mathbf{V}(y^3 - x^4 - x^3) \subset \mathbb{R}^2$.
   (b) $\mathbf{V}(2x^2 + y^2 - 5z^2) \subset \mathbb{R}^3$.
2. Find all integer solutions of the node curve $y^2 - x^3 - x^2 = 0$.

§4. Ideals for Algebraic Sets

Let k, $\mathbf{A}^n = \mathbf{A}^n_k$, and $k[\mathbf{x}] = k[x_1, ..., x_n]$ be as before.

For any subset $Y \subseteq \mathbf{A}^n$ it is easy to see that the set

$$\mathbf{I}(Y) = \left\{ f \in k[\mathbf{x}] \;\middle|\; f(P) = 0 \text{ for all } P \in Y \right\}$$

is an ideal of $k[\mathbf{x}]$. (Check this!)

4.1. Definition $\mathbf{I}(Y)$ is called the *ideal* of Y in $k[\mathbf{x}]$.

Before we can go further, in fact before we can even give any interesting examples, we need more deeply to explore the relationship between subsets of $\mathbf{A}^n$ and ideals in $k[\mathbf{x}]$ more deeply. More precisely, the real importance of the ideals is that they provide us with a language for computing with algebraic sets.

4.2. Lemma If J is an ideal of $k[\mathbf{x}]$, then $J \subseteq \mathbf{I}(\mathbf{V}(J))$, and $\mathbf{V}(\mathbf{I}(\mathbf{V}(J))) = \mathbf{V}(J)$.

Proof Exercise. □

4.3. Proposition (i) For any two subsets Y_1, Y_2 of $\mathbf{A}^n$, we have $\mathbf{I}(Y_1 \cup Y_2) = \mathbf{I}(Y_1) \cap \mathbf{I}(Y_2)$.
(ii) For any subset $Y \subseteq \mathbf{A}^n$, $\mathbf{V}(\mathbf{I}(Y)) = \overline{Y}$, the smallest affine algebraic set containing Y (i.e. the closure of Y with respect to the Zariski topology of $\mathbf{A}^n$, see Ch.II §1).

Proof Exercise. □

Example (i) If k is infinite then $\mathbf{I}(\mathbf{A}^n) = \{0\}$.

(ii) For $V = \mathbf{V}(y - x^2, z - x^3) \subset \mathbb{R}^3$, $\mathbf{I}(V) = \langle y - x^2, z - x^3 \rangle$.

Generally, J is not necessarily equal to $\mathbf{I}(\mathbf{V}(J))$. For instance, $\langle x^2, y^2 \rangle$ is properly contained in $\mathbf{I}(\mathbf{V}(x^2, y^2))$. Nevertheless, the ideal of an affine algebraic set always contains enough information to determine the affine algebraic set uniquely.

4.4. Proposition Let V and W be affine algebraic sets in $\mathbf{A}^n$. Then:

(i) $V \subseteq W$ if and only if $\mathbf{I}(V) \supseteq \mathbf{I}(W)$.
(ii) $V = W$ if and only if $\mathbf{I}(V) = \mathbf{I}(W)$.

Proof Exercise. □

Exercises for §4

1. Prove Lemma 4.2, Proposition 4.3 and Proposition 4.4.
2. Let V be an algebraic set in $\mathbf{A}^n$, $P \in \mathbf{A}^n$ a point not in V. Show that there exists a polynomial $f \in k[x_1, ..., x_n]$ such that $f(Q) = 0$ for all $Q \in V$, but $f(P) = 1$.
3. Let $\{P_1, ..., P_r\}$ be a finite set of points in $\mathbf{A}^n$. Show that there are polynomials $f_1, ..., f_r \in k[x_1, ..., x_n]$ such that $f_i(P_j) = 0$ if $i \neq j$ and $f_i(P_i) = 1$.
4. Let V be an algebraic set of $\mathbf{A}^n$, $P_1, P_2 \notin V$. Show that there exists a polynomial $f \in k[x_1, ..., x_n]$ such that $f(P_i) \neq 0$, $i = 1, 2$, but $f \in \mathbf{I}(V)$.

§5. Hilbert's Nullstellensatz

In §4 we observed that any affine algebraic set $V \subseteq \mathbf{A}^n = \mathbf{A}^n_k$ may be associated to an ideal of $k[\mathbf{x}] = k[x_1, ..., x_n]$,

$$\mathbf{I}(V) = \left\{ f \in k[\mathbf{x}] \;\middle|\; f(P) = 0 \text{ for all } P \in V \right\},$$

i.e., the ideal of all polynomials vanishing on V. That is, we have a map

$$\begin{array}{ccc} \{\text{affine algebraic sets}\} & \longrightarrow & \{\text{ideals}\} \\ V & \mapsto & \mathbf{I}(V) \end{array}$$

Conversely, given an ideal $I \subseteq k[\mathbf{x}]$, we can define the affine algebraic set $\mathbf{V}(I)$. Thus, we have a map

$$\begin{array}{ccc} \{\text{ideals}\} & \longrightarrow & \{\text{affine algebraic sets}\} \\ I & \mapsto & \mathbf{V}(I) \end{array}$$

These two maps define a correspondence between ideals and affine algebraic sets. The aim of this section is to explore the nature of this correspondence.

The first thing to note is that this correspondence (more precisely, the map **V**) is not injective: *different ideals can define the same algebraic sets.* For example, $\langle x \rangle$ and $\langle x^2 \rangle$ are different ideals in $k[x]$ which have the same algebraic set $\mathbf{V}(x) = \mathbf{V}(x^2) = \{0\}$. More serious problems may arise if the field k is not algebraically closed. Foe example, consider the three polynomials 1, $1+x^2$, and $1+x^2+x^4$ in $\mathbb{R}[x]$. These generate different ideals

$$I_1 = \langle 1 \rangle = \mathbb{R}[x], \quad I_2 = \langle 1+x^2 \rangle, \quad I_3 = \langle 1+x^2+x^4 \rangle,$$

but each polynomial does not have real roots, so that the corresponding algebraic sets are all empty: $\mathbf{V}(I_1) = \mathbf{V}(I_2) = \mathbf{V}(I_3) = \emptyset$. Examples of polynomials in two variables without real roots include $1+x^2+y^2$ and $1+x^2+y^4$. These define different ideals in $\mathbb{R}[x,y]$ that correspond to the empty algebraic set.

Does the problem of having different ideals representing the empty algebraic set disappear if the field k is algebraically closed? It does in the one-variable case, when the ring is $k[x]$. To see this, recall that any ideal I in $k[x]$ may be generated by a single polynomial because $k[x]$ is a PID(principal ideal domain). So we may write $I = \langle f \rangle$ for some polynomial $f \in k[x]$. Then $\mathbf{V}(I)$ is the set of roots of f, that is, the set of $a \in k$ such that $f(a) = 0$. But since k is algebraically closed, every nonzero nonconstant polynomial in $k[x]$ has a root. Hence the only way that we could have $\mathbf{V}(I) = \emptyset$ would be for f to be a nonzero constant, i.e. $I = \langle 1 \rangle = k[x]$. This shows that $I = k[x]$ is the only ideal of $k[x]$ that represents the empty algebraic set when k is algebraically closed; we do not consider $I = k[X]$ as a *proper* ideal.

The wonderful thing is that the same property holds for $k[x_1, ..., x_n]$ where $n > 1$. In any polynomial ring, algebraic closedness of the ground-field is enough to guarantee that the only ideal which represents the empty algebraic set is the entire polynomial ring itself. This is the *Weak Nullstellensatz*, which is the basis of (and is equivalent to) one of the most celebrated mathematical results of the late nineteenth century, Hilbert's Nullstellensatz. Even today, one customarily uses the original German name Nullstellensatz: A word formed, in typical German fashion, from three simpler words: Null(=Zero), Stellen(=Places), Satz(=Theorem).

5.1. Theorem (The Weak Nullstellensatz) Suppose that k is an algebraically closed field. Let I be an ideal of $k[\mathbf{x}]$ with $\mathbf{V}(I) = \emptyset$. Then $I = k[\mathbf{x}]$.

Proof If $I \neq k[\mathbf{x}]$, then we may assume that I is a maximal ideal, for there is a maximal ideal J containing I, and $\mathbf{V}(J) \subseteq \mathbf{V}(I)$. So $L = k[\mathbf{x}]/I$ is a field, and k may be regarded as a subfield of L. Suppose we knew that $k = L$. Then for each i there is an $a_i \in k$ such that the I-residue of x_i is a_i, or $x_i - a_i \in I$. But $\langle x_1 - a_1, ..., x_n - a_n \rangle$ is a maximal ideal (§1 Exercise 5), so $I = \langle x_1 - a_1, ..., x_n - a_n \rangle$, and $\mathbf{V}(I) = \{(a_1, ..., a_n)\} \neq \emptyset$.
Thus we have reduced the problem to showing:

- If an algebraically closed field k is a subfield of a field L, and there is a ring homomorphism from $k[\mathbf{x}]$ onto L (which is the identity on k), then $k = L$.

This fact is proved in Appendix I. □

Remark In the special case when $k = \mathbb{C}$, the Weak Nullstellensatz may be thought of as the "Fundamental Theorem of Algebra for multivariable polynomials"–every system of polynomials that generates an ideal smaller than $\mathbb{C}[\mathbf{x}]$ has a common zero in $\mathbf{A}^n_{\mathbb{C}}$. But then, we immediately have the following natural questions.

- (Consistency of Nullstellensatz) Suppose that we have $f_1, ..., f_s \in k[\mathbf{x}]$. How do we know whether $1 \in \langle f_1, ..., f_s \rangle$ or not?
- (Finiteness of zeros) Suppose that we have $f_1, ..., f_s \in k[\mathbf{x}]$. If we know that $\mathbf{V}(f_1, ..., f_s) \neq \emptyset$, then how do we know whether $\mathbf{V}(f_1, ..., f_s)$ is finite or infinite? If $\mathbf{V}(f_1, ..., f_s)$ is finite, can we predict the number of solutions before finding all of the solutions explicitly?

In CH.IV. §5, we will see that there exists a *consistency algorithm* to determine whether $1 \in \langle f_1, ..., f_s \rangle$ or not. For the finiteness problem, we will show in CH.V. §2 that there exists a *finiteness algorithm* to determine whether $\mathbf{V}(f_1, ..., f_s)$ is finite or not, and to predict the number of solutions provided $\mathbf{V}(f_1, ..., f_s)$ is finite. In the exercises the reader will be invited to prove a pure algebraic criterion that will be further explored for curves in CH.VI §7.
We will not discuss how to find the solutions of a system of polynomial equations, because that is another vast area of computational

mathematics. We refer the reader to [CLO'] or any book concerning symbolic solutions of polynomial equations by using the Groebner basis method or Wu's method, both of which have been very well developed in recent years.

Inspired by the Weak Nullstellensatz, one may hope that the correspondence between affine ideals and algebraic sets is one-to-one provided one restricts to algebraically closed fields. Unfortunately, our earlier example $\mathbf{V}(x) = \mathbf{V}(x^2) = \{0\}$ works over any field. Similarly, the ideals $\langle x^2, y\rangle$ and $\langle x, y\rangle$ (and, for that matter, $\langle x^n, y^m\rangle$ where n and m are integers > 1) are different but define the same algebraic set: namely, the single point $(0,0) \in \mathbf{A}^2_k$. These examples illustrate a basic reason why different ideals can define the same algebraic set (equivalently, that the map **V**can fail to be one-to-one): namely, a power of a polynomial vanishes on the same set as the original polynomial. The Hilbert Nullstellensatz states that, over an algebraically closed field, this is the *only* reason that different ideals may define the same algebraic set: if a polynomial f vanishes at all points of some algebraic set $\mathbf{V}(I)$, then some power of f must belong to I.

5.2. Theorem (Hilbert's Nullstellensatz) Let k be an algebraically closed field. If $f_1, ..., f_s \in k[\mathbf{x}]$ are such that $f \in \mathbf{I}(\mathbf{V}(f_1, ..., f_s))$, then there exists an integer $m \geq 1$ such that

$$f^m \in \langle f_1, ..., f_s\rangle$$

and conversely.

Proof Given a polynomial f which vanishes at every common zero of the polynomials $f_1, ..., f_s$, we must show that there exists an integer $m \geq 1$ and polynomials $A_1, ..., A_s$ such that

$$f^m = \sum_{i=1}^{s} A_i f_i.$$

The most direct proof is based on what looks as an ingenious trick. Consider the ideal

$$\tilde{I} = \langle f_1, ..., f_s, 1 - yf\rangle \subset k[x_1, ..., x_n, y],$$

where $f, f_1, ..., f_s$ are as above. We claim that

$$\mathbf{V}\left(\tilde{I}\right) = \emptyset.$$

To see this, let $(a_1, ..., a_n, a_{n+1}) \in \mathbf{A}^{n+1}$. Either $(a_1, ..., a_n)$ is a common zero of $f_1, .., f_s$, or $(a_1, ..., a_n)$ is not a common zero of $f_1, ..., f_s$. In the first case $f(a_1, ..., a_n) = 0$ since f vanishes at any common zero of $f_1, ..., f_s$. Thus, the polynomial $1 - yf$ takes the value $1 - a_{n+1}f(a_1, ..., a_n) = 1 \neq 0$ at the point $(a_1, ..., a_n, a_{n+1})$. In particular, $(a_1, ..., a_n, a_{n+1}) \notin \mathbf{V}(\tilde{I})$. In the second case, for some i, $1 \leq i \leq s$, we must have $f_i(a_1, ..., a_n) \neq 0$. Thinking of f_i as a function of $n+1$ variables which does not depend on the last variable, we have $f_i(a_1, ..., a_n, a_{n+1}) \neq 0$. In particular, we again conclude that $(a_1, ..., a_n, a_{n+1}) \notin \mathbf{V}(\tilde{I})$. Since $(a_1, ..., a_n, a_{n+1}) \in \mathbf{A}^{n+1}$ was arbitrary, we conclude that $\mathbf{V}(\tilde{I}) = \emptyset$ as claimed.
Apply the Weak Nullstellensatz to conclude that $1 \in \tilde{I}$. That is,

$$(1) \qquad 1 = \sum_{i=1}^{s} p_i(x_1, ..., x_n, y) f_i + q(x_1, ..., x_n, y)(1 - yf)$$

for some polynomials $p_i, q \in k[\mathbf{x}]$. Now set $y = 1/f(x_1, ..., x_n)$. Then relation (1) above implies that

$$(2) \qquad 1 = \sum_{i=1}^{s} p_i \left(x_1, ..., x_n, \frac{1}{f}\right) f_i.$$

Multiply both sides of this equation by a power f^m, where m is chosen sufficiently large to clear all denominators. This yields

$$(3) \qquad f^m = \sum_{i=1}^{s} A_i f_i,$$

for some polynomials $A_i \in k[\mathbf{x}]$, as desired.
Conversely, if $f^m \in \langle f_1, ..., f_s\rangle \subset \mathbf{I}(\mathbf{V}(f_1, ..., f_s))$, then $(f(P))^m = 0$ for all $P \in \mathbf{V}(f_1, ..., f_s)$. But this can happen only if $f(P) = 0$, i.e., $f \in \mathbf{I}(\mathbf{V}(f_1, ..., f_s))$. This completes the proof of the theorem. □

Exercises for §5

1. In the proof of Theorem 5.2, why is it reasonable to set $y = \dfrac{1}{f(x_1, ..., x_n)}$?

2. Let I be an ideal in $k[x_1, ..., x_n]$. Suppose $V = \mathbf{V}(I)$ is a finite set. Show that $|V| \leq \dim_k (k[x_1, ..., x_n]/I)$. Hence, if $\dim_k (k[x_1, ..., x_n]/I)$ is finite, then V is a finite set. (Hint: Use §4 Exericse 3 and prove that the residues $[f_i]$ of the f_i given there are k-linear independent in $k[x_1, ..., x_n]/I$.

§6. Radical Ideals and the Nullstellensatz

In this section we explore further the relation between ideals in $k[\mathbf{x}] = k[x_1, ..., x_n]$ and algebraic sets in $\mathbf{A}^n = \mathbf{A}_k^n$. From Theorem 5.2 we first derive the following fact.

6.1. Lemma Let V be an affine algebraic set. If $f^m \in \mathbf{I}(V)$, then $f \in \mathbf{I}(V)$.

□

Thus, an ideal consisting of *all* polynomials which vanish on an algebraic set V has the property that if some power of a polynomial belongs to the ideal , then the polynomial itself must belong to the ideal. This inspires the following definition.

6.2. Definition An ideal I is *radical* if $f^m \in I$ for any integer $m \geq 1$ implies that $f \in I$.

Now Lemma 6.1 may be rephrased in terms of radical ideals as follows.

6.3. Corollary $\mathbf{I}(V)$ is a radical ideal.

□

The foregoing suggests that there is a one-to-one correspondence between affine algebraic aets and radical ideals. To clarify this we first introduce the operation of taking the radical of an ideal.

6.4. Definition Let $I \subseteq k[\mathbf{x}]$ be an ideal. The *radiccal* of I, denoted

$\sqrt{I}$, is the set

$$\left\{f \in k[\mathbf{x}] \;\middle|\; f^m \in I \text{ for some integer } m \geq 1\right\}.$$

Note that we always have $I \subseteq \sqrt{I}$. But generally the equality may not hold. Consider, for example, the ideal $J = \langle x^2, y^3 \rangle \subset k[x, y]$. Although x nor y belongs to J, it is clear that $x \in \sqrt{J}$ and $y \in \sqrt{J}$. Note that $(x \cdot y)^2 = x^2y^2 \in J$ since $x^2 \in J$; thus, $x \cdot y \in \sqrt{J}$. It is less obvious that $x + y \in \sqrt{J}$. To see this, observe that

$$(x+y)^4 = x^4 + 4x^3y + 6x^2y^2 + 4xy^3 + y^4 \in J$$

because $x^4, 4x^3y, 6x^2y^2 \in J$ (they are all multiples of x^2) and $4xy^3, y^4 \in J$ (because they are multiples of y^3). Thus, $x + y \in \sqrt{J}$. However, neither xy nor $x + y$ is in J.

6.5. Lemma If I is an ideal in $k[\mathbf{x}]$, then $\sqrt{I}$ is an ideal of $k[\mathbf{x}]$ containing I. I is radical if and only if $I = \sqrt{I}$; in particular, $\sqrt{I}$ is a radical ideal.

Proof Exercise. □

We are now ready to state the ideal-theoretic form of the Nullstellensatz.

6.6. Theorem (The Nullstellensatz) Let k be an algebraically closed field. If I is an ideal in $k[\mathbf{x}]$, then

$$\mathbf{I}(\mathbf{V}(I)) = \sqrt{I}.$$

□

The most important consequence of the Nullstellensatz is that it allows us to set up a "dictionary" between geometry and algebra. The basis of the dictionary is contained in the following theorem.

6.7. Theorem (The ideal-algebraic set correspondence) Let k be an arbitrary field.

(i) The maps

$$\{\text{affine algebraic sets}\} \xrightarrow{\mathbf{I}} \{\text{ideals}\}$$

and

$$\{\text{ideals}\} \xrightarrow{\mathbf{V}} \{\text{affine algebraic sets}\}$$

are inclusion-reversing, i.e. if $I_1 \subseteq I_2$ are ideals, then $\mathbf{V}(I_1) \supseteq \mathbf{V}(I_2)$ and, similarly, if $V_1 \subseteq V_2$ are algebraic sets, then $\mathbf{I}(V_1) \supseteq \mathbf{I}(V_1)$. Furthermore, for any algebraic set V, we have

$$\mathbf{V}(\mathbf{I}(V)) = V.$$

So that I is always injective.
(ii) If k is algebraically closed, and if we restrict to radical ideals, then the maps

$$\{\text{affine algebraic sets}\} \xrightarrow{\mathbf{I}} \{\text{radical ideals}\}$$

and

$$\{\text{radical ideals}\} \xrightarrow{\mathbf{V}} \{\text{affine algebraic sets}\}$$

are inclusion-reversing bijections which are inverses of each other.

□

The power of this theorem is that a question about algebraic sets may be rephrased as a question about radical ideals (and conversely), provided that we are working over an algebraically closed field.
In view of the Nullstellensatz and the importance it assigns to radical ideals, one may try to compute generators for the radical if one knows a set of generators of the original ideal. In fact, there are three pertinent questions to ask about an ideal $I = \langle f_1, ..., f_s \rangle$:

- (Radical generators) Is there an algorithm which produces a set $\{g_1, ..., g_m\}$ of polynomials such that $\sqrt{I} = \langle g_1, ..., g_m \rangle$?
- (Radical ideal) Is there an algorithm to decide whether I is radical?
- (Radical membership) Given $f \in k[\mathbf{x}]$ is there an algorithm that will allow us to check whether $f \in \sqrt{I}$?

Before discussing some special cases for the first two questions, let us quote some comments from [CLO'] concerning the above questions.

The existence of these algorithms follows from work of Hermann (1926) (see also Mines, Richman, and Ruitenberg (1988) and Seidenberg (1974, 1984) for more modern expositions). Unfortunately, the algorithms given in these papers for the first two questions are not very practical and would not be suitable for calculating on a computer. However, recent work by Gianni, Trager, and Zacharias (1988) and Eisenbud, Huneke, and Vasconcelos (1990) have led to algorithms implemented in the computer algebra system SCRATCH-PAD and the computer algebra system *Macaulay*, respectively, for finding the radical of an ideal.
The algorithms mentioned above are rather sophisticated and involve concepts beyond the scope of this text, we refer to [CLO'] for the related references. Nevertheless, in CH.V. §4, we will quote (from [CLO']) the *radical membership algorithm*.

There is one case where we can compute the radical of an ideal, i.e., when we are dealing with a principal ideal $I = \langle f \rangle$. If we have f expressed as a product of irreducible polynomials, then it is easy to write down an explicit expression for the radical of the principal ideal generated by f.

6.8. Proposition Let $f \in k[\mathbf{x}]$ and $I = \langle f \rangle$ the principal ideal generated by f. If $f = f_1^{n_1} f_2^{n_2} \cdots f_r^{n_r}$ is the factorization of f into a product of distinct irreducible polynomials, then

$$\sqrt{I} = \sqrt{\langle f \rangle} = \langle f_1 f_2 \cdots f_r \rangle.$$

Proof We first show that $f_1 f_2 \cdots f_r \in \sqrt{I}$. Let N be an integer strictly greater than $\max\{n_1, ..., n_r\}$. Then

$$(f_1 f_2 \cdots f_r)^N = f_1^{N-n_1} f_2^{N-n_2} \cdots f_r^{N-n_r} f$$

is a polynomial multiple of f. This shows that $(f_1 f_2 \cdots f_r)^N \in I$, which implies that $f_1 f_2 \cdots f_r \in \sqrt{I}$. Thus, $\langle f_1 f_2 \cdots f_r \rangle \subseteq \sqrt{I}$. Conversely, suppose that $g \in \sqrt{I}$. Then there exists a positive integer M such that $g^M \in I$. This means that $g^M = h \cdot f$ for some polynomial h. Now suppose that $g = g_1^{m_1} g_2^{m_2} \cdots g_s^{m_s}$ is the factorization of g into a product of distinct irreducible polynomials. Then $g^M = g_1^{m_1 M} g_2^{m_2 M} \cdots g_s^{m_s M}$ is the factorization of g^M into a product of distinct irreducible polynomials. Thus,

$$g_1^{m_1 M} g_2^{m_2 M} \cdots g_s^{m_s M} = h \cdot f_1^{n_1} f_2^{n_2} \cdots f_r^{n_r}.$$

But, by unique factorization, the irreducible polynomials on both sides of the above equation must be the same (up to multiplication by constants). Since the $f_1, ..., f_r$ are irreducible, each f_i, $1 \leq i \leq r$ must be equal to a constant multiple of some g_j. This implies that g is a polynomial multiple of $f_1 f_2 \cdots f_r$. Therefore $g \in \langle f_1 f_2 \cdots f_r \rangle$, and this finishes the proof. □

In the general case, it is not easy to get the factorization of f into irreducible polynomials. However, there exists an algorithm to compute the generator of $\sqrt{\langle f \rangle}$ from f *without* factoring f first. We refer the reader to [CLO'] for the details.

Finally, let us point out that the intersection of two ideals corresponds to the same algebraic set as the product (Theorem 2.4), but usually the intersection is much more difficult to compute than the product. However, the intersection behaves much better with respect to the operation of taking radicals: The product of radical ideals need not be a radical ideal (consider $I \cdot J$ where $I = J$), but the intersection of radical ideals is always a radical ideal. The latter fact follows from application of the following proposition to radical ideals.

6.9. Proposition If I, J are any ideals, then

$$\sqrt{I \cap J} = \sqrt{I} \cap \sqrt{J}.$$

Proof Exercise. □

Exercises for §6

1. Show that an ideal I is radical if and only if $I = \sqrt{I}$.
2. Prove Lemma 6.5 and Proposition 6.9.
3. If I is a graded ideal of $k[x_1, ..., x_n]$ with respect to the natural gradation on $k[x_1, ..., x_n]$ (see §1), then $\sqrt{I}$ is a graded ideal.
4. Show that the intersection of any collection of prime ideals is radical.

5. Let I, J be ideals of $k[x_1, ..., x_n]$. Suppose $I \subset \sqrt{J}$. Show that $I^n \subset J$ for some $n > 0$. (Hint: As an ideal, I is finitely generated.)
6. Let R be a ring. Two ideals I, J of R is said to be *comaximal* if $I + J = R$, i.e., if $1 = a + b$, $a \in I$, $b \in J$. For example, two distinct maximal ideals are comaximal.
   a. Show that if I and J are comaximal, then $I \cdot J = I \cap J$. (Hint: $I \cap J = (I \cap J)R = (I \cap J)(I + J)$.)
   b. Show that $I + J^2 = R$. Show that I^m and J^n are comaximal for all m, n.
   c. Suppose $I_1, ..., I_N$ are ideals in R, and I_i and $J_i = \cap_{j \neq i} I_j$ are comaximal for all i. Show that $I_1^n \cap \cdots \cap I_N^n = (I_1 \cdots I_N)^n = (I_1 \cap \cdots \cap I_N)^n$ for all n.
7. Let R be a ring and I an ideal of R. Show that $\sqrt{I} = \cap p$, where p ranges over all prime ideals of R containing I.

CHAPTER II
Polynomial and Rational Functions

One of the unifying themes of modern geometry is that in order to understand any class of geometric objects, one should also study *functions* on those objects and *mappings* between those objects, especially the mappings which preserve some geometric property of interest. For instance, in topology one considers the class of continuous functions on topological spaces; in differential geometry one considers the class of infinitely differentiable functions (i.e., smooth functions) on manifolds (smooth mappings between manifolds); in (real) complex analytic geometry one considers the class of (real) complex analytic functions.

- **Question** To study the geometry of an algebraic set, for example, the singularity of points, what class of functions can we use?

Surprisingly simple, algebraic geometry only uses polynomial functions (plus rational functions which are quotients of polynomial functions) on algebraic sets and polynomial mappings (plus rational mappings) between algebraic sets. If we take an open set U in the complex metric space $\mathbb{C}^n$ and write

$C^o(U)$	for the ring of all continuous functions,
$C^\infty(U)$	for the ring of all smooth functions,
$C^\omega(U)$	for the ring of all analytic functions,
$\mathbb{C}[\mathbf{x}]$	for the ring of all polynomial functions,

where $\mathbb{C}[\mathbf{x}] = \mathbb{C}[x_1, ..., x_n]$, then we have the following strict inclusion relations

$$\mathbb{R}[x] \subset C^\omega(U) \subset C^\infty(U) \subset C^o(U).$$

In this chapter, we introduce the polynomial mappings between algebraic sets and start the study of polynomial functions and rational functions on algebraic sets. The results of this investigation will constitute another chapter of the "algebra-geometry dictionary" that we started in Chapter I. The algebraic properties of polynomial and rational functions on an algebraic set yield insight into the geometric properties of the algebraic set itself. So this chapter motivates the "local" study of algebraic sets in CH.VI.

§1. The Zariski Topology and Irreducible Algebraic Sets

To see why polynomial functions and rational mappings are the "right" choice for the geometric study of algebraic sets, we first endow every algebraic set with an appropriate topological structure, compatible with the natural structure of an algebraic set (e.g., the notion of irreducible subset, the closure of a subset, and the notion of dense subset, etc. in this topological space, should translate to natural geometric properties of the algebraic set considered).

We start by analyzing the structure of an algebraic set. Let k be a field, $k[\mathbf{x}] = k[x_1, ..., x_n]$, and let $\mathbf{A}^n = \mathbf{A}^n_k$ be the affine n-space over k. In CH.I Proposition 3.3 we have already established that the union of two algebraic sets is an algebraic set. For example, consider $\mathbf{V}(xz, yz)$, which is the union of a line and a plane. Intuitively, it is natural to think of the line and the plane as "more fundamental" than $\mathbf{V}(xz, yz)$. Intuition also tells us that a line or a plane is "irreducible" or "indecomposable" in some sense: they do not seem to be a union of finitely many simpler algebraic sets in any obvious way.

1.1. Definition The *Zariski topology* on $\mathbf{A}^n$ is defined by taking the open subsets to be the complements of affine algebraic sets. For any subset $X \subset \mathbf{A}^n$, the Zariski topology on X is the induced topology. (Note that the Zariski topology does not intrinsically depend on the field k, this is an important asset of this definition.)

Example (i) Let us consider the Zariski topology on the affine line $\mathbf{A}^1_k$. Every ideal in $k[x]$ is principle, so every algebraic set is the zero locus of a single polynomial. Thus the algebraic sets in $\mathbf{A}^1_k$ are just the finite subsets (including the empty set) and the whole space (corresponding to $f = 0$). Thus the open sets are the empty set and the complements of finite subsets. Observe that this topology is not a Hausdorff topology.

(ii) In general, the Zariski topology on $\mathbf{A}^n$ $(n \geq 1)$ is not the product topology of the Zariski topology on $\mathbf{A}^1_k$. For example, if we identify $\mathbf{A}^1 \times \mathbf{A}^1$ with A^2 and consider any curve $C = \mathbf{V}(y - f(x))$ in $\mathbf{A}^2$, where $f(x)$ is a polynomial in $k[x]$ of degree ≥ 1, then from (i) above it is easy to see that the open subset set $U = \mathbf{A}^2 - C$ in $\mathbf{A}^2$ cannot be a product of two open subsets of $\mathbf{A}^1$.

1.2. Definition A nonempty subset Y of a topological space X is *irreducible* if it cannot be expressed as the union $Y = Y_1 \cup Y_2$ of two proper subsets, each one of which is closed in Y(where Y has the induced topology). The *empty set is not considered to be irreducible.*

Example (iii) If k is infinite, then $\mathbf{A}^1_k$ is irreducible, because its only proper closed subsets are finite, yet it is infinite.

(iv) Similarly, $\mathbf{A}^n$ is irreducible provided k is infinite.

(v) $\mathbf{V}(xz, xy)$ is not irreducible.

Let X be a topological space, Y a subset of X. Recall that the *closure* of Y in X, denoted $\overline{Y}$, is the smallest closed subset in X containing Y, i.e., $\overline{Y} = \cap W$ where W runs over all closed subsets of X containing Y. Also recall that a subset Z of X is said to be *dense* in X if $Z \cap U \neq \emptyset$ for every nonempty open subset $U \subset X$, or equivalently, if $\overline{Z} = X$.

The following proposition contains two basic properties of irreducible subsets in a topological space. The first result will play a *key* role in later discussion (see Example (vi), Proposition 1.6, Proposition 1.7 below, later §6 and CH.VI).

1.3. Proposition (i) Any nonempty open subset of an irreducible space is dense and irreducible.
(ii) If Y is an irreducible subset of the topological space X, then its closure $\overline{Y}$ in X is also irreducible.

Proof (i) If U is a nonempty open subset in the irreducible space X and $V \neq \emptyset$ is another open subset of X then $U \cap V = \emptyset$ leads to $X = (X - U) \cup (X - V)$, a contradiction. Hence U is dense. If $U = U_1 \cup U_2$ for closed U_1, U_2 in U, then there are proper closed V_1, V_2 in X such that $V_1 \cap U = U_1$, $V_2 \cap U = U_2$. But then $X = (V_1 \cup V_2) \cup (X - U)$, both $V_1 \cup V_2$ and $X - U$ are proper closed, again a contradiction.
(ii) Easy. □

Now, it is natural to ask

- How do we know whether an algebraic set is irreducible or not?

If the definition of irreducibility is to correspond to our geometric intuition, it is clear that a point, a line, and a plane ought to be irreducible. The twisted cubic $\mathbf{V}(y - x^2, z - x^3)$ in $\mathbb{R}^3$ appears to be irreducible. But how do we prove this? The key is to treat the problem algebraically: If we can characterize ideals that correspond to irreducible algebraic sets, then we may also establish whether an algebraic set is irreducible.

1.4. Theorem Let $V \subset \mathbf{A}^n$ be an affine algebraic set. Then V is irreducible if and only if $\mathbf{I}(V)$ is a prime ideal.

Proof First, assume that V is irreducible and let $fg \in \mathbf{I}(V)$. Set $V_1 = V \cap \mathbf{V}(f)$ and $V_2 = V \cap \mathbf{V}(g)$. Then $fg \in \mathbf{I}(V)$ implies that $V = V_1 \cup V_2$. Since V is irreducible, we have either $V = V_1$ or $V = V_2$.

Say the former holds, so that $V = V_1 = V \cap \mathbf{V}(f)$. This implies that f vanishes on V, so that $f \in \mathbf{I}(V)$. Thus, $\mathbf{I}(V)$ is prime.

Next, assume that $\mathbf{I}(V)$ is prime and let $V = V_1 \cup V_2$. Suppose that $V \neq V_1$. We claim that $\mathbf{I}(V) = \mathbf{I}(V_2)$. To prove this, note that $\mathbf{I}(V) \subset \mathbf{I}(V_2)$ since $V_2 \subset V$. For the opposite inclusion, first note that $\mathbf{I}(V)$ is properly contained in $\mathbf{I}(V_1)$ since V_1 is properly contained in V. Thus, we can pick $f \in \mathbf{I}(V_1) - \mathbf{I}(V)$. Now take any $g \in \mathbf{I}(V_2)$. Since $V = V_1 \cup V_2$, it follows that fg vanishes on V, and, hence, $fg \in \mathbf{I}(V)$. But $\mathbf{I}(V)$ is prime, so that f or g lies in $\mathbf{I}(V)$. We know that $f \notin \mathbf{I}(V)$ and, thus, $g \in \mathbf{I}(V)$. This proves $\mathbf{I}(V) = \mathbf{I}(V_2)$, whence $V = V_2$ because $\mathbf{I}$ is one-to-one. Thus V is an irreducible algebraic set. □

Of course, every prime ideal is radical. Then, using the correspondence between radical ideals and algebraic sets, we obtain the following corollary of Theorem 1.4.

1.5. Corollary When k is algebraically closed, the functions $\mathbf{I}$ and $\mathbf{V}$ (see CH.I §6) induce a bijection between irreducible algebraic sets in $\mathbf{A}^n$ and prime ideals in $k[\mathbf{x}]$.

□

Example (vi) As an example of how to use Theorem 1.4., let us prove that the ideal $\mathbf{I}(V)$ of the twisted cubic $V = \mathbf{V}(y-x^2, z-x^3) \subset \mathbb{R}^3$ is prime. Suppose that $fg \in \mathbf{I}(V)$. We have to show that $f \in \mathbf{I}(V)$ or $g \in \mathbf{I}(V)$. Since the curve is parametrized by (t, t^2, t^3), i.e., for any $P = (a, b, c) \in V$ there is some $t \in \mathbf{A}^1 = \mathbb{R}$ such that $a = t$, $b = t^2$ and $c = t^3$, and conversely any $t \in \mathbf{A}^1 = \mathbb{R}$ determines a point of V, we have $f(t, t^2, t^3)g(t, t^2, t^3) = 0$ for all $t \in \mathbf{A}^1$. Putting $f'(t) = f(t, t^2, t^3)$, $g'(t) = g(t, t^2, t^3)$, $f'(t)$ and $g'(t)$ may be viewed as polynomials in one variable t. Now let $V_{f'} = \{t \in \mathbf{A}^1 \mid f'(t) = f(t, t^2, t^3) = 0\}$. If $V_{f'} = \mathbf{A}^1$ then f vanishes on V and hence $f \in \mathbf{I}(V)$. If $V_{f'} \neq \mathbf{A}^1$, then g' vanishes on the nonempty open subset $U = \mathbf{A}^1 - V_{f'}$ of $\mathbf{A}^1$. It follows from Proposition 1.3(i) that $g'(t)$ vanishes on $\mathbf{A}^1$ since $\mathbb{R}$ is infinite and $\mathbf{A}^1$ is irreducible. But this means that g vanishes on V and hence $g \in \mathbf{I}(V)$, as desired. By the theorem, the twisted cubic is an irreducible algebraic set in $\mathbb{R}^3$.

In the same way as above, one may prove that a straight line is irreducible: first parametrize it, then apply the above argument. In fact, by using the observation in CH.I §3 and the above Proposition 1.3 the above argument holds much more generally. We leave the proofs of the following two results as exercises for the consequnet §3.

1.6. Proposition If k is an infinite field and $V \subset \mathbf{A}^n$ is an algebraic set defined parametrically

$$\begin{cases} x_1 = f_1(t_1, ..., t_m), \\ \vdots \\ x_n = f_n(t_1, ..., t_m), \end{cases}$$

where $f_1, ..., f_n$ are polynomials in $k[t_1, ..., t_m]$, then V is irreducible.

□

1.7. Proposition If k is an infinite field and V is an algebraic set defined by the rational parametrization

$$\begin{cases} x_1 = \dfrac{f_1(t_1, ..., t_m)}{g_1(t_1, ..., t_m)}, \\ \vdots \\ x_n = \dfrac{f_n(t_1, ..., t_m)}{g_n(t_1, ..., t_m)}, \end{cases}$$

where $f_1, ..., f_n, g_1, ..., g_n \in k[t_1, ..., t_m]$, then V is irreducible.

□

The simplest algebraic set in $\mathbf{A}^n$ is a single point $\{P = (a_1, ..., a_n)\}$, namely, $\mathbf{V}(x_1 - a_1, ..., x_n - a_n) = \{P\}$. It has been shown that $\langle x_1 - a_1, ..., x_n - a_n\rangle$ is a maximal ideal and $\mathbf{I}(\{P\}) = \langle x_1 - a_1, ..., x_n - a_n\rangle$ (CH.I.§1, Exercise 5). Hence every point $P = (a_1, ..., a_n) \in \mathbf{A}^n$ corresponds to a maximal ideal even if k is not infinite. The converse does not hold if k is not algebraically closed. For example, the ideal $\langle x^2 + 1\rangle$ is a maximal ideal in $\mathbb{R}[x]$, but it does not correspond to a point in $\mathbb{R}$. Over an algebraically closed field, it turns out that every maximal ideal corresponds to some point of $\mathbf{A}^n$.

1.8. Theorem If k is an algebraically closed field, then every maximal ideal of $k[\mathbf{x}]$ is of the form $\langle x_1 - a_1, \ldots, x_n - a_n\rangle$ for some $a_1, \ldots, a_n \in k$.

Proof Let $I \subset k[\mathbf{x}]$ be maximal. Since $I \neq k[\mathbf{x}]$, we have $\mathbf{V}(I) \neq \emptyset$ by the Weak Nullstellensatz. Hence, there is some point $P = (a_1, .., a_n) \in \mathbf{V}(I)$. Passing to ideals, we have

$$\mathbf{I}(\mathbf{V}(I)) \subseteq \mathbf{I}(\{P\}).$$

But $\mathbf{I}(\mathbf{V}(I)) = \sqrt{I}$ by the Nullstellensatz. Now I is maximal by hypothesis, hence $\sqrt{I} = I$. Thus, we can write

$$I \subseteq \mathbf{I}(\{P\}) = \langle x_1 - a_1, \ldots, x_n - a_n\rangle \neq k[\mathbf{x}].$$

It follows again from the maximality of I that $I = \langle x_1 - a_1, \ldots, x_n - a_n\rangle$. □

Note that the proof of Theorem 1.8. relies heavily on the Nullstellensatz. It is not difficult to see that it is, in fact, equivalent to the Nullstellensatz. We have the following easy corollary of Theorem 1.8.

1.9. Corollary If k is an algebraically closed field, then there is a bijection between points of $\mathbf{A}^n$ and maximal ideals of $k[\mathbf{x}]$.

□

Thus, we have extended our algebra-geometry dictionary:

- Over an algebraically closed field, every nonempty irreducible algebraic set corresponds to a prime ideal, and conversely. Every point corresponds to a maximal ideal, and conversely.

Exercises for §1

1. Show that $\mathbf{V}(y - x^2) \subset \mathbf{A}^2_{\mathbb{C}}$ is irreducible; in fact $\mathbf{I}(\mathbf{V}(y - x^2)) = \langle y - x^2\rangle$.
2. Show that $f = y^2 + x^2(x-1)^2 \in \mathbb{R}[x, y]$ is an irreducible polynomial, but that $\mathbf{V}(f)$ is reducible.
3. Show that if $f \in \mathbb{C}[x_1, \ldots, x_n]$ is irreducible, then $V(f)$ is irreducible. Also show that if $V = \mathbf{V}(g)$ is an irreducible hypersurface in $\mathbf{A}^n$, there is no irreducible algebraic set W such that $V \subset W \subset \mathbf{A}^n$, $W \neq V$, $W \neq \mathbf{A}^n$.

4. Show that any linear space of $k^n = \mathbf{A}_k^n$, where k is a field, is irreducible.
5. Show that $\mathbf{V}(y^2 - (x-1)(x-\lambda))$ is an irreducible curve in $\mathbf{A}_{\mathbb{C}}^2$, where $\lambda \in \mathbb{C}$.
6. Show that if I is any proper ideal in $\mathbb{C}[x_1, ..., x_n]$, then $\sqrt{I}$ is the intersection of all maximal ideals containing I. (Hint: Use Theorem 1.8.)

§2. Decomposition of an Algebraic Set

In the last section, we have built-up some theory concerning irreducible algebraic sets. We now try to build algebraic sets from irreducible ones. As before we put $k[\mathbf{x}] = k[x_1, ..., x_n]$.

By Hilbert's basis theorem (Theorem 8 and Exercise 6 of CH.I §1), if we pass to the ideal $\mathbf{I}(V)$ of any algebraic set V it follows immediately that $\mathbf{A}^n = \mathbf{A}_k^n$ is a *Noetherian space* in the sense that every collection of algebraic sets in $\mathbf{A}^n$ has a minimal member. This observation enables us to prove the following decomposition theorem for affine algebraic sets.

2.1. Theorem Let V be an algebraic set in $\mathbf{A}^n$. Then there are unique irreducible algebraic sets $V_1, ..., V_m$ such that $V = V_1 \cup \cdots \cup V_m$ and $V_i \not\subset V_j$ for all $i \neq j$.

Proof Let $S = \{$ algebraic sets $V \subset \mathbf{A}^n \mid V$ is not the union of a finite number of irreducible algebraic sets $\}$. We want to show that S is empty. If not, let V be a minimal member of S. Since $V \in S$, V is not irreducible, so $V = V_1 \cup V_2$, V_i are proper closed subsets of V. Then $V_i \notin S$, so $V_i = V_{i1} \cup \cdots \cup V_{im_i}$ with V_{ij} irreducible. But then $V = \cup_{i,j} V_{ij}$, a contradiction.

So any algebraic set V may be written as $V = V_1 \cup \cdots \cup V_m$, where each V_i is irreducible. To get the second condition, simply throw away any V_i such that $V_i \subset V_j$ for $i \neq j$. To show uniqueness, let $V = W_1 \cup \cdots \cup W_m$ be another such decomposition. Then $V_i = \cup_j (W_j \cap V_i)$, so $V_i \subset W_{j(i)}$ for some $j(i)$. Similarly, $W_{j(i)} \subset V_k$ for some k. But $V_i \subset V_k$ implies $i = k$, so $V_i = W_{j(i)}$. Likewise each W_j

is equal to some $V_{i(j)}$. □

The V_i appearng in the theorem are called the *irreducible components* of V; $V = V_1 \cup \cdots \cup V_m$ is the *minimal decomposition* (or sometimes, the *irredundant union*) of V into irreducible components.

Example (i) As a simple example of Theorem 2.1, consider the algebraic set $\mathbf{V}(xz, yz)$ which is a union of a line (the z-axis) and a plane (the xy-plane), both of which are irreducible.

(ii) Let $f \in k[\mathbf{x}]$ and $f = f_1^{n_1} f_2^{n_2} \cdots f_r^{n_r}$ the factorization of f into a product of distinct irreducible polynomials. Then it is not hard to see that $\mathbf{V}(f) = \mathbf{V}(f_1) \cup \cdots \cup \mathbf{V}(f_r)$ is the decomposition of $\mathbf{V}(f)$ into irreducible components. Moreover, if k is algebraically closed, then $\mathbf{I}(\mathbf{V}(f)) = \langle f_1 f_2 \cdots f_r \rangle$ (CH.I Proposition 6.8).

We remark that the uniqueness part of Theorem 2.1 is false if one does not insist that the union be finite. (A plane π is the union of all the points on it. It is also the union of some line in π and all the points not on the line–these are infinitely many lines in π with which one could start.) This should remind the reader of the fact that although the proof of Theorem 2.1 is easy, it is far from vacuous: One makes subtle use of finiteness (which follows from the Hilbert Basis Theorem).

Theorem 2.1 can also be expressed purely algebraically using the one-to-one correspondence between radical ideals and algebraic sets.

2.2. Theorem If k is algebraically closed, then every radical ideal I in $k[\mathbf{x}]$ may be written uniquely as a finite intersection of prime ideals, $I = P_1 \cap \cdots \cap P_r$, where $P_i \not\subset P_j$ for $i \neq j$. (As in the case of algebraic sets, we often call such a presentation of a radical ideal a *minimal decomposition* or an *irredundant intersection*).

Proof This follows immediately from Theorem 2.1 and the ideal-algebraic set correspondence. □

Remark (i) We should mention that Theorem 2.2. holds for any field k, although the proof in the general case is different (see any commutative algebra textbook).

(ii) Based on what we have seen in the last section and the foregoing dicussion, the following questions arise naturally

- (Primality) Is there an algorithm for deciding if a given ideal is prime?
- (Irreducibility) Is there an algorithm for deciding if a given affine algebraic set is irreducible?
- (Decomposition) Is there an algorithm for finding the minimal decomposition of a given algebraic set or radical ideal?

The answer to all three questions above is yes. We suggest the reader go back to the remark given before Proposition 6.8 in CH.I §6 for the references where the algorithmic solution to the questions raised above can also be found.

We finish this section by taking a closer look at the affine plane $\mathbf{A}^2$, and find all its algebraic sets. By theorem 2.1 it is enough to find the irreducible algebraic sets.

2.3. Proposition Let f and g be polynomials in $k[x,y]$ with no common factors, and let $V = \mathbf{V}(f,g)$ be the algebraic set defined by f and g. Then $V = \mathbf{V}(f) \cap \mathbf{V}(g)$ is a finite set of points.

Proof f and g have no common factors in $k[x][y]$, so they also have no common factors in $k(x)[y]$ (why?). Since $k(x)[y]$ is a PID (principal ideal domain), $(f,g) = 1$ in $k(x)[y]$, so $pf + qg = 1$ for some $p,q \in k(x)[y]$. There is a nonzero $d \in k[x]$ such that $dp = a$, $dq = b \in k[x,y]$. Therefore $af + bg = d$. If $P = (u,v) \in V$, then $d(P) = d(u) = 0$. But d has only a finite number of zeros. This shows that only a finite number of x-coordinates appear among the points of V. Since the same reasoning applies to the y-coordinates, there can be only a finite number of points. □

2.4. Corollary If f is an irreducible polynomial in $k[x,y]$, and if $\mathbf{V}(f)$ is infinite, then $\mathbf{I}(\mathbf{V}(f)) = \langle f \rangle$, and $\mathbf{V}(f)$ is irreducible.

Proof If $g \in \mathbf{I}(\mathbf{V}(f))$, $\mathbf{V}(f,g)$ is infinite, so by the proposition f divides g, i.e., $g \in \langle f \rangle$. The fact that $\mathbf{V}(f)$ is irreducible then follows from Theorem 1.4. □

2.5. Corollary Suppose k is infinite. The irreducible algebraic subsets of $\mathbf{A}^2$ are: $\mathbf{A}^2$, $\emptyset$, points, and irreducible plane curves $\mathbf{V}(f)$, where f is an irreducible polynomial and $\mathbf{V}(f)$ is infinite.

Proof Let V be an irreducible algebraic set in $\mathbf{A}^2$. If V is finite or $\mathbf{I}(V) = (0)$, V is of the required type. Otherwise, $\mathbf{I}(V)$ contains a nonconstant polynomial f; since $\mathbf{I}(V)$ is prime, some irreducible factor of f belongs to $\mathbf{I}(V)$, so we may assume f is irreducible. Then we claim $\mathbf{I}(V) = \langle f \rangle$; for if $g \in \mathbf{I}(V)$, $g \notin \langle f \rangle$, then $V \subset \mathbf{V}(f, g)$ is finite, contradicting the infiniteness of V. □

Exercises for §2

1. Let V, W be algebraic sets in $\mathbf{A}^n$, $V \subset W$. Show that each irreducible component of V is contained in some irreducible component of W.
2. Find the irreducible components of $\mathbf{V}(y^2 - xy - x^2y + x^3)$ in $A^2_{\mathbb{R}}$, and also in $A^2_{\mathbb{C}}$.
3. Let $I = \langle xz - y^2, z^3 - x^5 \rangle$. Express $\mathbf{V}(I)$ as a finite union of irreducible algebraic sets. (Hint; Use the parametrization (t^3, t^4, t^5) and $(t^3, -t^4, t^5)$.)
4. Let $f \in \mathbb{C}[x_1, ..., x_n]$ and let $f = f_1^{n_1} \cdots f_r^{n_r}$ be the decomposition of f into irreducible factors. Show that $\mathbf{V}(f) = \mathbf{V}(f_1) \cup \cdots \cup \mathbf{V}(f_r)$ is the decomposition of $\mathbf{V}(f)$ into irreducible components and $\mathbf{I}(\mathbf{V}(f)) = \langle f_1 f_2 \cdots f_r \rangle$. (See CH.I Proposition 6.8.)

§3. Polynomial Mappings and Polynomial Functions

In this section we introduce a most important class of mappings, the polynomial mappings between affine algebraic sets, and start the study of polynomial functions.

Since we are working over an *arbitrary* field k, if we look at the definition of an affine space $\mathbf{A}^n = \mathbf{A}^n_k$ and recall from CH.I §1 the fact that we started algebraic geometry by considering every polynomial f over k as a function $\mathbf{A}^n \to k$, the first natural choice of finding a mapping ϕ from $\mathbf{A}^m$ to $\mathbf{A}^n$ is to take n polynomials $f_1, ..., f_n \in k[y_1, ..., y_m]$ and put

$$\phi(P) = (f_1(P), ..., f_n(P)) \in \mathbf{A}^n, \quad \text{for } P \in \mathbf{A}^m.$$

This leads to the more general

3.1. Definition Let $V \subset \mathbf{A}^m$, $W \subset \mathbf{A}^n$ be affine algebraic sets. A function ϕ: $V \to W$ is said to be a *polynomial mapping* (or *regular mapping*) if there exist polynomials $f_1, ..., f_n \in k[x_1, ..., x_m]$ such that

$$\phi(a_1, ..., a_m) = (f_1(a_1, ..., a_m), ..., f_n(a_1, ..., a_m))$$

for all $(a_1, ..., a_m) \in V$. We say that the n-tuple of polynomials $(f_1, ..., f_n)$ *represents* ϕ, and call $f_1, ..., f_n$ the *coordinate functions* of ϕ.

To say that ϕ is a polynomial mapping from $V \subset \mathbf{A}^m$ to $W \subset \mathbf{A}^n$ represented by $(f_1, ..., f_n)$ means that $(f_1(a_1, ..., a_m), ..., f_n(a_1, ..., a_m))$ must satisfy the defining equations of W for all $(a_1, ..., a_m) \in V$.

Example (i) Any linear mapping ϕ from k^m to k^n defines a polynomial mapping $\overline{\phi}$ from $\mathbf{A}^m$ to $\mathbf{A}^n$: Let $\{e_i\}_{i=1}^m$ be a basis of $\mathbf{A}^m$ and $\{\eta_j\}_{j=1}^n$ a basis of $\mathbf{A}^n$. If $A = (a_{ij})$ is the matrix of ϕ, then $\phi(e_i) = \sum_{j=1}^n a_{ij}\eta_j$ for $i = 1, ..., m$. Put $f_j = \sum_{i=1}^m a_{ij}x_i$, $j = 1, ..., n$, then for every $(c_1, ..., c_m) \in \mathbf{A}^m$ we have $\overline{\phi}(c_1, ..., c_m) = (f_1(c_1, ..., c_m), ..., f_n(c_1, ..., c_m))$.

(ii) Let $V = \mathbf{V}(y - x^2, z - x^3)$ be the twisted cubic curve in $\mathbb{R}^3$ and $W = \mathbf{V}(y^3 - z^2) \subset \mathbb{R}^2$. Then the projection π_1: $\mathbb{R}^3 \to \mathbb{R}^2$ represented by (y, z) yields a polynomial mapping π_1: $V \to W$. This is true because every point in $\pi_1(V) = \{(x^2, x^3),\ x \in \mathbb{R}\}$ satisfies the defining equation of W.

(iii) Indeed, any polynomial parametric representation of an algebraic set (see CH.I §3) determines a polynomial mapping.

Of particular interest is the case $W = k$, where ϕ simply becomes a *scalar polynomial function* defined on the algebraic set V. One reason to consider polynomial functions from V to k is that a general polynomial mapping ϕ: $V \to \mathbf{A}^n$ is constructed by using any n

polynomial functions ϕ_i: $V \to k$ for the components. Hence, if we understand functions ϕ: $V \to k$, we understand how to construct all mappings ϕ: $V \to \mathbf{A}^n$ as well.

To begin the study of polynomial functions, note that, for $V \subset \mathbf{A}^n$, Definition 3.1. states that a mapping ϕ: $V \to k$ is a polynomial function if *there exists* a polynomial $f \in k[x_1, ..., x_m]$ representing ϕ. In fact, we usually specify a polynomial function by giving an explicit polynomial representative. Thus, finding a representative is not actually the essential issue. What we will see next, however, is that the cases where a representative is uniquely determined are very rare. For example, consider the algebraic set $V = \mathbf{V}(y - x^2) \subset I\!R^2$. The polynomial $f = x^3 + y^3$ represents a polynomial function from V to $I\!R$. However, $g = x^3 + y^3 + (y - x^2)$, $h = x^3 + y^3 + (x^4y - x^6)$, and $F = x^3 + y^3 + A(x, y)(y - x^2)$ for any $A(x, y) \in I\!R[x, y]$ define *the same polynomial function* on V. Indeed, since $\mathbf{I}(V)$ is the set of polynomials which are zero at every point of V, adding any element of $\mathbf{I}(V)$ to f does not change the values of the polynomial at the points of V. The general case is summed up in the following proposition.

3.2. Proposition Let $V \subset \mathbf{A}^m$ be an affine algebraic set. Then
(i) f and $g \in k[x_1, ..., x_m]$ represent the same polynomial function on V if and only if $f - g \in \mathbf{I}(V)$.
(ii) $(f_1, ..., f_n)$ and $(g_1, ..., g_n)$ represent the same polynomial mapping from V to $\mathbf{A}^n$ if and only if $f_i - g_i \in \mathbf{I}(V)$ for each i, $1 \le i \le m$.

Proof Exercise. □

Thus, the correspondence between polynomials in $k[x_1, ..., x_m]$ and polynomial functions is one-to-one only in the case that $\mathbf{I}(V) = \{0\}$. It is not difficult to prove that $\mathbf{I}(V) = \{0\}$ if and only if k is infinite and $V = \mathbf{A}^m$. (What about the case where k is not infinite?)

3.3. Definition We denote by $k[V]$ the collection of polynomial functions ϕ: $V \to k$.

Indeed, since k is a field, we may define a sum and product operation for any pair of functions ϕ, ψ: $V \to k$ by adding and multiplying pointwise. For each $P \in V$,

$$\begin{aligned}(\phi + \psi)(P) &= \phi(P) + \psi(P) \\ (\phi \cdot \psi)(P) &= \phi(P) \cdot \psi(P).\end{aligned}$$

Furthermore, if we pick specific representatives $f, g \in k[x_1, ..., x_m]$ for ϕ, ψ, respectively, then, by definition, the polynomial sum $f + g$ represents $\phi + \psi$ and the polynomial product $f \cdot g$ represents $\phi \cdot \psi$, It follows that $\phi + \psi$ and $\phi \cdot \psi$ are polynomial functions on V. Thus, $k[V]$ has sum and product operations constructed from the sum and product operations in $k[x_1, ..., x_m]$. In this way $k[V]$ becomes a commutative ring.

Now we are ready to start investigating what $k[V]$ can tell us about the geometric properties of an algebraic set. First, recall that an algebraic set is *reducible* if it can be written as the union of two nonempty proper subalgebraic sets: $V = V_1 \cup V_2$, where $V_1 \neq V$ and $V_2 \neq V$. For example, the algebraic set $V = \mathbf{V}(x^3 + xy^2 - xz, yx^2 + y^3 - yz)$ in $\mathbf{A}^3$ is reducible because we may decompose V as $V = \mathbf{V}(x^2 + y^2 - z) \cup \mathbf{V}(x, y)$. We would like demonstrate that geometric properties such as reducibility can be derived from algebraic properties of $k[V]$. To see this, let

$$(*) \qquad f = x^2 + y^2 - z, \qquad g = 2x^2 - 3y^4 z \in k[x, y, z]$$

and let ϕ, ψ be the corresponding elements of $k[V]$.
Note that neither ϕ nor ψ is identically zero on V. For example, at $(0, 0, 5) \in V$, $\phi(0, 0, 5) = f(0, 0, 5) = -5 \neq 0$. Similarly, at $(1, 1, 2) \in V$, $\psi(1, 1, 2) = g(1, 1, 2) = -4 \neq 0$. However, the product function $\phi \cdot \psi$ is zero at every point of V. The reason is that

$$\begin{aligned}f \cdot g &= (x^2 + y^2 - z)(2x^2 - 3y^4 z) \\ &= 2x(x^3 + xy^2 - xz) - 3y^3 z(x^2 y + y^3 - yz) \\ &\in \langle x^3 + xy^2 - xz, x^2 y + y^3 - yz \rangle.\end{aligned}$$

Hence, $f \cdot g \in \mathbf{I}(V)$, so the corresponding polynomial function $\phi \cdot \psi$ on V is identically zero. This fact tells us at least two things:

- $k[V]$ is not a commutative domain.
- The existence of $\phi \neq 0$ and $\psi \neq 0$ in $k[V]$ such that $\phi \cdot \psi = 0$ is a direct consequence of the reduciblity of V: f in $(*)$ is zero

on $V_1 = \mathbf{V}(x^2+y^2-z)$, but not on $V_2 = \mathbf{V}(x, y)$, and similarly g is zero on V_2, but not on V_1. This is why $f \cdot g = 0$ at every point of $V = V_1 \cup V_2$.

The general case of the relation showing by the example given above can be stated as follows.

3.4. Proposition Let $V \subset \mathbf{A}^n$ be an affine algebraic set. The following statements are equivalent:
(i) V is irreducible.
(ii) $\mathbf{I}(V)$ is a prime ideal.
(iii) $k[V]$ is a domain.

Proof (i) $\Leftrightarrow$ (ii) follows from Theorem 1.4, and (ii) $\Leftrightarrow$ (iii) may be directly checked. □

Thus, the collection of polynomial functions on an affine algebraic set can detect geometric properties such as reducibility or irreducibility. In addition, knowing the structure of $k[V]$ can also provide information leading toward the beginings of a *classification* of algebraic sets, a topic we are going to start exploring from the next section and further in CH.VI once we have developed several different tools to analyze the algebraic properties of $k[V]$.

Exercises for §3

1. Prove §1 Proposition 1.6 and Proposition 1.7.
2. Give a detailed proof of Proposition 3.2.
3. If ϕ: $V \to W$ is an onto polynomial map, and X is an algebraic subset of W, show that $\phi^{-1}(X)$ is an algebraic subset of V. If $\phi^{-1}(X)$ is irreducible, show that X is irreducible. This gives a useful test for irreducibility.
4. Show that $\mathbf{V}(xz-y^2, yz-x^3, z^2-x^2y) \subset \mathbf{A}^3_{\mathbb{C}}$ is an irreducible algebraic set. (Hint: $y^3 - x^4$, $z^3 - x^5$, $z^4 - y^5 \in \mathbf{I}(V)$. Find a polynomial map from $\mathbf{A}^1_{\mathbb{C}}$ onto V).
5. Show that the *projection map* π: $\mathbf{A}^n \to \mathbf{A}^r$, $n \geq r$, defined by $\pi(x_1, ..., x_n) = (x_1, ..., x_r)$ is a polynomial map.
6. Let ϕ: $\mathbf{A}^1 \to V = \mathbf{V}(y^2 - x^2(x+1))$ be defined by $\phi(t) = (t^2-1, t(t^2-1))$. Show that ϕ is one-to-one and onto, except that $\phi(\pm 1) = (0, 0)$.

§4. The Coordinate Ring of an Algebraic Set

Set $\mathbf{A}^n = A^n_k$, $k[\mathbf{x}] = k[x_1, ..., x_n]$. Let $V \subset \mathbf{A}^n$ be an affine algebraic set, $k[V]$ the ring of polynomial functions on V as defined in §3. Concerning the algebraic structure of $k[V]$, we have the following important fact.

4.1. Theorem Let $V \subset \mathbf{A}^n$ be an affine algebraic set and $\mathbf{I}(V)$ its ideal. Then there is a ring isomorphism $k[\mathbf{x}]/\mathbf{I}(V) \cong k[V]$. Hence $k[V]$ is a Noetherian ring (in the sense of Appendix I §1), in particular, $k[V]$ is ring-finite over k in the sense of Appendix I.

Proof Construct a map π: $k[\mathbf{x}] \to k[V]$ by mapping a polynomial in $k[\mathbf{x}]$ to the polynomial function it defines on V. By definition of polynomial functions, this map is surjective; by definition of the ring structure on $k[V]$, π is a ring homomorphism (in fact a k-linear ring homomorphism). A polynomial $f \in k[\mathbf{x}]$ defines the zero polynomial function on V if and only if it is in $\mathbf{I}(V)$ by Proposition 3.2. Hence $\text{Ker}\pi = \mathbf{I}(V)$. The Noetherian property follows from the fact that $k[\mathbf{x}]$ is Noetherian (Hilbert basis theorem, CH.I Theorem 1.8) and hence $k[\mathbf{x}]/\mathbf{I}(V)$ is Noetherian. □

Thus, given a polynomial $f \in k[\mathbf{x}]$, we let $\pi(f) = [f]$ denote the polynomial function in $k[V]$ represented by f. In particular, each variable x_i gives a polynomial function $[x_i]$: $V \to k$ whose value at a point $P \in V$ is the i-th coordinate of P. We call $[x_i] \in k[V]$ the i-th *coordinate function* on V. Then the isomorphism $k[\mathbf{x}]/\mathbf{I}(V) \cong k[V]$ shows that the coordinate functions generate $k[V]$ in the sense that any polynomial function on V is a k-linear combination of products of the $[x_i]$. This explains the following terminology.

4.2. Definition The *coordinate ring* of an affine algebraic set $V \subset \mathbf{A}^n$ is the ring $k[V]$, which is usually identified with $k[\mathbf{x}]/\mathbf{I}(V)$ by Theorem 4.1.

From now on, many results concerning an algebraic set will be phrased in terms of the coordinate ring. For example:

- An algebraic set is irreducible if and only if its coordinate ring is a domain.

In the "algebra-geometry" dictionary of Chapter I, we related algebraic sets in $\mathbf{A}^n$ to ideals in $k[\mathbf{x}]$. One theme of this chapter is that this dictionary still works if we replace $\mathbf{A}^n$ and $k[\mathbf{x}]$ by a general algebraic set V and its coordinate ring $k[V]$. This can be done by using the ideal correspondence through the ring homomorphism $k[\mathbf{x}] \to k[\mathbf{x}]/\mathbf{I}(V) \cong k[V]$, namely we have the following one-to-one and onto mappings between the given sets:

(a) $\{$ ideals of $k[\mathbf{x}]$ containing $\mathbf{I}(V)\}$ and $\{$ ideals of $k[\mathbf{x}]/\mathbf{I}(V)\}$.
(b) $\{$ radical ideals of $k[\mathbf{x}]$ containing $\mathbf{I}(V)\}$ and $\{$ radical ideals of $k[\mathbf{x}]/\mathbf{I}(V)\}$.
(c) $\{$ prime ideals of $k[\mathbf{x}]$ containing $\mathbf{I}(V)\}$ and $\{$ prime ideals of $k[\mathbf{x}]/\mathbf{I}(V)\}$.
(d) $\{$ maximal ideals of $k[\mathbf{x}]$ containing $\mathbf{I}(V)\}$ and $\{$ maximal ideals of $k[\mathbf{x}]/\mathbf{I}(V)\}$.

Consequently, we obtain the following important properties of the above correspondence.

4.3. Theorem Let k be an algebraically closed field and let $V \subset \mathbf{A}^n$ be an affine algebraic set.
(i) The correspondences

$$\left\{ \begin{array}{c} \text{affine algebraic subsets} \\ W \subset V \end{array} \right\} \begin{array}{c} \xrightarrow{\mathbf{I}_V} \\ \xleftarrow[\mathbf{V}_V]{} \end{array} \left\{ \begin{array}{c} \text{radical ideals} \\ J \subset k[V] \end{array} \right\}$$

are inclusion-reversing bijections and are inverses of each other, where $\mathbf{I}_V$ denotes taking ideal $\mathbf{I}(W)$ with W as idicated above and $\mathbf{V}_V$ denotes taking algebraic set $\mathbf{V}(J)$ with J as indicated above; we use $\mathbf{I}_V$ resp. $\mathbf{V}_V$ just for the restrictions of $\mathbf{I}$ and $\mathbf{V}$.
(ii) Under the correspondence given in (i), irreducible algebraic subsets resp. points of V correspond to prime ideals resp. to maximal ideals of $k[V]$.

Proof Exercise. □

Next, we return to the problem of finding a *classification* of algebraic sets posed in the end of §3. What does it mean for two affine algebraic sets to be "isomorphic"? One reasonable answer is given in the following definition.

4.4. Definition Let $V \subset \mathbf{A}^m$ and $W \subset \mathbf{A}^n$ be algebraic sets. We say that V and W are *isomorphic* if there exist polynomial mappings $\alpha: V \to W$ and $\beta: W \to V$ such that $\alpha \circ \beta = \mathrm{id}_W$, and $\beta \circ \alpha = \mathrm{id}_V$. (For any algebraic set V, we write id_V for the identity mapping from V to itself. This is always a polynomial mapping.)

Intuitively, algebraic sets that *are* isomorphic should share properties such as irreducibility, etc. In addition, algebraic subsets of V should correspond to algebraic subsets of W, and so forth. For instance, saying that an algebraic set $W \subset \mathbf{A}^n$ is isomorphic to $V = \mathbf{A}^m$ implies that there is a one-to-one and onto polynomial mapping α: $\mathbf{A}^m \to W$ with a polynomial inverse. Thus we have a polynomial *parametrization* of W with especially nice properties!

Example (i) This example is inspired by a technique used in geometric modeling, which illustrates the usefulness of the above idea. Let us consider the two surfaces

$$\begin{aligned} Q_1 &= \mathbf{V}\left(x^2 - xy - y^2 + z^2\right) = \mathbf{V}(f_1), \\ Q_2 &= \mathbf{V}\left(x^2 - y^2 + z^2 - z\right) = \mathbf{V}(f_2) \end{aligned}$$

in $\mathbb{R}^3$. (These might be boundary surfaces of a solid region in a shape we were designing, for example.) To study the *intersection curve* $C = \mathbf{V}(f_1, f_2)$ of the two surfaces, we could proceed as follows. Neither Q_1 nor Q_2 is an especially simple surface, so the intersection curve is fairly difficult to visualize directly. However, usually we are *not limited* to using the particular equations f_1, f_2 to define the curve! It is easy to check that $C = \mathbf{V}(f_1, f_1 + cf_2)$, where $c \in \mathbb{R}$ is any nonzero real number. Hence, the surfaces $F_c = \mathbf{V}(f_1 + cf_2)$ also contain C. These surfaces, together with Q_2, are often called the elements of the *pencil* of surfaces determined by Q_1 and Q_2. (A pencil of algebraic sets is a one-parameter family of algebraic sets, parametrized by the points of k. In the above case, the parameter is $c \in k$.)

If we can find a value of c making the surface F_c particularly simple, then understanding the curve C will be correspondingly easier. Here, if we take $c = -1$, then F_{-1} is defined by

$$\begin{aligned} 0 &= f_1 - f_2 \\ &= z - xy. \end{aligned}$$

The surface $F = F_{-1}$ is much easier to understand because it is *isomorphic as an algebraic set* to $\mathbb{R}^2$. To see this, note that we have polynomial mappings:

$$\begin{array}{ccccccc} \mathbb{R}^2 & \xrightarrow{\alpha} & Q = F_{-1}, & \qquad & Q & \xrightarrow{\pi} & \mathbb{R}^2, \\ (x,y) & \mapsto & (x,y,xy), & & (x,y,z) & \mapsto & (x,y), \end{array}$$

which satisfy $\alpha \circ \pi = \mathrm{id}_Q$ and $\pi \circ \alpha = \mathrm{id}_{\mathbb{R}^2}$.
Hence, curves on Q may be *viewed* as plane curves in the following way. To study C, we can project to the curve $\pi(C) \subset \mathbb{R}^2$, and we obtain the equation

$$x^2y^2 + x^2 - xy - y^2 = 0$$

for $\pi(C)$ by substituting $z = xy$ in either f_1 or f_2. Note that π and α restrict to isomorphisms between C and $\pi(C)$, so we have not really lost anything by projecting in this case. In particular, each point (a,b) on $\pi(C)$ corresponds to exactly one point (a,b,ab) on C. It may be shown that $\pi(C)$ can also be parametrized as

$$\begin{cases} x = \dfrac{-t^2+t+1}{t^2+1}, \\ \\ y = \dfrac{-t^2+t+1}{t(t+2)}. \end{cases}$$

From this we also obtain a parametrization of C via the mapping α.

Given the above example, there are two natural questions to raise.

- Given a polynomial mapping ϕ: $V \to W$ with coordinate functions $f_1, \ldots, f_n$, is there an algorithmic way to check whether ϕ is an isomorphism or not?
- Given two affine algebraic sets $V \subset \mathbf{A}^m$ and $W \subset \mathbf{A}^n$, what can their coordinate rings tell us about a possible isomorphism between V and W?

The answer to the first question is definitely yes. We will mention the algorithmic result in CH.IV §5. Here we would like to point out that this question is closely related to the famous *Jacobian Conjecture* which was posed by O.H. Keller in 1939, and has resisted all attempts of proof or refutation. (A survey paper about this conjecture was written by H. Bass, E.H. Connell and D. Wright: " The Jacobian conjecture: Reduction of degree and formal expansion of the inverse", *Bull. Amer. Math. Soc.*, 7(1982).)

- **Jacobian Conjecture** Let k be a field of characteristic 0 (but not necessarily algebraically closed), and let ϕ: $\mathbf{A}_k^n \to \mathbf{A}_k^n$ be a polynomial mapping with coordinate functions $f_1, ..., f_n \in k[x_1, ..., x_n]$. Then ϕ is a polynomial isomorphism if and only if the determinant of the Jacobi matrix

$$\begin{pmatrix} \frac{\partial f_1}{\partial x_1} & \frac{\partial f_1}{\partial x_2} & \cdots & \frac{\partial f_1}{\partial x_n} \\ \vdots & \vdots & \cdots & \vdots \\ \frac{\partial f_n}{\partial x_1} & \frac{\partial f_n}{\partial x_2} & \cdots & \frac{\partial f_n}{\partial x_n} \end{pmatrix}$$

is in $k - \{0\}$.

There is another way of looking at the same problem. Every ring endomorphism ψ of $k[x_1, ..., x_n]$ which is also k-linear is of the form

$$h \mapsto h(f_1, ... f_n)$$

where $f_i = \psi(x_i)$ for $1 \leq i \leq n$. The existence of $g_1, ..., g_n$ with

$$g_i(f_1, ..., f_n) = x_i, \quad 1 \leq i \leq n$$

is equivalent to the surjectivity of the endomorphism in question (see Exercise 7).

For the second question raised above, it is better to consider the relation between the two coordinate rings

$$k[V] \cong k[x_1, ..., x_m]/\mathbf{I}(V) \quad \text{and} \quad k[W] \cong k[y_1, ..., y_n]/\mathbf{I}(W)$$

as follows.

If we have a polynomial mapping α: $V \to W$, then every polynomial function ϕ: $W \to k$ in $k[W]$ yields a polynomial function $\phi \circ \alpha$: $V \to k$ in $k[V]$, i.e. we have the following commutative diagram:

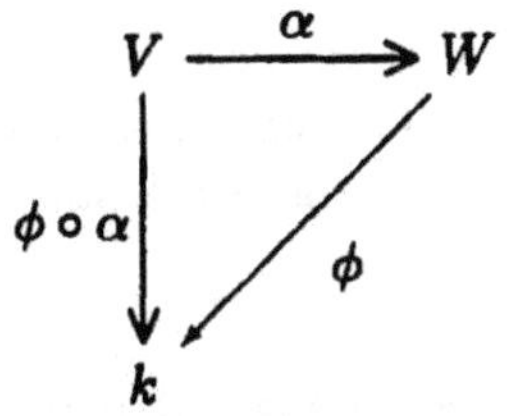

This leads to the existence of a map from $k[W]$ to $k[V]$ with the following properties.

4.5. Proposition Let V and W be algebraic sets (possibly in different affine spaces).
(i) Let α: $V \to W$ be a polynomial mapping. Then α induces a ring homomorphism α_*: $k[W] \to k[V]$ with $\alpha_*(\phi) = \phi \circ \alpha$ which is k linear. (Note that α_* "goes in the opposite direction" when compared to α since α_* maps functions on W to functions on V. For this reason we call α_* the ***pullback mapping*** on functions.)
(ii) Conversely, let β: $k[W] \to k[V]$ be a ring homomorphism which is k-linear. Then there is a unique polynomial mapping α: $V \to W$ such that $\beta = \alpha_*$.

Proof (i) Easy.
(ii) Since β: $k[W] = k[y_1, ..., y_n]/\mathbf{I}(W) \to k[x_1, .., x_m]/\mathbf{I}(V) = k[V]$ is a ring homomorphism, suppose that $\beta(y_i + \mathbf{I}(W)) = g_i + \mathbf{I}(V) \in k[V]$ with $g_i \in k[x_1, ..., x_m]$, $i = 1, ..., n$, then for $F + \mathbf{I}(W) \in k[W]$ we have $\beta(F + \mathbf{I}(W)) = F(g_1, ..., g_n) + \mathbf{I}(V)$. Thus, if $F \in \mathbf{I}(W)$ then $0 = \beta(F + \mathbf{I}(W)) = F(g_1, ..., g_n) + \mathbf{I}(V)$, i.e. $F(g_1, ..., g_n) \in \mathbf{I}(V)$. Now, if we define α: $V \to W$ by putting $\alpha(a_1, ..., a_m) = (g_1(a_1, ..., a_m), ..., g_n(a_1, ..., a_m))$, then it is easy to see that α is the desired unique polynomial mapping. □

4.6. Lemma Let α: $V \to W$ and β: $W \to K$ be polynomial mappings between algebraic sets. Then
(i) $\beta \circ \alpha$: $V \to K$ is a polynomial mapping.
(ii) $(\beta \circ \alpha)_* = \alpha_* \circ \beta_*$.

Proof Straightforward verification. □

4.7. Theorem Two affine algebraic sets $V \subset \mathbf{A}^m$ and $W \subset \mathbf{A}^n$ are isomorphic if and only if there is a k-linear ring isomorphism from $k[V]$ onto $k[W]$.

□

We conclude with several examples in order to illustrate isomorphisms of algebraic sets and the corresponding ring isomorphisms of their coordinate rings.

Example (ii) Let A be an invertible $n \times n$ matrix with entries in k and consider the linear mapping L_A: $\mathbf{A}^n \to \mathbf{A}^n$ defined by $L_A(x) = Ax$, where Ax is the matrix product. From the example in §3 we retain that L_A is a polynomial mapping, and if $A = (a_{ij})$ then L_A has coordinate functions $f_j = \sum_{i=1}^n a_{ij}x_i$, $j = 1, ..., n$, and the induced k-linear ring homomorphism L_{A*}: $k[x_1, ..., x_n] \to k[x_1, ..., x_n]$ is defined by $L_{A*}(g) = g(f_1, ..., f_n)$ for $g \in k[x_1, ..., x_n]$. Since A is invertible, L_A is an isomorphism of algebraic sets. Hence L_{A*} is a ring automorphism of $k[x_1, ..., x_n]$.

(iii) Let V be an algebraic set in $\mathbf{A}^n$ and $f \in k[V]$. Define

$$G(f) = \left\{ (a_1, ..., a_n, a_{n+1}) \in \mathbf{A}^{n+1} \;\middle|\; \begin{array}{l} (a_1, ..., a_n) \in V, \\ \text{and } a_{n+1} = f(a_1, ..., a_n) \end{array} \right\},$$

the *graph* of f. Then

(a) $G(f)$ is an affine algebraic set in $\mathbf{A}^{n+1}$; $G(f) = \mathbf{V}(\mathbf{I}(V), x_{n+1} - f)$;

(b) if V is irreducible then $G(f)$ is irreducible; and

(c) the map $(a_1, ..., a_n) \mapsto (a_1, ..., a_n, f(a_1, ..., a_n))$ defines an isomorphism of V and $G(f)$. (Projection gives the inverse.)

(iv) Consider the curve $V = \mathbf{V}(y^5 - x^2)$ in $\mathbb{R}^2$. We claim that V is not isomorphic to $\mathbb{R}$ as an algebraic set, even though there is a one-to-one polynomial mapping from V to $\mathbb{R}$ given by projecting V onto the x-axis. The obstruction is hiding in the coordinate ring of V: $\mathbb{R}[V] = \mathbb{R}[x, y]/\langle y^5 - x^2 \rangle$. If there were an isomorphism α: $\mathbb{R} \to V$, then the "pullback" α_*: $\mathbb{R}[V] \to \mathbb{R}[u]$ would be a ring isomorphism given by

$$\alpha_*([x]) = c(u), \qquad \alpha_*([y]) = d(u),$$

where $c(u), d(u) \in I\!R[u]$ are polynomials. Since $y^5 - x^2$ represents the zero function on V, we must have $\alpha_*(y^5 - x^2) = (d(u))^5 - (c(u))^2 = 0$ in $I\!R[u]$.

We may assume that $c(0) = d(0) = 0$ since the parametrization α can be "arranged" so that $\alpha(0) = (0,0) \in V$. But then, let us examine the possible polynomial solutions

$$c(u) = c_1 u + c_2 u^2 + \cdots, \qquad d(u) = d_1 u + d_2 u^2 + \cdots$$

of the equation $(c(u))^2 = (d(u))^5$. Since $(d(u))^5$ contains no power of u lower than u^5, the same must be true of $(c(u))^2$. However,

$$(c(u))^2 = c_1^2 u^2 + 2c_1c_2u^3 + (c_2^2 + 2c_1c_3)u^4 + \cdots.$$

The coefficient of u^2 must be zero, which implies $c_1 = 0$. The coefficient of u^4 must also be zero, which implies $c_2 = 0$ as well. Since $c_1, c_2 = 0$, the smallest power of u that can appear in c^2 is u^6, which implies that $d_1 = 0$.

It follows that u cannot be in the image of α_* since the image of α_* consists of polynomials in $c(u)$ and $d(u)$. This is a contradiction since α_* was supposed to be a ring isomorphism *onto* $I\!R[u]$. Thus, V and $I\!R$ are not isomorphic.

Exercises for §4

1. Give a detailed proof of Theorem 4.3.
2. Let $V_1 = \mathbf{V}(y - x^2) \subset \mathbf{A}_k^2$. Show that $k[V]$ is isomorphic to a polynomial ring in one variable over k.
3. Let $V_2 = \mathbf{V}(xy - 1) \subset \mathbf{A}_k^2$. Show that $k[V]$ is not isomorphic to a polynomial ring in one variable over k.
4. Let f be an irreducible quadratic polynomial in $k[x,y]$, and let W be the conic defined by f. Show that $k[W]$ is isomorphic to $k[V_1]$ or $k[V_2]$. Which one is it when?
5. Let ϕ: $\mathbf{A}^1 \to V = \mathbf{V}(y^2 - x^3) \subset \mathbf{A}^2$ be defined by $\phi(t) = (t^2, t^3)$. Show that although ϕ is one-to-one, onto polynomial map, ϕ is not an isomorphism. (Hint: $\phi_*(k[V]) = k[T^2, T^3] \subset k[T] = k[\mathbf{A}^1]$.)
6. Let ϕ: $\mathbf{A}^1 \to V = \mathbf{V}(y^2 - x^3 - x^2) \subset \mathbf{A}^2$ be defined by $\phi(t) = (t^2 - 1, t^3 - t)$, Is ϕ an isomorphism?
7. Let α be the ring homomorphism from the polynomial ring $k[x,y,z]$ to the polynomial ring $k[T]$ defined by $\alpha(x) = T^9$, $\alpha(y) = T^6$, $\alpha(z) = T^4$. Let $V = \mathbf{V}(x^2 - y^3, y^2 - z^3) \subset \mathbf{A}^3$.

(a) Show that α induces a ring homomorphism $\overline{\alpha}$: $k[V] \to k[T]$.

(b) What is the polynomial map ϕ: $\mathbf{A}^1 \to V$ such that $\phi_* = \overline{\alpha}$?

(c) Show that ϕ is one-to-one and onto, but not an isomorphism.

§5. Affine Change of Coordinates

Recall from the last section that if α: $\mathbf{A}^m = \mathbf{A}_k^m \to \mathbf{A}_k^n = \mathbf{A}^n$ is a polynomial mapping with coordinate functions $f_1, ..., f_n \in k[y_1, ..., y_m]$, then the induced ring homomorphism α_*: $k[x_1, ..., x_n] \to k[y_1, ..., y_m]$ is given by $\alpha_*(g) = g(f_1, ..., f_n)$ for every $g \in k[x_1, ..., x_n]$.

Notation

- If I is an ideal of $k[x_1, .., x_n]$, we write I^α for the ideal of $k[y_1, ..., y_m]$ generated by $\alpha_*(I) = \{\alpha_*(g) \mid \text{all } g \in I\}$; (is $\alpha_*(I)$ an ideal of $k[y_1, ..., y_m]$?)
- If V is an algebraic set of $\mathbf{A}^n$ and $I = \mathbf{I}(V)$, we put $V^\alpha = \mathbf{V}(I^\alpha) = \mathbf{V}(\alpha_*(g) \mid g \in \mathbf{I}(V)) \subset \mathbf{A}^m$.

5.1. Lemma With notation as above, we have
(i) $V^\alpha = \alpha^{-1}(V)$;
(ii) If $g \in k[x_1, ..., x_n]$ and $V = \mathbf{V}(g)$ is the hypersurface of g in $\mathbf{A}^n$, then V^α is the hypersurface defined by $\alpha_*(g)$ in $\mathbf{A}^m$ (in case $\alpha_*(g)$ is not a constant).

Proof Direct from foregoing observations. □

5.2. Definition An *affine change of coordinates* on $\mathbf{A}^n$ is a polynomial mapping α: $\mathbf{A}^n \to \mathbf{A}^n$ with coordinate functions $f_1, ..., f_n \in k[x_1, ..., x_n]$, such that each f_i has degree 1, and such that α is one-to-one and onto.

Let α be an affine change of coordinates on $\mathbf{A}^n$ with coordinate functions

$$f_i = \sum_{j=1}^{n} a_{ij}x_j + a_{i0}, \quad i = 1, ..., n.$$

then α may be regarded as a composition of two polynomial mappings:

$$\mathbf{A}^n \xrightarrow{\alpha'} \mathbf{A}^n \xrightarrow{\alpha''} \mathbf{A}^n$$

where α' is a *linear map* with coordinate functions

$$f_i' = \sum_{j=1}^{n} a_{ij}x_j, \quad i = 1, ..., n,$$

and α'' is a *translation* with coordinate functions

$$f_i'' = x_i + a_{i0}, \quad i = 1, ..., n.$$

Since any translation has an inverse (also a translation), it follows that α will be one-to-one (and onto) if and only if α' is invertible.

5.3. Proposition If α and β are affine changes of coordinates on $\mathbf{A}^n$, then so are $\alpha \circ \beta$ and α^{-1}; α is an isomorphism of $\mathbf{A}^n$ with itself. □

5.4. Theorem Let V be an affine algebraic set in $\mathbf{A}^n$. If α: $\mathbf{A}^n \to \mathbf{A}^n$ is an affine change of coordinates on $\mathbf{A}^n$, then V is isomorphic to $\mathbf{V}(I^\alpha)$, where $I = \mathbf{I}(V)$. Consequently,

$$k[V] = \frac{k[x_1, ..., x_n]}{\mathbf{I}(V)} \xrightarrow{\cong} \frac{k[x_1, ..., x_n]}{\mathbf{I}(\mathbf{V}(I^\alpha))} = k[\mathbf{V}(I^\alpha)].$$

Proof By Lemma 5.1. we have $\alpha^{-1}(V) = \mathbf{V}(I^\alpha)$, it follows that $V = \alpha(\mathbf{V}(I^\alpha))$. □

Exercises for §5

1. A set $V \subset \mathbf{A}^n$ is called a *linear subvariety* of $\mathbf{A}^n$ if $V = \mathbf{V}(f_1, ..., f_r)$ for some polynomials f_i of degree 1.
   (a) Show that if α is an affine change of coordinates on $\mathbf{A}^n$, then V^α is also a linear subvariety of $\mathbf{A}^n$.

(b) If $V \neq \emptyset$, show that there is an affine change of coordinates α of $\mathbf{A}^n$ such that $V^\alpha = \mathbf{V}(x_{m+1}, ..., x_n)$. (Hint: Use induction on r). So V is irreducible.

(c) Show that the m which appears in part (b) is independent of the choice of α. V is then isomorphic (as an algebraic set) to $\mathbf{A}^m$. (Hint: Suppose there were an affine change of coordinates β such that $\mathbf{V}(x_{m+1}, ..., x_n)^\beta = \mathbf{V}(x_{s+1}, ..., x_n)$, $m < s$; show that $x_{m+1}, ..., x_n$ would be dependent.)

2. Give a detailed proof of Lemma 5.1.

§6. Rational Functions and Local Rings

In this section, we introduce the notion of rational functions on an irreducible algebraic set V and associate to every point $P \in V$ a local ring (see the definition below) consisting of rational functions. The importance of these objects will be more evident in CH.VI.

Let V be an affine algebraic set in $\mathbf{A}^n = \mathbf{A}^n_k$. From §2 we know that V may be decomposed as a union of a finite number irreducible algebraic sets.

6.1. Definition An irreducible affine algebraic set is called an *affine variety*. Any nonempty open subset of a variety is called a *quasi-affine variety*.

Let V be an affine variety in $\mathbf{A}^n$. Then we have seen in §3 that the coordinate ring $k[V]$ of V is an integral domain. Thus, the field of fractions of $k[V]$ exists and is denoted $k(V)$.

6.2. Definition Let V be an affine variety in $\mathbf{A}^n$. We call $k(V)$ the *function field* (or *field of rational functions*) of V. An element of $k(V)$ is called a *rational function* on V.

Note the consistency of our notation. We use $k[x_1, ..., x_n]$ for a polynomial ring and $k[V]$ for the coordinate ring of V. Similarly, we use

$k(x_1, ..., x_n)$ for a rational function field and $k(V)$ for the function field of V.
As with any rational function, we must be careful to avoid zeros of the denominator if we want a well-defined function value in k. Below we explain in some detail that a rational function is indeed defined "locally" on V.

Let V be a variety in $\mathbf{A}^n$, and let $\frac{\phi}{\psi}$ be a rational function on V where ϕ and ψ are polynomial functions in $k[V]$.

6.3. Definition If $P \in V$, we say that $\frac{\phi}{\psi}$ is *defined at P* if for some $[h], [s] \in k[V]$ with $h, s \in k[x_1, ..., x_n]$ such that $\frac{\phi}{\psi} = \frac{[h]}{[s]}$ we have $s(P) \neq 0$.

Note that there may be many different ways to write $\frac{\phi}{\psi}$ as a ratio of polynomial functions; $\frac{\phi}{\psi}$ is defined at P if it is possible to find a "denominator" for $\frac{\phi}{\psi}$ which doesn't vanish at P. If $k[V]$ is a UFD (unique factorization domain), however, there is an essentially unique representation $\frac{\phi}{\psi} = \frac{[h]}{[s]}$, where $[h]$ and $[s]$ have no common factors (check it!), and then $\frac{\phi}{\psi}$ is defined at P if and only if $s(P) \neq 0$.

Example (i) $V = \mathbf{V}(xw - yz) \subset \mathbf{A}_k^4$. $k[V] = k[x, y, z, w]/\langle xw - yz \rangle$. If $P = (x, y, z, w) \in V$ with $y \neq 0$, then $\frac{\phi}{\psi} = \frac{[x]}{[y]} \in k(V)$ is defined at P. If $Q = (x, y, z, w) \in V$ with $w \neq 0$, then $\frac{\phi}{\psi} = \frac{[z]}{[w]}$ is defined at Q.

Let $\frac{\phi}{\psi} \in k(V)$. Put

$$J_{\frac{\phi}{\psi}} = \left\{ g \in k[x_1, ..., x_n] \;\middle|\; [g]\frac{\phi}{\psi} \in k[V] \right\}.$$

Then $J_{\frac{\phi}{\psi}}$ is an ideal of $k[\mathbf{x}]$ containing $\mathbf{I}(V)$. (Check this!)

6.4. Proposition $\mathbf{V}\left(J_{\frac{\phi}{\psi}}\right) = \left\{ P \in V \;\middle|\; \frac{\phi}{\psi} \text{ is not defined at } P \right\}.$

Proof Let $\frac{[h]}{[s]}$ be any representation of $\frac{\phi}{\psi}$ in $k(V)$. Then $[s] \in J_{\frac{\phi}{\psi}}$ implies $s(Q) = 0$ for every $Q \in \mathbf{V}(J_{\frac{\phi}{\psi}})$, i.e., $\frac{\phi}{\psi}$ is not defined at $Q \in \mathbf{V}(J_{\frac{\phi}{\psi}})$. Conversely, let $P \in V$ be such that $\frac{\phi}{\psi}$ is not defined at P. Then for any representation $\frac{[h]}{[s]} \in k(V)$, $s(P) = 0$. Now if $g \in J_{\frac{\phi}{\psi}}$, then $[g]\frac{\phi}{\psi} \in k[V]$, say $[g]\frac{\phi}{\psi} = [u]$. If $[g] = 0$, then $g \in \mathbf{I}(V)$ and $g(P) = 0$; if $[g] \neq 0$, then $\frac{[u]}{[g]}$ is a representation of $\frac{\phi}{\psi}$ in $k(V)$ and $g(P) = 0$ by the preceding argument. Hence the proposition holds. □

6.5. Definition The algebraic set $\mathbf{V}\left(J_{\frac{\phi}{\psi}}\right)$ is called the *pole set* of $\frac{\phi}{\psi}$ in V.

Example (ii) Again take $V = \mathbf{V}(xw - yz) \subset \mathbf{A}^4$. Then for $\frac{\phi}{\psi} = \frac{[x]}{[y]} = \frac{[z]}{[w]} \in k(V)$, $\mathbf{V}\left(J_{\frac{\phi}{\psi}}\right) = \{(x, y, z, w) \in \mathbf{A}^4 \mid y = w = 0\}$.

Let $\frac{\phi}{\psi} \in k(V)$ be a rational function on an affine variety V. From Proposition 6.4 we infer that, although $\frac{\phi}{\psi}$ need not define a function on V, it does define a function on the *Zariski dense open subset*

$$U_{\frac{\phi}{\psi}} = \left\{P \in V \;\middle|\; \frac{\phi}{\psi} \text{ is defined at } P\right\} = V - \mathbf{V}\left(J_{\frac{\phi}{\psi}}\right)$$

of V, namely, it is "almost everywhere defined". (Why $U_{\frac{\phi}{\psi}} \neq \emptyset$?) More precisely, we have

6.6. Observations (i) $\frac{\phi}{\psi}$ defines a function $\vartheta \colon U_{\frac{\phi}{\psi}} \to k$ with $\vartheta(P) = \frac{\phi}{\psi}(P) = \frac{[h]}{[s]}(P)$ for *some representation* $\frac{[h]}{[s]}$ of $\frac{\phi}{\psi}$.

(ii) For any $P \in U_{\frac{\phi}{\psi}}$, there is an open neighborhood of P say, $U_P \subset U_{\frac{\phi}{\psi}}$, and $[h], [s] \in k[V]$ such that $s(Q) \neq 0$ for all $Q \in U_P$ and $\frac{\phi}{\psi}(Q) = \frac{[h]}{[s]}(Q)$ on U_P.

Warning Note that $\frac{\phi}{\psi}$ may not be represented by a unique $\frac{[h]}{[s]}$ with the property $s(P) \neq 0$ for all $P \in U_{\frac{\phi}{\psi}}$. For example, consider $V = \mathbf{V}(xw - yz) \subset \mathbf{A}^4$, and $\frac{\phi}{\psi} = \frac{[x]}{[y]} = \frac{[z]}{[w]} \in k(V)$. It is impossible to write $\frac{\phi}{\psi} = \frac{[h]}{[s]}$, where $s(P) \neq 0$ for all $P \in U_{\frac{\phi}{\psi}}$. Indeed, if $\frac{\phi}{\psi} = \frac{[h]}{[s]}$ on $U_{\frac{\phi}{\psi}} = \{(x, y, z, w) \in V \mid y \neq 0, \text{ or } w \neq 0\}$, then since $[x] \neq 0$, we have

$$\frac{\phi}{\psi} = \frac{[h]}{[s]} = \frac{[h][x]}{[s][x]} = \frac{[hx]}{[sx]}$$

which is not defined on $(0, 0, 1, 1) \in U_{\frac{\phi}{\psi}}$, a contradiction.

However, based on the above observations, we claim that $\frac{\phi}{\psi}$ is *uniquely* determined by the restriction to some nonempty open set.

6.7. Lemma Let V and $k(V)$ be as before, and let $\frac{\phi}{\psi}, \frac{\phi'}{\psi'} \in k(V)$ be two rational functions. Suppose that $\frac{\phi}{\psi}$ and $\frac{\phi'}{\psi'}$ are defined on some open subset $U \subset V$. If $\frac{\phi}{\psi} = \frac{\phi'}{\psi'}$ on U, then $\frac{\phi}{\psi} = \frac{\phi'}{\psi'}$ in $k(V)$.

Proof By the assumption it is obvious that $U \subset U_{\frac{\phi}{\psi}} \cap U_{\frac{\phi'}{\psi'}}$. Let $P \in U$. By Proposition 1.3(i) and the above Observation 6.9(ii), we may assume that $\frac{\phi}{\psi}$ and $\frac{\phi'}{\psi'}$ are represented by $\frac{[h]}{[s]}$ and $\frac{[h']}{[s']}$, respectively, on some open neighborhood $U_P \subset U \subset U_{\frac{\phi}{\psi}} \cap U_{\frac{\phi'}{\psi'}}$ containing P. Thus, by the assumption we have $(hs' - h's)(P) = 0$ for all $P \in U_P$. Put $g = hs' - h's$. Then g defines a polynomial function $V \to k$ and $U \subset g^{-1}(0)$. It follows that $g^{-1}(0) = V$ because $g^{-1}(0)$ is closed and U is dense in V. This shows that $g \in \mathbf{I}(V)$, i.e., $[hs'] = [h's]$ and $\frac{\phi}{\psi} = \frac{[h]}{[s]} = \frac{[h']}{[s']} = \frac{\phi'}{\psi'}$ in $k(V)$. □

The "local" property of a rational function discussed above may be generalized to define the regular functions on a quasi-variety in CH.VI, in order to introduce a local theory in algebraic geometry.

Now, let V be an affine variety and $P \in V$; to P we associate a

subring of $K(V)$ as follows.
Put

$$\mathcal{O}_{P,V} = \left\{ \frac{\phi}{\psi} \in k(V) \;\middle|\; \frac{\phi}{\psi} \text{ is defined at } P \right\}$$

6.8. Lemma $\mathcal{O}_{P,V}$ is a subring of $k(V)$. Moreover, $k \subset k[V] \subset \mathcal{O}_{P,V} \subset k(V)$.

Proof Straightforward. □

6.9. Definition $\mathcal{O}_{P,V}$ is called the *local ring* of V at P (this name will be algebraically qualified in Proposition 6.11 and geometrically qualified in CH.VI §5).

In what follows, we always regard $k[V]$ and $\mathcal{O}_{P,V}$ as subrings of $k(V)$.

6.10 Proposition If k is algebraically closed, then

$$k[V] = \bigcap_{P \in V} \mathcal{O}_{P,V},$$

i.e., the rational functions that are defined at every $P \in V$ are nothing but all polynomial functions on V.

Proof If $\frac{\phi}{\psi} \in \cap_{P\in V}\mathcal{O}_{P,V}$, then $\mathbf{V}(J_{\frac{\phi}{\psi}}) = \emptyset$, so $1 \in J_{\frac{\phi}{\psi}}$ by Nullstellensatz, i.e., $1 \cdot \frac{\phi}{\psi} = \frac{\phi}{\psi} \in k[V]$, which proves the proposition. □

Let $\frac{\phi}{\psi} \in \mathcal{O}_{P,V}$. We define the *value of* $\frac{\phi}{\psi}$ *at* P, written $\frac{\phi}{\psi}(P)$, by taking any representation of $\frac{\phi}{\psi}$, $\frac{[h]}{[s]}$ say, with $s(P) \neq 0$, putting

$$\frac{\phi}{\psi}(P) = \frac{h(P)}{s(P)}$$

(one checks that this is independent of the choice of $[h]$ and $[s]$ for which $\frac{\phi}{\psi} = \frac{[h]}{[s]}$, $s(P) \neq 0$.) Thus, we have a ring epimorphism (check it!):

$$\begin{array}{ccc} \mathcal{O}_{P,V} & \longrightarrow & k \\ \\ \dfrac{\phi}{\psi} & \mapsto & \dfrac{\phi}{\psi}(P) \end{array}$$

with kernel

$$\mathcal{M}_{P,V} = \left\{ \frac{\phi}{\psi} \in \mathcal{O}_{P,V} \;\middle|\; \frac{\phi}{\psi}(P) = 0 \right\}$$

which is called the *maximal ideal of V at P*.

A commutative ring is called a *local ring*, if it has a unique maximal ideal. (See Exercise 3 for the characterization of a local ring in terms of non-invertible elements.)

6.11. Proposition With notation as above, $\mathcal{O}_{P,V}$ is a local Noetherian domain (in the sense of the above definition and Appendix I §1) with unique maximal ideal $\mathcal{M}_{P,V}$.

Proof The fact that $\mathcal{O}_{P,V}$ is a local ring with unique maximal ideal $\mathcal{M}_{P,V}$ is obvious from the definition. The Noetherian property of $\mathcal{O}_{P,V}$ follows if we show that any ideal I of $\mathcal{O}_{P,V}$ is finitely generated. By the Hilbert basis theorem (Ch.I Theorem 1.8) and the foregoing Theorem 4.1 we know that $k[V]$ is Noetherian. Choosing generators $\xi_1, ..., \xi_r$ for the ideal $I \cap k[V]$ of $k[V]$. We claim that $\xi_1, ..., \xi_r$ generate I as an ideal in $\mathcal{O}_{P,V}$. For if $f \in I \subset \mathcal{O}_{P,V}$, there is $h, s \in k[V]$ with $s(P) \neq 0$ and $sf \in k[V]$; then $sf \in k[V] \cap I$, so $sf = \sum u_i \xi_i$, $u_i \in k[V]$, so $f = \sum \frac{u_i}{s} \xi_i$, as desired. □

The above proposition motivates the following natural question.

- For a nonempty affine variety V, is the correspondence

$$\begin{array}{ccc} V & \longrightarrow & \{\mathcal{O}_{P,V}\}_{P \in V} \\ P & \mapsto & \mathcal{O}_{P,V} \end{array}$$

  a one-to-one correspondence?

Let p be any prime ideal of $k[V]$. We leave it as an exercise for the reader to check that the set

$$S_p^{-1}k[V] = \left\{ \frac{[g]}{[h]} \in k(V) \;\middle|\; [h] \notin p \right\}$$

is a subring of $K(V)$, and if I is an ideal of $k[V]$ then the set

$$S_p^{-1}I = \left\{ \frac{[g]}{[h]} \in S_p^{-1}k[V] \;\middle|\; [g] \in I \right\}$$

is an ideal of $S_p^{-1}k[V]$. The reader may look up in Appendix II §1 that the ring $S_p^{-1}k[V]$ coincides with the *localization* of $k[V]$ at the prime ideal p defined there. Localization theory plays a very important role in modern algebraic geometry and commutative algebra.

6.12. Lemma With notation as above,
(i) $S_p^{-1}k[V]$ is a local ring with the unique maximal ideal $M_p = S_p^{-1}p$
(ii) $M_p \cap k[V] = p$
(iii) If p and q are distinct prime ideals of $k[V]$, then $S_p^{-1}k[V] \neq S_q^{-1}k[V]$.

Proof By Exercise 3 this is straightforward. □

6.13. Lemma Let $P \in V$ and $\mathcal{O}_{P,V}$ the local ring of V at P with maximal ideal $\mathcal{M}_{P,V}$. Let $\mathbf{I}(P)$ be the ideal of P in $k[V]$, i.e.,

$$\mathbf{I}(P) = \left\{[g] \in k[V] \,\middle|\, g(P) = 0\right\}.$$

Then $\mathbf{I}(P)$ is a maximal ideal of $k[V]$, and $S_{\mathbf{I}(P)}^{-1}k[V] = \mathcal{O}_{P,V}$, $\mathcal{M}_{P,V} = M_{\mathbf{I}(P)}$ where $M_{\mathbf{I}(P)} = S_{\mathbf{I}(P)}^{-1}\mathbf{I}(P)$.

Proof If $P = (a_1, ..., a_n)$ then we know that $\Omega = \langle x_1 - a_1, ..., x_n - a_n\rangle$ is a maximal ideal of $k[x_1, ..., x_n]$ containing $\mathbf{I}(V)$ (CH.I §1 Exercise 5). It follows from the definition of $\mathbf{I}(P)$ that it is a maximal ideal of $k[V]$, indeed, $\mathbf{I}(P) = \Omega/\mathbf{I}(V)$.
By the definition $S_{\mathbf{I}(P)}^{-1}k[V]$, the inclusion $S_{\mathbf{I}(p)}^{-1}k[V] \subset \mathcal{O}_{P,V}$ is also clear. Now let $\frac{\phi}{\psi} \in \mathcal{O}_{P,V}$. Then $\frac{\phi}{\psi}$ is defined at P and $\frac{\phi}{\psi}$ is represented by some $\frac{[g']}{[h']} \in K(V)$ with $[h'](P) \neq 0$. Thus $[h'] \notin \mathbf{I}(P)$, and it follows that $\frac{[g']}{[h']} \in S_{\mathbf{I}(P)}^{-1}k[V]$. This shows that $\mathcal{O}_{P,V} \subset S_{\mathbf{I}(p)}^{-1}k[V]$.
Hence $\mathcal{O}_{P,V} = S_{\mathbf{I}(p)}^{-1}k[V]$. The equality $\mathcal{M}_{P,V} = M_{\mathbf{I}(P)}$ follows from Lemma 6.12. □

Combining Theorem 4.3 and the above results, we obtain:

6.14. Theorem Let k be algebraically closed. If V is an affine variety, then there is a one-to-one correspondence between V and $\{\mathcal{O}_{P,V}\}_{P \in V}$.

□

Theorem 6.14 tells us that if k is algebraically closed, different points corresponds to different local rings. In CH.VI, we will extend this result to a more general setting.
The local rings $\mathcal{O}_{P,V}$ play an prominent role in the modern study of algebraic varieties. All the properties of V which depend only on a "neighborhood" of P (the "local" properties) are reflected in the ring $\mathcal{O}_{P,V}$ (see Exercise 2 below for one indication of this, and see CH.VI §§5–6 for more).

The finall result of this section shows that for an affine variety V and a point $P \in V$, $k[V]$, $k(V)$ and $\mathcal{O}_{P,V}$ are invariants up to polynomial isomorphism.
Let V and W be affine varieties and α: $V \to W$ a polynomial mapping. If α_*: $k[W] \to k[V]$ is the ring homomorphism induced by α (Proposition 4.5), then it is easy to see that α_* induces a field homomorphism $k(W) \to k(V)$, and hence a local ring homomorphism $\mathcal{O}_{P,V} \to \mathcal{O}_{\phi(P),W}$.

6.15. Theorem If ϕ: $W \to V$ is a polynomial isomorphism between affine varieties, then ϕ induces ring isomorphisms: $k[V] \cong k[W]$, $k(V) \cong k(W)$, $\mathcal{O}_{P,V} \cong \mathcal{O}_{\phi(P),W}$ which are the identity on k (i.e. k-linear). □

6.16. Corollary (Comparing with Theorem 5.4) With notation as in §5, let α: $\mathbf{A}^n \to \mathbf{A}^n$ be an affine change of coordinates on $\mathbf{A}^n$, and let V an affine variety in $\mathbf{A}^n$. If $P \in V$, then $\mathcal{O}_P \cong \mathcal{O}_{\alpha^{-1}(P)}$.

□

Comparing to Theorem 4.7, the following question arises naturally:

- Suppose that V, W are varieties such that $k(V) \cong k(W)$. Can we derive an isomorphism between V and W? If not, what shoud the relation between V and W be?

If we pay a little attention to the proof of Theorem 4.7 and §4 Exercises 3, 5, and 6, we will find that the answer to the above question is not so easy to give immediately; indeed, this makes the theme of CH.VI.

Exercises for §6

1. Let $V = \mathbf{V}(y^2 - x^2(x+1)) \subset \mathbf{A}^2$, and $[x], [y]$ the residues of x, y in $k[V]$; let $z = \frac{[y]}{[x]} \in k(V)$. Find the pole sets of z and of z^2. (Note that V is a variety by Exercise 5 of §3).
2. Let $\mathcal{O}_{P,V}$ be the local ring of a variety V at a point P. Show that there is a natural one-to-one correspondence between the prime ideals in $\mathcal{O}_{P,V}$ and the subvarieties of V containing P. (Hint: If I is prime in $\mathcal{O}_{P,V}$, $I \cap k[V]$ is prime in $k[V]$, and I is generated by $I \cap k[V]$; use Theorem 4.3.)
3. Let R be a ring. Then the following are equivalent:
   (a) R is a local ring;
   (b) The set of non-units in R forms an ideal.
4. Prove Propositions 6.11–6.12 and Corollary 6.13.
5. Let V be an affine variety in $\mathbf{A}^n$. Let $k(\mathbf{x}) = k(x_1, ..., x_n)$ be the rational function field in n variables. Put

$$\mathcal{O}_V = \left\{ f = \frac{F}{G} \in k(\mathbf{x}) \;\middle|\; G \not\subset \mathbf{I}(V) \right\}$$

$$\mathcal{M}_V = \left\{ f = \frac{F}{G} \in \mathcal{O}_V \;\middle|\; F \in \mathbf{I}(V) \right\}$$

   Show that $\mathcal{O}_V$ is a local ring with the unique maximal ideal $\mathcal{M}_V$, and moreover $\mathcal{O}_V/\mathcal{M}_V \cong k(V)$.
6. Let V be an affine variety in $\mathbf{A}^n$ and p a prime ideal of $k[V]$. Let notation be as in the present section, $k[\mathbf{x}] = k[x_1, ..., x_n]$.
   a. Show that $S_p^{-1}k[V]$ is a subring of $k(V)$, and for any ideal $I \subset k[V]$, $S_p^{-1}I$ is an ideal of $S_p^{-1}k[V]$.
   b. Show that if Ω is a prime ideal of $k[\mathbf{x}]$ containing $\mathbf{I}(V)$ such that $p = \Omega/\mathbf{I}(V)$, then $S_\Omega^{-1}k[\mathbf{x}]/S_\Omega^{-1}\mathbf{I}(V) \cong S_p^{-1}k[V]$, and $S_\Omega^{-1}\Omega/S_\Omega^{-1}\mathbf{I}(V) \cong M_p$, where $M_p = S_p^{-1}p$.
   c. Let V be an affine variety in $\mathbf{A}^n$, $I = \mathbf{I}(V) \subset k[\mathbf{x}]$, $P \in V$, and let J be an ideal of $k[\mathbf{x}]$ which contains I. Let J' be the image of J in $k[V]$. Show that there is a natural homomorphism φ from $\mathcal{O}_{P,\mathbf{A}^n}/J\mathcal{O}_{P,\mathbf{A}^n}$ to $\mathcal{O}_{P,V}/J'\mathcal{O}_{P,V}$, and φ is an

isomorphism. In particular, $\mathcal{O}_{P,\mathbf{A}^n}/I\mathcal{O}_{P\mathbf{A}^n}$ is isomorphic to $\mathcal{O}_{P,V}$.

d. Show that if $V = \mathbf{V}(I) = \{P\}$, then $k[\mathbf{x}]/I \cong \mathcal{O}_{P,\mathbf{A}^n}/I\mathcal{O}_{P,\mathbf{A}^n}$.

CHAPTER III
Projective Algebraic Sets

So far, all of the algebraic sets (varieties) we have studied are subsets of affine space $\mathbf{A}^n = \mathbf{A}_k^n$ over a field k. Suppose that we want to study *all* the points of an intersection of two curves in $\mathbb{R}^2$, for example:

(i) Consider the lines L_1: $y = -x$ and L_2: $y = \alpha x + 1$. If α varies continuously and $\alpha \neq -1$, then we "see" that L_1 and L_2 always intersects at one point P; when α reaches -1, we "see" that L_1 and L_2 are parallel and it seems that the point P of intersection disappeared. But our geometric intuition tells us that the point P of intersection does not really disappear, it just "runs" to some sort of point at ∞, i.e., the parallel lines L_1 and L_2 still meet at ∞. Moreover, as indicated by the following picture, there should be different points at ∞, depending on the direction of the lines.

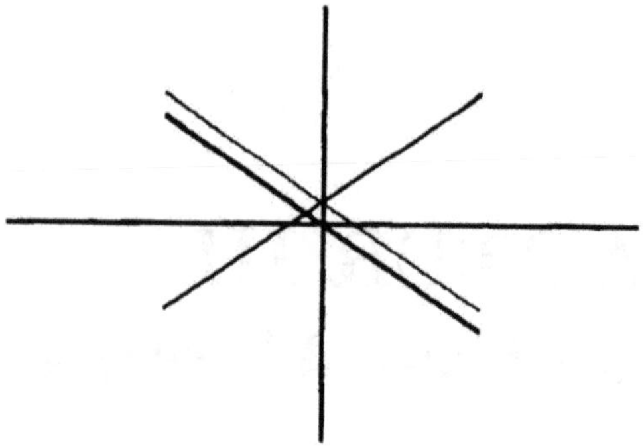

Figure 3

(ii) Consider the hyperbola C: $y^2 = x^2 + 1$ and the line L: $y = \alpha x$, $\alpha \in \mathbb{R}$, the parametrization of the first curve is $x = \dfrac{1+t^2}{1-t^2}$, $y = \dfrac{2t}{1-t^2}$.

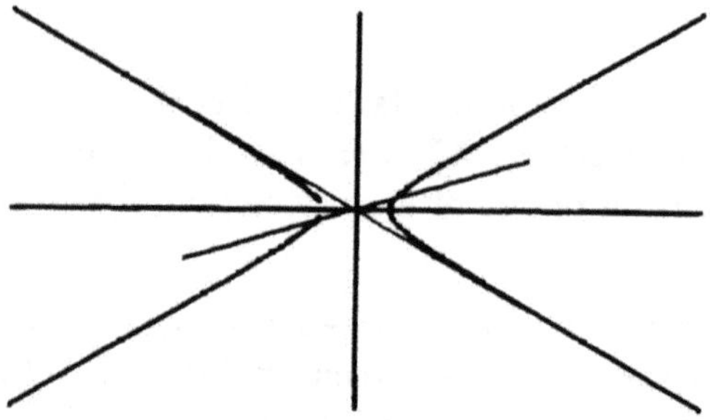

Figure 4

If α varies continuously and $\alpha \neq \pm 1$, C and L intersect in two points; when α reaches ± 1, we "see" that the points of intersection "run" to some sort of points at ∞, namely, the curve is asymptotic to the line, and they meet at ∞.

Other points which seem to be missing from certain pictures or descriptions are given by parametrizations of algebraic sets, for example:

(iii) If we look at the parametric representations of the unit circle $x^2 + y^2 = 1$ over $\mathbb{R}$ (CH.I §3), we have to deal separately with the exception $(-1, 0)$, because in the picture given by the rational parametric equations we cannot see this "smooth" point. To understand why this point is "missing", let us recall the rational parametric representation of the circle again:

$$(*) \qquad \begin{cases} x = \dfrac{1-t^2}{1+t^2} \\[2ex] y = \dfrac{2t}{1+t^2} \end{cases}$$

If we are working over the real numbers, we may talk about limits, in particular, the equation $(*)$ allows us to compute the limit of (x, y) as t increases. The limit is $(-1, 0)$, the missing point. Thus, $(-1, 0)$ corresponds to an "infinite" value of t in $\mathbf{A}^1_{\mathbf{R}}$. Of course, we are not allowed to say that we "obtain" this point by putting $t = \infty$.

- **Question** Is it possible to enlarge $\mathbf{A}^n$ to an appropriate geometric space in which we can "see" all points (including points at ∞) of an affine algebraic set of $\mathbf{A}^n$?

The answer to the above question is definitely *yes*—this is the topic of the present chapter. Actually, after enlarging an affine n-space $\mathbf{A}^n$ to a projective n-space $\mathbf{P}^n$ by adding to $\mathbf{A}^n$ all points at ∞ and constructing the projective algebra-geometry dictionary as in the affine case, we obtain a suitable solution to the problems. The extra advantage will be that projective sets may be covered by a collection of affine algebraic sets, so that the projective theory is "covered" by affine methods up to some modifications.

§1. Projective Space

In this section, we construct a projective space by adding to an affine space points at ∞. If we start with $\mathbf{A}^2_{\mathbf{R}}$, one way to achieve this is to identify each point $(a, b) \in \mathbb{R}^2$ with the point $(a, b, 1) \in \mathbb{R}^3$. Every point $(x, y, 1)$ determine a line in $\mathbb{R}^3$ which passes through $(0, 0, 0)$ and $(x, y, , 1)$. Every line through $(0, 0, 0)$, except those lying in the plane $z = 0$, corresponds to exactly one such point. The lines through $(0,0,0)$ in the plane $z = 0$ may be thought of as corresponding to the "points at infinity". This leads to the following definition.

1.1. Definition Let k be any field, and let $\mathbf{A}^{n+1} = \mathbf{A}_k^{n+1}$ be the affine $n+1$-space over k.
If $P = (a_1, ..., a_n, a_{n+1})$, $Q = (b_1, ..., b_n, b_{n+1})$ are two *different* points in $\mathbf{A}^{n+1}$, then the *line* L passing through P and Q is parametrically defined as

$$L = \left\{(\lambda a_1 + tb_1, \lambda a_2 + tb_2, ..., \lambda a_{n+1} + tb_{n+1}) \;\middle|\; \lambda,\ t \in k\right\}.$$

The *projective n-space* over k, denoted $\mathbf{P}_k^n$, or simply $\mathbf{P}^n$, is defined to be the set of all lines through $(0, 0, ..., 0)$ in $\mathbf{A}^{n+1}$.

Note that any point $(a_1, ..., a_{n+1}) \neq (0, 0, ..., 0)$ determines a *unique* line L through $(0, 0, ..., 0)$ in $\mathbf{A}^{n+1}$:

$$L = \left\{(\lambda a_1, ..., \lambda a_{n+1}) \;\middle|\; \lambda \in k\right\}.$$

Two such point $(a_1, ..., a_{n+1})$ and $(b_1, ..., b_{n+1})$ determine the same line if and only if there is a nonzero $\lambda \in k$ such that $b_i = \lambda a_i$ for $i = 1, ..., n+1$; let us say that $(a_1, ..., a_{n+1})$ and $(b_1, ..., b_{n+1})$ are equivalent, denoted $(a_1, ..., a_{n+1}) \sim (b_1, ..., b_{n+1})$, if this is the case. Then

$$\mathbf{P}^n \cong \left(\mathbf{A}^{n+1} - \{(0, ..., 0)\}\right) / \sim.$$

Elements of $\mathbf{P}^n$ will be called *points*. If a point $P \in \mathbf{P}^n$ is determined as above by some $(a_1, ..., a_{n+1}) \in \mathbf{A}^{n+1}$, we say that $(a_1, ..., a_{n+1})$ is a set of *homogeneous coordinates* for P. We often write $P = (a_1, ..., a_{n+1})$ to indicate that $(a_1, ..., a_{n+1})$ are homogeneous coordinates for P, or that $(a_1, ..., a_{n+1})$ represents P.

Note that the i-th coordinate a_i of a point P in $\mathbf{P}^n$ is not well-defined, but it is well-defined to say whether the i-th coordinate is zero or nonzero. Moreoevr, if $a_i \neq 0$, the ratios a_j/a_i are well-defined (since they are unchanged under equivalence).

Let

$$U_i = \left\{P = (a_1, ..., a_{n+1}) \in \mathbf{P}^n \;\middle|\; a_i \neq 0\right\}.$$

Each $P \in U_i$ has a unique representative (the set of homogeneous coordinates) of the form

$$P = \left(\frac{a_1}{a_i}, \ldots, \frac{a_{i-1}}{a_i}, 1, \frac{a_{i+1}}{a_i}, \ldots, \frac{a_{n+1}}{a_i}\right).$$

The coordinates of P which are not equal to 1 are called the *non-homogeneous coordinates* with respect to U_i (or a_i, or i). Putting

$$P_* = \left(\frac{a_1}{a_i}, \ldots, \frac{a_{i-1}}{a_i}, \frac{a_{i+1}}{a_i}, \ldots, \frac{a_{n+1}}{a_i}\right),$$

We may define a map

$$\begin{array}{rccc} \varphi_i : & U_i & \xrightarrow{\varphi_i} & \mathbf{A}^n \\ & P & \mapsto & P_* \end{array}$$

It is easy to see that φ_i sets up a bijective correspondence between the points of $\mathbf{A}^n$ and the points of U_i, and the iverse of φ_i is

$$\begin{array}{rccc} \phi_i : & \mathbf{A}^n & \longrightarrow & U_i \\ & (a_1, \ldots, a_n) & \mapsto & (a_1, \ldots, a_{i-1}, 1, a_i, a_{i+1}, \ldots, a_n). \end{array}$$

It is clear that

$$\mathbf{P}^n = \bigcup_{i=1}^{n+1} U_i,$$

so $\mathbf{P}^n$ is covered by $n+1$ sets each of which looks just like affine n-space (we shall further exploit this observation later in this chapter and in CH.VI).

From now on we let U_{n+1} be a priviliged affine part.

Let

$$H_\infty^{n+1} = \mathbf{P}^n - U_{n+1} = \left\{P = (a_1, \ldots, a_n, a_{n+1}) \;\middle|\; a_{n+1} = 0\right\}.$$

then

$$(\Delta) \qquad \mathbf{P}^n = U_{n+1} \cup H_\infty^{n+1}.$$

H_{∞}^{n+1} is often called the ***hyperplane at infinity.*** Since we obtain a bijective

$$\begin{array}{ccc} H_{\infty}^{n+1} & \longrightarrow & \mathbf{P}^{n-1} \\ (a_1, ..., a_n, 0) & \mapsto & (a_1, ..., a_n), \end{array}$$

H_{∞}^{n+1} may be identified with $\mathbf{P}^{n-1}$. To see what $H_{\infty}^{n+1} = \mathbf{P}^{n-1}$ means geometrically, note that a point $P \in \mathbf{P}^{n-1}$ is given by a line $L \subset \mathbf{A}^n$ going through the origin. Consequently, in the decomposition (Δ), one should think of P as representing the asymptotic direction of all lines in $\mathbf{A}^n$ parallel to L. This allows us to view P as a point at ∞.

Example (i) $\mathbf{P}_k^0$ is a point.

(ii) $\mathbf{P}_k^1 = \{(x, 1) \mid x \in k\} \cup \{(1, 0)\}$. $\mathbf{P}_k^1$ is the affine line plus one ***point at infinity.*** $\mathbf{P}_k^1$ is the ***projective line*** over k.
When $k = \mathbb{R}$, note that $(-1, 0)$ also represents the point of $\mathbf{P}_{\mathbb{R}}^1$ at ∞. But we have seen that $(-1, 0)$ is just the point missed by the parametric representation $(*)$ of the unit circle given in previous Example (iii). If we view $(-1, 0)$ as the limit point obtained by letting t grow to infinity, we may intuitively interprete this as the limit for t^{-1} going to zero. This suggests to introduce new variables u and v in $(*)$ where $t = \dfrac{u}{v}$. It is assumed for the moment that $v \neq 0$. The equations $(*)$ then take the form

$$(**) \qquad \begin{cases} x = \dfrac{v^2 - u^2}{u^2 + v^2} \\[2ex] y = \dfrac{2uv}{u^2 + v^2} \end{cases}$$

To each ordered pair (u, v) with $v \neq 0$ corresponds a point on the circle, and (su, sv) will determine the same point if s is a nonzero real number. If we we now drop the restriction $v \neq 0$, we see that the ordered pair $(1, 0)$, indeed $(s, 0)$ for any real number $s \neq 0$, corresponds to the point $(-1, 0)$. This situation finds its most appropriate setting in the language of projective spaces. What we have constructed is a mapping

$$\begin{array}{ccc} P^1_{\mathbb{R}} & \longrightarrow & C \subset \mathbf{A}^2_{\mathbb{R}} \\ P = (u, v) & \mapsto & (x, y) \end{array}$$

where C denotes the circle and (x, y) is given by the rational functions appearing in (**), indeed we have a bijective correspondence between $\mathbf{P}^1_{\mathbb{R}}$ and C. Thus we have constructed a model for the real projective 1-space.

(iii) $\mathbf{P}^2_k = \{(x, y, 1) \mid (x, y) \in \mathbf{A}^2\} \cup \{(x, y, 0) \mid (x, y) \in \mathbf{P}^1\}$. Here H^3_∞ is called the *line at infinity*. $\mathbf{P}^2_k$ is called the *projective plane* over k.

(iv) Consider a line L: $y = mx + b$ in $\mathbf{A}^2 = \mathbf{A}^2_{\mathbb{R}}$. To find the the corresponding points of L in $\mathbf{P}^2 = \mathbf{P}^2_{\mathbb{R}}$, let us take a point (x_0, y_0) on L. Then the line L' in $\mathbf{A}^3$ through $(0, 0, 0)$ and $(x_0, y_0, 1)$ is

$$\frac{X_0}{X} = \frac{mX_0 + b}{Y} = \frac{1}{Z},$$

i.e., any point (X, Y, Z) on L' should satisfy $Z \neq 0$ and $Y = mX + bZ$. Thus, if we identify $\mathbf{A}^2$ with $\phi_3(\mathbf{A}^2) = U_3 \subset \mathbf{P}^2$, then the points on the line L correspond to the points $(X, Y, Z) \in \mathbf{P}^2$ with $Y = mX + bZ$, $Z \neq 0$. Furthermore, let $\mathcal{L} = \{(X, Y, Z) \in \mathbf{P}^2 \mid Y = mX + bZ\}$ be the projective line defined by $Y = mX + bZ$ in $\mathbf{P}^2$. Then $\mathcal{L} \cap U_3 = \phi_3(L)$ and $\mathcal{L} \cap H^3_\infty = \{(1, m, 0)\}$. So all lines with the same slope, when extended to $\mathbf{P}^2$ in this way, pass through the same point at infinity.

(v) Similarly, consider again the hyperbola C: $y^2 = x^2 + 1$ in $\mathbf{A}^2 = \mathbf{A}^2_{\mathbb{R}}$. The points on the curve correspond to the points (X, Y, Z) in $\mathbf{P}^2 = \mathbf{P}^2_{\mathbb{R}}$ with $Y^2 = X^2 + Z^2$, $Z \neq 0$. Let $\mathcal{C}$ be the corresponding quadratic curve $\mathcal{C}$: $\{(X, Y, Z) \in \mathbf{P}^2 \mid Y^2 = X^2 + Z^2\}$ in $\mathbf{P}^2$. Then $\mathcal{C} \cap U_3 = \phi_3(C)$ and $\mathcal{C} \cap H^3_\infty = \{P = (1, 1, 0), Q = (1, -1, 0)\}$. Clearly, P, Q are the points where the lines $y = x$ and $y = -x$ intersect the curve. Furthermore, the y-axis intersects $\mathcal{C}$ at $(0, 1, 1)$ in $\mathbf{P}^2$. (If we work over $\mathbb{C}$, what will $\mathcal{C} \cap H^3_\infty$ be?)

(vi) If we consider the circle $x^2 + y^2 = 1$ over $\mathbb{R}$, then the points on the circle correspond to the points (X, Y, Z) in $\mathbf{P}^2_{\mathbb{R}}$ with $X^2 + Y^2 =$

Z^2, $Z \neq 0$. In this case we see that $\{(X,Y,Z) \in \mathbf{P}^2_{\mathbb{R}} \mid X^2+Y^2 = Z^2\} \cap H^3_\infty = \emptyset$. This shows that the corresponding projective curve has no points at ∞ at all, i.e., it lies entirely in U_3. (What will happen if we work over $\mathbb{C}$?)

Finally, take note of the following prospects.

- Except for the important fact of having constructed a new model of real projective 1-space, the conclusion resulting from Example(ii) is the following: to obtain parametrizations of curves in which no points are missing, it is necessary to consider projective space, even though the original curve lies in the affine plane. For instance, every nonhomogeneous conic

  $$C: \quad ax^2+bxy+cy^2+dx+ey+f, \quad a,b,c,d,e,f \in \mathbb{R},$$

  in $\mathbf{A}^2_{\mathbb{R}}$ corresponds to a quadratic curve

  $$\mathcal{C}: \quad \begin{array}{l} AX^2+2BXY+CY^2+2DXZ+2EYZ+FZ^2 \\ \text{where } A,B,C,D,E,F \in \mathbb{R}, \end{array}$$

  in $\mathbf{P}^2_{\mathbb{R}}$. In CH.VI §2 we will see that if C is nondegenerate, then it has a rational (indeed polynomial) parametric representation through a mapping from $\mathbf{P}^1_{\mathbb{R}}$ to $\mathbf{P}^2_{\mathbb{R}}$ which recovers the conic C without losing any information of C.
- Later in §6, we will see how the affine-projective transfer principle may be set-up in general.
- In CH.V §5, we will see that if k is algebraically closed and $V \subset \mathbf{P}^n_k$ is any projective algebraic set (see the definition in the next section) of dimension > 0, then V meets every hypersurface (see the definition in the next section) in $\mathbf{P}^n_k$, in particular, it contains points at infinity. In particular, we will see in CH.VII §3 (Exercise 2) that any two projective curves of $\mathbf{P}^2_k$ have nonempty intersection.

Exercises for §1

1. Which points in $\mathbf{P}^2$ do not belong to two of the three sets U_1, U_2, U_3?
2. In this exercise, we will study how lines in $\mathbf{A}^n$ relate to points at infinity in $\mathbf{P}^n$. We will use the decomposition $\mathbf{P}^n = U_{n+1} \cup$

H_∞^{n+1} where U_{n+1} resp. H_∞^{n+1} is identified with $\mathbf{A}^n$ resp. with $\mathbf{P}^{n-1}$. Given a line L in $\mathbf{A}^n$, we may parametrize L by the formula $a + bt$, where $a \in L$ and b is a nonzero vector parallel to L. In coordinates, we write this parametrization as $(a_1 + b_1 t, ..., a_n + b_n t)$.

a. We may embed L in $\mathbf{P}^n$ using the homogeneous coordinates

$$(a_1 + b_1 t, ..., a_n + b_n t, 1).$$

To find out what happens as $t \to \pm\infty$, divide by t to obtain

$$\left(\frac{a_1}{t} + b_1, ..., \frac{a_n}{t} + b_n, \frac{1}{t}\right).$$

When $t \to \pm\infty$, which point of $H_\infty^{n+1} = \mathbf{P}^{n-1}$ do you obtain this way?

b. The line L will have many parametrizations. Show that the points of $\mathbf{P}^{n-1}$ given by part a are the same for all parametrization of L. (Hint: Two nonzero vectors are parallel if and only if one is a scalar multiple of the other..)

c. Parts a and b show that a line L in $\mathbf{A}^n$ has a well-defined point at infinity in $H_\infty^{n+1} = \mathbf{P}^{n-1}$. Show that two lines in $\mathbf{A}^n$ are paralell if and only if they have the same point at infinity.

3. Consider the twisted cubic $W = \mathbf{V}(y - x^2, z - x^3) \subset \mathbf{A}^3_{\mathbf{R}}$. If we parametrize W by (t, t^2, t^3) in $\mathbf{A}^n_{\mathbf{R}}$, show that when $t \to \pm\infty$, the point $(t, t^2, t^3, 1)$ in $\mathbf{P}^3_{\mathbf{R}}$ approaches $(0, 0, 1, 0)$. Thus, we expect W to have one point at infinity.

4. Let $F \in k[x_1, ..., x_{n+1}]$ (k infinite). Write $F = \sum F_i$, F_i a homogeneous element of degree i (see CH.I §1). Let $P \in \mathbf{P}^n$, and suppose $F(a_1, ..., a_{n+1}) = 0$ for every choice of homogeneous coordinates $(a_1, ..., a_{n+1})$ for P. Show that each $F_i(a_1, ..., a_{n+1}) = 0$ for all homogeneous coordinates for P. (Hint: Fix $(a_1, ..., a_{n+1})$, and consider $g(\lambda) = F(\lambda a_1, ..., \lambda a_{n+1}) = \sum \lambda^i F_i(a_1, ..., a_{n+1})$.)

§2. The Projective Algebra-Geometry Dictionary

In this section we develop the projective algebra-geometry dictionary with respect to $\mathbf{P}^n = \mathbf{P}^n_k$. Since the concepts and most of the proofs are entirely similar to those for affine algebraic sets, many easy verifications will be left to the reader.

A point $P = (a_1, ..., a_n, a_{n+1}) \in \mathbf{P}^n$ is said to be a *zero* of a polynomial $F \in k[x_1, ..., x_n, x_{n+1}]$ if $F(a_1, ..., a_{n+1}) = 0$ *for every choice of homogeneous coordinates* $(a_1, ..., a_n, a_{n+1})$ for P; we then write $F(P) = 0$.
If we view $k[x_1, ..., x_n, x_{n+1}]$ as a graded ring with the natural gradation (see CH.I. §1) and write F as a sum of homogeneous elements:

$$F = F_m + F_{m-1} + \cdots + F_1 + F_0,$$

where each F_i is a homogeneous element of degree i, then $F(P) = 0$ means that $F_i(a_1, ..., a_n, a_{n+1}) = 0$ for any choice of homogeneous coordinates for P, i.e. $F_i(P) = 0$, and $F_0 = 0$ (§1 Exercise 4).
Thus, in $\mathbf{P}^n$, only the sets of zeros of homogeneous polynomials are considered, i.e., for any set $T \subset k[x_1, ..., x_n, x_{n+1}]$ consisting of *homogeneous elements*, we let

$$\mathbf{V}(T) = \left\{ P \in \mathbf{P}^n \;\middle|\; P \text{ is a common zero of all } F \in T \right\}.$$

If $I = \langle T \rangle$ is the *graded* ideal generated by T, then, comparing with CH.I §2 Proposition 2.1, we have the following basic facts concerning $\mathbf{V}(T)$.

- $\mathbf{V}(I) = \mathbf{V}(T)$;
- There exist homogeneous elements $F_{m_1}, ..., F_{m_r}$ of I such that $I = \langle F_{m_1}, ..., F_{m_r} \rangle$ (the Hilbert basis theorem for graded ideal) and consequently $\mathbf{V}(I) = \mathbf{V}(F_{m_1}, ..., F_{m_r})$, where each F_{m_i} has the total degree m_i.
- If $T_1 \subset T_2$ are subsets consisting of homogeneous polynomials, then $\mathbf{V}(T_1) \supset \mathbf{V}(T_2)$.

2.1. Definition (a) A subset $V \subset \mathbf{P}^n$ is a *projective algebraic set* if $V = \mathbf{V}(T)$ for a subset of homogeneous polynomials in $k[x_1, ..., x_n, x_{n+1}]$.

From the definition (a) it is easy to see that the projective versions of CH.I Proposition 2.3 and Theorem 2.4 hold. Thus, we may also define:

(b) The *Zariski topology* on $\mathbf{P}^n$ is defined by taking projective algebraic sets as closed sets. And for any subset $X \subset \mathbf{P}^n$, the Zariski topology on X is the induced topology.
(c) An algebraic set $V \subset \mathbf{P}^n$ is *irreducible* if it is not the union of two smaller algebraic sets. If V is irreducible, we call V a *projective variety*. An open set of a projective variety is called a *quasi-projective variety*.

As in the affine case (CH.II Proposition 1.3), any nonempty open subset of a projective variety V is necessarily *dense* in V with respect to the Zariski topology on V.

For any subset $V \subset \mathbf{P}^n$, the *ideal* of V in $k[x_1, ..., x_n, x_{n+1}]$ is defined to be

$$\mathbf{I}(V) = \left\{F \in k[x_1, ..., x_n, x_{n+1}] \;\middle|\; F(P) = 0 \text{ for all } P \in V\right\}.$$

2.2. Proposition (i) $\mathbf{I}(V)$ is a graded ideal of $k[x_1, ..., x_n, x_{n+1}]$.
(ii) An algebraic set V is irreducible if and only if $\mathbf{I}(V)$ is a prime ideal.
(iii) Any projective algebraic set can be written uniquely as a union of finitely many projective varieties, its *irreducible components*.

Proof (i) Exercise.
To prove (ii), note that a graded ideal is a prime ideal if and only if it is a *graded* prime ideal (see CH.I. §1).
(iii) This may be proved as in the affine case. □

Note that we have used the same notation for projective algebraic sets and the ideal of a subset of $\mathbf{P}^n$ as in the affine case. In practice it should always be clear which one is meant; if there is any danger of confusion, we will write $\mathbf{V}_p$, $\mathbf{I}_p$ for the projective operations, $\mathbf{V}_a$, $\mathbf{I}_a$ for the affine ones.

2.3. Definition If V is a projective algebraic set in $\mathbf{P}^n$, we define

$$\mathbf{C}(V) = \left\{(a_1, ..., a_{n+1}) \in \mathbf{A}^{n+1} \;\middle|\; \begin{array}{l}(a_1, ..., a_{n+1}) \in V \\ \text{or } (a_1, ..., a_{n+1}) = (0, ..., 0)\end{array}\right\}$$

to be the *affine cone* over V. (Indeed, if $V = \mathbf{V}_p(F_{m_1}, ..., F_{m_s})$ then $\mathbf{C}(V) = \mathbf{V}_a(F_{m_1}, ..., F_{m_s})$.)

2.4. Lemma (i) For any projective algebraic set V, if $V \neq \emptyset$, then

$$\mathbf{I}_a(\mathbf{C}(V)) = \mathbf{I}_p(V).$$

(ii) If I is a graded ideal in $k[x_1, ..., x_n, x_{n+1}]$ such that $\mathbf{V}_p(I) \neq \emptyset$, then

$$\mathbf{C}(\mathbf{V}_p(I)) = \mathbf{V}_a(I).$$

Proof Exercise. □

The above lemma reduces many questions about $\mathbf{P}^n$ to questions about $\mathbf{A}^{n+1}$. For example:

2.5. Theorem (Projective Nullstellensatz) Let k be algebraically closed, and let I be a graded ideal in $k[x_1, ..., x_n, x_{n+1}]$. Then
(i) $\mathbf{V}_p(I) = \emptyset$ if and only if there is an integer N such that I contains all homogeneous elements of degree $\geq N$. (This is equivalent to say that I contains $(k[x_1, ..., x_{n+1}]^+)^N$, where $k[x_1, ..., x_{n+1}]^+ = \oplus_{n\geq 1} k[x_1, ..., x_{n+1}]_n = \langle x_1, ..., x_n, x_{n+1}\rangle$ is the unique maximal graded ideal of $k[x_1, ..., x_{n+1}]$ which is usually called the *positive cone* of $k[x_1, ..., x_{n+1}]$.)
(ii) If $\mathbf{V}_p(I) \neq \emptyset$, then $\mathbf{I}_p(\mathbf{V}_p(I)) = \sqrt{I}$.

Proof (i) The following are equivalent:

$\mathbf{V}_p(I) = \emptyset$;
$\mathbf{V}_a(I) \subset \{(0, ..., 0, 0)\}$
$\sqrt{I} = \mathbf{I}_a(\mathbf{V}_a(I)) \supset \langle x_1, ..., x_n, x_{n+1}\rangle$ (by the affine Nullstellensatz);
$\langle x_1, ..., x_n, x_{n+1}\rangle^N \subset I$.

(ii) $\mathbf{I}_p(\mathbf{V}_p(I)) = \mathbf{I}_a(\mathbf{C}(\mathbf{V}_p(I))) = \mathbf{I}_a(\mathbf{V}_a(I)) = \sqrt{I}$. □

Because of the projective Nullstellensatz, we may phrase the projective ideal-algebraic set correspondence as follows.

2.6. Theorem (i) Let k be an infinite field. Then the maps

$$\left\{\begin{array}{c}\text{graded ideals}\\ \text{in } k[x_1, ..., x_n, x_{n+1}]\end{array}\right\} \underset{\mathbf{I}_p}{\overset{\mathbf{V}_p}{\rightleftarrows}} \left\{\begin{array}{c}\text{projective algebraic sets}\\ \text{in } \mathbf{P}^n\end{array}\right\}$$

are inclusion-reversing. Furthermore, for any projective algebraic set V, we have

$$\mathbf{V}_p(\mathbf{I}_p(V)) = V,$$

so that $\mathbf{I}_p$ is always one-to-one.
(ii) Let k be an algebraically closed field. If we restrict the correspondence of (i) to nonempty projective algebraic sets and radical homogeneous ideals properly contained in $\langle x_1, ..., x_n, x_{n+1}\rangle$, then

$$\left\{ \begin{array}{c} \text{radical graded ideals properly} \\ \text{contained in } \langle x_1, ..., x_n, x_{n+1}\rangle \end{array} \right\}$$

$$\mathbf{V}_p \Big\downarrow \quad \Big\uparrow \mathbf{I}_p$$

$$\left\{ \begin{array}{c} \text{nonempty projective algebraic} \\ \text{sets in } \mathbf{P}^n \end{array} \right\}$$

are inclusion-reversing bijections which are inverses of each other.

□

There is a one-to-one correspondence between *projective hypersurfaces* $V = \mathbf{V}(F) \subset \mathbf{P}^n$ and the (nonconstant) homogeneous polynomials F which define V provided F has no multiple factors (F is determined up to multiplication by a nonzero $\lambda \in k$). Irreducible hypersurfaces correspond to irreducible homogeneous polynomials. If $n = 2$, $V = \mathbf{V}(F)$ is called a *projective plane curve*.
A *hyperplane* is a hypersurface defined by a homogeneous element of degree 1. Hyperplanes in $\mathbf{P}^2$ are called *projective lines*. The hyperplanes $\mathbf{V}(x_i)$, $i = 1, ..., n+1$, may be called the *coordinate hyperplanes*, or the *hyperplanes at* ∞ with respect to U_i. If $n = 2$, the $\mathbf{V}(x_i)$ are the three *coordinate axes*. We also have a correspondence between *irreducible* projective algebraic sets and graded *prime* ideals.

Exercises for §2

1. Prove Proposition 2.2(i), Lemma 2.4.
2. Show that any two distinct lines in $\mathbf{P}^2$ intersect in one point.
3. Let H_1, ..., H_m be hyperplanes in $\mathbf{P}^n$, $m \leq n$. Show that $H_1 \cap \cdots \cap H_m \neq \emptyset$.

4. Show that $V = \mathbf{V}(yw - z^2, xz - y^2, xw - yz) \subset \mathbf{P}^3$ is irreducible. (V is called a twisted cubic curve in $\mathbf{P}^3$.)
5. Decompose the algebraic set defined by two of the equations in Exercise 4 into irreducible components.
6. Show that every irreducible projective hypersurface is determined by an irreducible homogeneous polynomial.

§3. Homogeneous Coordinate Ring and Function Field

3.1. Definition Let V be a projective algebraic set in $\mathbf{P}^n = \mathbf{P}^n_k$ and $\mathbf{I}(V)$ its ideal in $k[x_1, ..., x_n, x_{n+1}]$. Then the quotient ring $k[x_1, ..., x_n, x_{n+1}]/\mathbf{I}(V)$ is called the *homogeneous coordinate ring* of V.

Since $\mathbf{I}(V)$ is a graded ideal (Proposition 2.2), it follows from CH.I §1 Proposition 1.9 that

$$k[x_1, ..., x_n, x_{n+1}]/\mathbf{I}(V) = \oplus_{n \geq 0}(k[x_1, ..., x_n, x_{n+1}]/\mathbf{I}(V))_n$$

is a graded ring with the gradation

$$(k[x_1, ..., x_n, x_{n+1}]/\mathbf{I}(V))_n = (k[x_1, ..., x_n, x_{n+1}]_n + \mathbf{I}(V))/\mathbf{I}(V),\ n \geq 0.$$

So we usually write $k^g[V]$ for the coordinate ring of V.

Remark Let V be a projective algebraic set in $\mathbf{P}^n$. Then, in contrast with the case of affine algebraic sets, no elements of $k^g[V]$ except the constants determine functions on V. That was why we did not introduce $k^g[V]$ by considering the polynomial functions on V, and as a consequence we also do not have a projective analogue of the polynomial mappings on affine algebraic sets. To construct effective mappings between projective algebraic sets later in CH.VI, we have to consider the rational functions below.

If V is a projective variety in $\mathbf{P}^n$, then $\mathbf{I}(V)$ is a prime ideal, so $k^g[V]$ is a domain. Let $Q(V)$ be the quotient field of $k^g[V]$. Then, most elements of $Q(V)$ connot be regarded as functions. However, if $[F]$, $[G]$ are both homogeneous elements in $k^g[V]$ of the same degree d, then $\frac{[F]}{[G]}$ does define a function, at least when $[G]$ is not zero: for

$$\frac{[F]}{[G]}(\lambda(a_1,...,a_n,a_{n+1})) = \frac{\lambda^d F(a_1,...,a_{n+1})}{\lambda^d G(a_1,...,a_{n+1})} = \frac{F(a_1,...,a_{n+1})}{G(a_1,...,a_{n+1})},$$

so the value of $\frac{[F]}{[G]}$ is *independent of the choice of homogeneous coordinates.*

3.2. Definition Let V be a projective variety in $\mathbf{P}^n$. The *function field* of V, denoted $k(V)$, is defined to be

$$k(V) = \left\{ \begin{array}{l} \frac{[F]}{[G]} \in Q(V), \\ \text{for some homogeneous elements } [F],[G] \in k^g[V] \\ \text{with the same degree} \end{array} \right\}.$$

Elements of $k(V)$ are called *rational functions* on V.

One may directly verify that $k(V)$ is a subfield of $Q(V)$, and $k \subset k(V) \subset Q(V)$, but $k^g[V] \not\subset k(V)$.

$k(V)$ may also be obtained by considering another interesting graded ring associated to $k^g[V]$, which is useful in understanding the structure of the more general function field on a variety in CH.VI §2. For each $n \in \mathbb{Z}$, consider the additive subgroup of $Q(V)$

$$Q^g(V)_n = \left\{ \frac{[F]}{[G]} \;\middle|\; \begin{array}{l} [F],[G] \text{ are homogeneous elements} \\ \text{with } \deg([F]) - \deg([G]) = n \end{array} \right\},$$

and define $Q^g(V) = \oplus_{n\in\mathbb{Z}} Q^g(V)_n$, then

- $Q^g(V)$ is a $\mathbb{Z}$-graded ring with the n-th homogeneous component $Q^g(V)_n$ as defined above, and
- $Q^g(V)$ is a *graded field* in the sense that every nonzero homogeneous element is invertible.

3.3. Proposition With the notation as above, $k(V) = Q^g(V)_0$. □

As in the affine case, the elements in $k(V)$ are "locally" defined on V in the sense of the following definition.

3.4. Definition Let V be a projective variety in $\mathbf{P}^n$. Let $P \in V$, $\frac{\phi}{\psi} \in k(V)$. We say that $\frac{\phi}{\psi}$ is *defined* at P if $\frac{\phi}{\psi}$ can be written as $\frac{\phi}{\psi} = \frac{[F]}{[G]}$ for some homogeneous elements $[F], [G]$ of the same degree, and $[G](P) = G(P) \neq 0$.

Using the same notation as in the affine case, for any point $P \in V$, where V is a projective variety, we also let

$$\mathcal{O}_{P,V} = \left\{ \frac{\phi}{\psi} \in k(V) \;\middle|\; \frac{\phi}{\psi} \text{ is defined at } P \right\}.$$

Then the *value* $\frac{\phi}{\psi}(P) = \frac{[F]}{[G]}(P) = \frac{F(P)}{G(P)}$ of a function $\frac{\phi}{\psi} \in \mathcal{O}_{P,V}$ is well-defined. Moreover, by considering the ring homomorphism π: $\mathcal{O}_{P,V} \to k$ with $\pi(z) = z(P)$ we have

3.5. Proposition $\mathcal{O}_{P,V}$ is a subring of $k(V)$; it is a local ring with the unique maximal ideal

$$\mathcal{M}_P = \left\{ \frac{\phi}{\psi} \in k(V) \;\middle|\; \frac{\phi}{\psi} = \frac{[F]}{[G]},\ [G](P) \neq 0,\ [F](P) = 0 \right\}.$$

Proof Exercise. □

It is easy to rephrase the discussion of rational functions given in CH.II §6 for the rational functions defined in this section. We omit repetition here, but a uniform theory including both the affine and the projective cases will be given in CH.VI.

Exercises for §3

1. Show that $Q^g(V)$ is a $\mathbb{Z}$-graded ring in the sense of CH.I §1.
2. Prove Proposition 3.5.

3. Let $\frac{\phi}{\psi} \in k(V)$ where V is a projective variety. Show that the pole set of $\frac{\phi}{\psi}$ (as defined in the affine case, CH.II Definition 6.7) is an algebraic subset of V.

§4. Projective Change of Coordinates

As in the affine case (CH.II §5), we may also study the projective algebraic sets by choosing an *appropriate coordinate system*. This can be seen as follows.

We start with another interpretation of projective space $\mathbf{P}^n = \mathbf{P}^n_k$. Namely, the set of $(n+1)$-tuples over k may be thought of as a vector space V_{n+1} of dimension $n+1$ over k, with the usual addition and scalar multiplication. In terms of the vector space structure, a point of $\mathbf{P}^n$ may be identified with a one dimensional subspace of V_{n+1}. Thus,

- lines in $\mathbf{P}^n$ are defined as two-dimensional subspaces (comparing with Definition 1.1), and
- hyperplanes are defined as n-dimensional subspaces.

Moreover, the linear transformations of V_{n+1} induce the "appropriate" transformations on $\mathbf{P}^n$ as explained hereafter.

Consider $\mathbf{P}^n$ as the set of one-dimensional subspaces of the vector space V_{n+1} of $n+1$-tuples over k. Let ϕ: $V_{n+1} \to V_{n+1}$ be an *invertible* linear transformation. If $(a_1, ..., a_n, a_{n+1}) \in V_{n+1}$ with some $a_i \neq 0$, then $\phi(a_1, ..., a_n, a_{n+1}) = (b_1, ..., b_n, b_{n+1}) \in V_{n+1}$ with some $b_j \neq 0$. If we view $(a_1, ..., a_n, a_{n+1})$ as a point in $\mathbf{P}^n$, then $(b_1, ..., b_n, b_{n+1})$ is also a point in $\mathbf{P}^n$; since ϕ is linear, we have $\phi(\lambda(a_1, ..., a_n, a_{n+1})) = \lambda(b_1, ..., b_n, b_{n+1})$. Thus, ϕ induces a transformation, denoted Φ, from $\mathbf{P}^n$ into itself.

4.1. Definition We say that Φ is a *projective change of coordinates.*

Since ϕ is an invertible linear transformation, it follows that a projective change of coordinates is invertible.

Now let Φ be a projective change of coordinates induced by an invertible linear transformation ϕ: $V_{n+1} \to V_{n+1}$, and let A_ϕ be the matrix of ϕ. Then for any $P = (a_1, ..., a_n, a_{n+1}) \in \mathbf{P}^n$, we have

$$\Phi(a_1, ..., a_n, a_{n+1}) = (a_1, ..., a_n, a_{n+1})A_\phi.$$

From the above equation one infers that a projective change of coordinates Φ preserves incidence properties in the sense that

- points are mapped to points, planes to planes, etc.,
- intersection properties are preserved. For example, if L_1 and L_2 are two lines of $\mathbf{P}^n$ intersecting in a point P then $\Phi(L_1)$ and $\Phi(L_2)$ intersect in $\Phi(P)$.

All these properties are just translations into "projective" language of the fact that *the corresponding linear transformation ϕ of V_{n+1} maps r-dimensional subspaces to r-dimensional subspaces, and their intersections to the intersections of the images.* More generally, we have the following

4.2. Theorem With notation as before, the following properties hold.
(i) If $V = \mathbf{V}(F)$ is a hyperplane in $\mathbf{P}^n$, then $\Phi(V)$ is a hyperplane defined by $G = F\left((x_1, ..., x_n, x_{n+1})A_\phi^{-1}\right)$.
(ii) If V is an algebraic set in $\mathbf{P}^n$, then $\Phi^{-1}(V)$ is also an algebraic set in $\mathbf{P}^n$. More precisely, if $V = \mathbf{V}(F_1, ..., F_r)$, then

$$\Phi^{-1}(V) = \mathbf{V}\left(F_1((x_1, ..., x_n, x_{n+1})A_\phi), ..., F_r((x_1, ..., x_n, x_{n+1})A_\phi)\right)$$

(iii) V is a variety in $\mathbf{P}^n$ if and only if $\Phi^{-1}(V)$ is a variety.
(iv) Φ induces ring isomorphisms

$$k^g[V] \xrightarrow{\cong} k^g[\Phi^{-1}(V)]$$

$$k(V) \xrightarrow{\cong} k(\Phi^{-1}(V))$$

$$\mathcal{O}_P \xrightarrow{\cong} \mathcal{O}_Q, \text{ if } \Phi(Q) = P.$$

Proof Exercise. □

Let V be an algebraic set in $\mathbf{P}^n$ and Φ a projective change of coordinates. In view of Theorem 4.2, we say that V and $\Phi(V)$ are *projectively equivalent*.

Example (i) All hyperplanes $\mathcal{H} \subset \mathbf{P}^n$ are projectively equivalent. Indeed, if $\mathcal{H}$ is defined by the linear equation of the form

$$F = a_1x_1 + \cdots + a_nx_n + a_{n+1}x_{n+1} = 0, \ a_i \in k,$$

where the a_i are not all zero, then by considering the linear transformation $\phi(x_1, ..., x_n, x_{n+1}) \mapsto (X_1, ..., X_n, X_{n+1})$ with

$$\begin{aligned} X_1 &= a_1x_1 + \cdots + a_nx_n + a_{n+1}x_{n+1}, \\ X_2 &= x_2, \\ &\vdots \\ X_n &= x_n, \\ X_{n+1} &= x_{n+1}, \end{aligned}$$

which corresponds to the matrix

$$A_\phi = \begin{pmatrix} a_1 & a_2 & \cdots & a_{n+1} \\ 0 & 1 & \cdots & 0 \\ \vdots & \vdots & \ddots & \vdots \\ 0 & 0 & \cdots & 1 \end{pmatrix}$$

Supppose $a_1 \neq 0$. Then A_ϕ is invertible, and one checks directly that $\mathbf{V}(F)$ and $\mathbf{V}(X_1)$ are projectively equivalent in the sense of Theorem 4.2. A similar argument shows that if $a_i \neq 0$ for some i, then $\mathbf{V}(F)$ is projectively equivalent to $\mathbf{V}(X_i)$ (exercise).
In §1 we observed that $\mathbf{V}(X_1)$ may be viewed as a copy of the projective space $\mathbf{P}^{n-1}$. It follows that all hyperplanes in $\mathbf{P}^n$ look like $\mathbf{P}^{n-1}$.

(ii) Let $\mathbf{V}(F)$ be the quadric in $\mathbf{P}^n$ given by the equation

$$(*) \qquad F = \sum_{i,j=1}^{n+1} a_{ij}x_ix_j = 0, \quad a_{ij} \in k.$$

Assume that char$k \neq 2$. Then from linear algebra we know that $a_{ij} = a_{ji}$, $i,j = 1, ..., n+1$, and we also know that $\mathbf{V}(F)$ is projectively equivalent to a canonical quadric defined by an equation of the form

$$(**) \qquad G = \sum_{i=1}^{n+1} a_ix_i^2 = 0, \quad a_i \in k.$$

In linear algebra, this is done by translating the problem into matrix theory and using certain elementary results on the diagonalization of symmetric matrices. Here we give another proof based on an elementary application of completing squares.
First of all, we reduce the problem to the case $a_{11} \neq 0$:

(a) If $a_{11} = 0$ and $a_{hh} \neq 0$ for some h, then we set $X_1 = x_h$, $X_h = x_1$, and $X_i = x_i$ for $i \neq 0, h$. This, of course, amounts to a projective change of coordinates.

(b) If all $a_{ii} = 0$ then some $a_{hl} \neq 0$ with $h \neq l$ and by a transformation of the previous type, we reduce to the case $a_{12} \neq 0$. Set $X_1 = x_1$, $X_2 = x_2 - x_1$, and $X_i = x_i$ for $i \neq 1, 2$. Then (*) transform into

$$\sum_{i,j=1}^{n+1} a'_{ij} X_i X_j \text{ where } a'_{11} = 2a_{12} \neq 0 \ (\text{char} k \neq 2).$$

Now we arrive at the case where in the original equation $a_{11} \neq 0$.

(c) The next step consists of "completing the square" in x_1. We start from the observation that

$$a_{11}^{-1}\left[a_{11}x_1 + \sum_{i=2}^{n+1} a_{i1}x_i\right]^2 = a_{11}x_1^2 + 2\sum_{i=2}^{n+1} a_{i1}x_1x_i + $$
$$+ \sum_{i,j=2}^{n+1} a_{11}^{-1} a_{i1} a_{1j} x_i x_j.$$

Now set

$$\begin{cases} X_1 = x_1 + a_{11}^{-1} \sum_{i=1}^{n+1} a_{i1} x_i, \\ X_i = x_i, \quad \text{for } i \geq 2. \end{cases}$$

Making this substitution in (*) we obtain

$$a_{11}X_1^2 + H, \text{ where } H = \sum_{i,j=2}^{n+1} b_{ij} X_i X_j, \ b_{ij} = b_{ji} \in k.$$

(d) We now repeat the whole process from the beginning for H and continue in this way until the reduction is complete.

The quadric now has the desired form (**). Observe that if k is algebraically closed then each coefficient a_i in (**) has a square root and so a further transformation of the form

$$\begin{cases} X_i = \sqrt{a_i} \cdot x_i \text{ for } a_i \neq 0, \\ X_i = x_i \text{ for } a_i = 0 \end{cases}$$

plus a very trivial transformation changes (**) to the form

$$(***) \qquad \sum_{i=1}^{r} x_i^2 = 0, \quad r \leq n+1.$$

(iii) Let k be a infinite field and $V = \mathbf{V}(F)$ a curve in $\mathbf{P}_k^2$ defined by a homogeneous polynomial of $k[x_1, x_2, x_3]$. Assume that this curve contains the line L defined by $P = \sum_{i=1}^{3} a_i x_i = 0$. Then there is a polynomial $G \in k[x_1, x_2, x_3]$ such that

$$F = P \cdot G.$$

To see this, suppose $a_1 \neq 0$ (the other case is similar). Consider the projective change of coordinates defined by putting

$$X_1 = P, \quad X_2 = x_2, \quad X_3 = x_3.$$

Then the polynomial $F(x_1, x_2, x_3)$ becomes a polynomial $H(X_1, X_2, X_3)$ and the hypothesis implies that $H(0, X_2, X_3)$ is identically zero as a polynomial function. Since k is infinite, this implies that X_1 divides $H(X_1, X_2, X_3)$. Now the desired result follows from an interpretation of this statement in terms of the original variables.

Exercises for §4

1. Give a detailed proof of Theorem 4.2.
2. In Example (i), if $a_i \neq 0$ for some i, show that $\mathbf{V}(F)$ is projectively equivalent to $\mathbf{V}(x_i)$.

§5. Dehomogenization and Homogenization of Polynomials

Now we investigate the relation between affine algebraic sets(varieties) and projective algebraic sets(varieties) by studying how the ideals in $k[x_1, ..., x_n]$ are related to the *graded* ideals in $k[x_1, ..., x_n, x_{n+1}]$.

Recall that affine n-space $\mathbf{A}^n = \mathbf{A}^n_k$ can be imbedded into the projective n-space $\mathbf{P}^n$ by the map ϕ: $(a_1, ..., a_n) \mapsto (a_1, ..., a_n, 1) \in U_{n+1}$. This ϕ defines a polynomial mapping from $\mathbf{A}^n$ to $\mathbf{A}^{n+1}$, where the latter is by definition the affine cone of $\mathbf{P}^n$, i.e., the coordinate functions are $f_1 = x_1, ..., f_n = x_n, f_{n+1} = 1 \in k[x_1, ..., x_n]$. Then we have the induced ring homomorphism (CH.II §4)

$$\begin{array}{rccc}\phi_* : & k[x_1, ..., x_n, x_{n+1}] & \longrightarrow & k[x_1, ..., x_n] \\ & F & \mapsto & F(x_1, ..., x_n, 1)\end{array}$$

Consider the natural gradation on $k[x_1, ..., x_n]$ resp. on $k[x_1, ..., x_n, x_{n+1}]$ given by total degree of polynomials (CH.I §1). If $f = F_0 + F_1 + \cdots F_d$ is the decomposition of f into homegeneous components in $k[x_1, ..., x_n]$, where each F_i has degree i, then we see that $F = x_{n+1}^d f_0 + x_{n+1}^{d-1} f_1 + \cdots + x_{n+1} f_{d-1} + f_d = x_{n+1}^d f(x_1/x_{n+1}, ..., x_n/x_{n+1})$ is a homogeneous element of degree d in $k[x_1, ..., x_n, x_{n+1}]$ such that $\phi_*(F) = f$. This shows that

- Under the ring homomorphism ϕ_*, every polynomial in $k[x_1, ..., x_n]$ is an image of some homogeneous polynomial in $k[x_1, ..., x_n, x_{n+1}]$.

On the other hand, it is an easy exercise to see that every polynomial $f \in k[x_1, ..., x_n, x_{n+1}]$ can be written as $H + D$, where H is a homogeneous polynomial and $D \in \langle 1 - x_{n+1} \rangle$. Hence the following properties hold:

(i) ϕ_* is surjective with $\ker\phi_* = \langle 1 - x_{n+1} \rangle$. Hence

$$k[x_1, ..., x_n] \cong k[x_1, ..., x_n, x_{n+1}]/\langle 1 - x_{n+1} \rangle.$$

(ii) Considering the natural gradation on $k[x_1, ..., x_n, x_{n+1}]$ given by total degree of polynomials (CH. I §1), we have

$$\frac{k[x_1,...,x_n,x_{n+1}]_p + \langle 1 - x_{n+1}\rangle}{\langle 1 - x_{n+1}\rangle} \subset \frac{k[x_1,...,x_n,x_{n+1}]_{p+1} + \langle 1 - x_{n+1}\rangle}{\langle 1 - x_{n+1}\rangle}$$

for all $p \geq 0$, and

$$\begin{aligned}\bigcup_{p\geq 0} \frac{k[x_1,...,x_n,x_{n+1}]_p + \langle 1 - x_{n+1}\rangle}{\langle 1 - x_{n+1}\rangle} &= \frac{k[x_1,...,x_n,x_{n+1}]}{\langle 1 - x_{n+1}\rangle} \\ &\cong k[x_1,...,x_n].\end{aligned}$$

So far, we have seen how polynomials in $k[x_1,...,x_n]$ and homogeneous polynomials in $k[x_1,...,x_n,x_{n+1}]$ are related via the ring homomorphism ϕ_*. To formalize all this we need:

5.1. Definition (i) For any $F \in k[x_1,...,x_n,x_{n+1}]$, write $F_* = \phi_*(F)$. F_* is called the *dehomogenization* of F with respect to x_{n+1}. (Similarly, one may define the dehomogenization of F with respect to any x_i.)
(ii) For any $f \in k[x_1,...,x_n]$ with $\deg(f) = d$, the homogeneous polynomial of degree d in $k[x_1,...,x_n,x_{n+1}]$, denoted

$$f^* = x_{n+1}^d f(x_1/x_{n+1},...,x_n/x_{n+1}),$$

is called the *homogenization* of f with respect to x_{n+1}. (Similarly, one may define the homogenization of f with respect to any x_i.)

5.2. Proposition (i) $(F+G)_* = F_* + G_*$, $(FG)_* = F_*G_*$.
(ii) $(fg)^* = f^*g^*$, $x_{n+1}^t(f+g)^* = x_{n+1}^r f^* + x_{n+1}^s g^*$, where $r = \deg(g)$, $s = \deg(f)$, and $t = r + s - \deg(f+g)$.
(iii) If F is homogeneous in $k[x_1,...,x_n,x_{n+1}]$ and r is the highest power of x_{n+1} which divides F, then $x_{n+1}^r(F_*)^* = F$; and for any $f \in k[x_1,...,x_n]$, $(f^*)_* = f$.

Proof Exercise. □

5.3. Corollary Up to powers of x_{n+1}, factoring a homogeneous element $F \in k[x_1,...,x_n,x_{n+1}]$ is the same as factoring $F_* \in k[x_1,...,x_n]$. In particular, if $F \in k[X,Y]$ is a homogeneous element, k algebraically closed, then F factors into a product of linear factors.

Proof The first claim follows directly from the proposition above. For the second, write $F = Y^r G$, where Y does not divide G. $F_* = G_* = \lambda\Pi(x - \lambda_i)$ since k is algebraically closed, so $F = \lambda Y^r \Pi(X - \lambda_i Y)$. □

Now, let I be an ideal of $k[x_1, ..., x_n]$. With notation as above, let I^* be the ideal in $k[x_1, ..., x_n, x_{n+1}]$ generated by $\{f^* \mid f \in I\}$. We call I^* the *homogenization* of I with respect to x_{n+1}.

5.4. Lemma (i) I^* is a graded ideal of $k[x_1, ..., x_n, x_{n+1}]$.
(ii) If I is an ideal of $k[x_1, ..., x_n]$, then every homogeneous element of I^* is of the form $x_{n+1}^r f^*$ where $r \geq 0$, $f \in I$.
(iii) Let F be a polynomial in $k[x_1, ..., x_n, x_{n+1}]$. Then $F \in I^*$ if and only if $F_* \in I$.

Proof Exercise. □

Coversely, Let I be an ideal of $k[x_1, ..., x_n, x_{n+1}]$. With notation as in the last section, then $\phi_*(I) = I_*$ is an ideal in $k[x_1, ..., x_n]$ since ϕ_* is a ring epimorphism. We call I_* the *dehomogenization* of I with respect to x_{n+1}.

5.5. Proposition (i) If I is an ideal of $k[x_1, ..., x_n]$, then $(I^*)_* = I$, and moreover $\left(\sqrt{I^*}\right)_* = \sqrt{I}$.
(ii) If I is a radical ideal of $k[x_1, ..., x_n]$, i.e., $I = \sqrt{I}$, then I^* is a radical ideal in $k[x_1, ..., x_n, x_{n+1}]$, i.e., $I^* = \sqrt{I^*}$. (Note that $\sqrt{I^*}$ is a graded ideal, see CH.I §6 Exercise 1.)
(iii) If I is a graded ideal of $k[x_1, ..., x_n, x_{n+1}]$, then $I \subset (I_*)^*$.

Proof Exercise. □

If I is a graded ideal in $k[x_1, ..., x_n, x_{n+1}]$, then generally $(I_*)^* \neq I$. For example, let $I = \langle zx, zy \rangle$ in $k[x, y, z]$. Taking the dehomogenization of I with respect to z, we see that $x, y \in (I_*)^*$ but $x, y \notin I$. Hence $(I_*)^* \neq I$.

5.6. Definition A graded ideal $I \subset k[x_1, ..., x_n, x_{n+1}]$ is said to be $(\phi_*)^*$*-closed*, if $(I_*)^* = I$.

From the above definition and Proposition 5.5(i) we immediately derive the following inclusion-preserving correspondence:

$$\left\{\begin{array}{c}(\phi_*)^*-\text{closed graded ideals}\\ \text{of } k[x_1,...,x_n,x_{n+1}]\end{array}\right\} \underset{I^*}{\overset{\phi_*}{\rightleftarrows}} \left\{\begin{array}{c}\text{ideals of}\\ k[x_1,...,x_n]\end{array}\right\}$$

5.7. Theorem Let I be a graded ideal of $k[x_1,...,x_{n+1}]$. Then the following are equivalent.
(i) I is $(\phi_*)^*$-closed;
(ii) The factor ring $k[x_1,...,x_{n+1}]/I$, viewed as a $k[x_1,...,x_{n+1}]$-module (see Appendix 1), is x_{n+1}-torsinfree, i.e., x_{n+1} does not annihilate any nonzero element of $k[x_1,...,x_{n+1}]/I$;
(iii) $x_{n+1}k[x_1,...,x_{n+1}]\cap I = x_{n+1}I$.

Proof (i) $\Rightarrow$ (ii) Since the natural gradation on $k[x_1,...,x_{n+1}]$ induces a gradation on $k[x_1,...,x_{n+1}]/I$ that makes $k[x_1,...,x_{n+1}]/I$ into a positively graded ring (CH.I §1), i.e.,

$$\begin{cases} k[x_1,...,x_{n+1}]/I = \oplus_{p\geq 0}(k[x_1,...,x_{n+1}]/I)_p \\ \\ \text{with } (k[x_1,...,x_{n+1}]/I)_p = (k[x_1,...,x_{n+1}]_p + I)/I,\ p\geq 0. \end{cases}$$

It is sufficient to show that x_{n+1} does not annihilate any nonzero homogeneous element of $k[x_1,...,x_{n+1}]/I$. Let $F\in k[x_1,...,x_{n+1}]_p$ be a homogeneous polynomial of degree p. If $x_{n+1}F\in I$, then $F_* = (x_{n+1}F)_*\in I_*$ by Proposition 5.2(i). Since F is homogeneous and I is $(\phi_*)^*$-closed, it follows from Proposition 5.2(iii) that there is $r\geq 0$ such that $F = x^r_{n+1}(F_*)^*\in (I_*)^* = I$. This proves (ii).
(ii) $\Rightarrow$ (iii) we only need to show that $x_{n+1}k[x_1,...,x_{n+1}]\cap I\subset x_{n+1}I$. But this is clear from (ii).
(iii) $\Rightarrow$ (i) In view of Proposition 5.5(ii) it is sufficient to establish that $(I_*)^*\subset I$. Let F be any homogeneous element in $(I_*)^*$. Then by Lemma 5.4(ii) we have $F = x^r_{n+1}f^*$ for some $r\geq 0$ and $f\in I_*$. Suppose $f = G_*$ with $G\in I$. Since I is a graded ideal, we may assume that G is a homogenmeous element of I. Then by Proposition 5.2(iii) there exists some $l\geq 0$ such that $x^l_{n+1}(G_*)^* = G$. Thus, $F = x^r_{n+1}f^* = x^r_{n+1}(G_*)^*$, and $x^l_{n+1}F = x^r_{n+1}G\in I$. It follows from (iii) that $F\in I$. This completes the proof. □

5.8. Corollary Let I be a graded ideal of $k[x_1, ..., x_{n+1}]$. Then

$$(I_*)^* = \left\{F \in k[x_1, ..., x_{n+1}] \;\middle|\; x_{n+1}^r F \in I \text{ for some } r \geq 0\right\},$$

i.e, every element in $(I_*)^*/I$ is x_{n+1}-torsion, and $k[x_1, ..., x_{n+1}]/(I_*)^*$ is x_{n+1}-torsionfree.

□

The next proposition provides us with examples of $(\phi_*)^*$-closed ideals that will be used in CH.VII §2.

5.9. Proposition Let F, G be nonconstant homogeneous polynomials in $k[x_1, ..., x_{n+1}]$ and $\mathbf{V}(F)$, $\mathbf{V}(G)$ the corresponding projective hypersurfaces in $\mathbf{P}^n$. Suppose that $\mathbf{V}(x_{n+1}) \cap \mathbf{V}(F) \cap \mathbf{V}(G) = \emptyset$. Then $I = \langle F, G \rangle$ is $(\phi_*)^*$-closed.

Proof As a consequence of Theorem 5.7 it is sufficient to prove that x_{n+1} does not annihilate any nonzero element of $k[x_1, ..., x_{n+1}]/I$. Let us denote $F(x_1, ..., x_n, 0)$ by F_0 for every $F \in k[x_1, ..., x_{n+1}]$. Since $\mathbf{V}(F)$, $\mathbf{V}(G)$ and $\mathbf{V}(x_{n+1})$ have no common zeros, we may write

$$F = F_0 + x_{n+1}F', \quad F_0 \neq 0,$$
$$G = G_0 + x_{n+1}G', \quad G_0 \neq 0.$$

We claim that F_0 and G_0 have no common factors (check this!). Now if $x_{n+1}H = AF + BG \in I$, then $A_0F_0 = -B_0G_0$, so

$$\begin{matrix} B_0 = F_0C, \\ A_0 = -G_0C \end{matrix} \quad \text{for some } C \in k[x_1, ..., x_n].$$

Let $A_1 = A + CG$, $B_1 = B - CF$. Since $(A_1)_0 = (B_1)_0 = 0$, we have $A_1 = x_{n+1}A'$, $B_1 = x_{n+1}B'$ for some $A', B' \in k[x_1, ..., x_n, x_{n+1}]$. Since $x_{n+1}H = A_1F + B_1G$, it follows that $H = A'F + B'G \in I$, as desired. □

Certain related problems wll be considered later, e.g.:

- We will see in the next section (Proposition 6.9) that when k is algebraically closed, $\mathbf{V}((I_*)^*)$ is the smallest projective algebraic set containing $\mathbf{V}(I_*)$ (identified with $\phi(\mathbf{V}(I_*))$).
- In CH.IV, we will exhibit an algorithm that may be used to check whether a given graded ideal is $(\phi_*)^*$-closed or not.

Exercises for §5

1. If $F \in k[x_1, ..., x_n, x_{n+1}]$, prove $F = H + D$, where H is a homogeneous polynomial and $D \in \langle 1 - x_{n+1} \rangle$.
2. Prove Proposition 5.2, Lemma 5.4, and Proposition 5.5.
3. Factor $y^3 - 2xy^2 + 2x^2y + 4x^3$ into linear factors in $\mathbb{C}[x, y]$.
4. Suppose $F, G \in k[x_1, ..., x_n]$ are homogeneous elements of degree r, $r + 1$ respectively, with no common factors (k a field). Show that $F + G$ is irreducible.
5. Let $F, G \in k[x_1, ..., x_{n+1}]$ be homogeneous elements. Suppose that F and G have no common factors. Show that F_* and G_*, the dehomogenizations of F and G with respect to some x_i, do not have common factors. Is the same true for homogenizations of two polynomials?
6. Find all subvarieties in $\mathbf{P}^1$ and $\mathbf{P}^2$.
7. Let $L_1, L_2, ...$ and $M_1, M_2, ...$ be sequences of linear forms in $k[x, y]$, and assume no $L_i = \lambda M_j$, $\lambda \in k$. Let

$$F_{ij} = L_1 L_2 \cdots L_i \cdot M_1 M_2 \cdots M_j, \quad i, j \geq 0, (F_{00} = 1).$$

Show that $\{F_{ij} \mid i + j = d\}$ forms a bsis for the k-subspace $k[x, y]_d$ consisting of all homogeneous polynomials of degree d.

§6. Affine-Projective Transfer of Algebraic Sets

As before, we let ϕ: $\mathbf{A}^n = \mathbf{A}^n_k \to U_{n+1} \subset \mathbf{P}^n = \mathbf{P}^n_k$ be the $n+1$-st imbedding of $\mathbf{A}^n$ in $\mathbf{P}^n$ with $\phi(a_1, ..., a_n) = (a_1, ..., a_n, 1)$, and let ϕ_* is the dehomogenization map induced by ϕ. Inspired by the correspondence between the ideals of $k[x_1, ..., x_n]$ and the $(\phi_*)^*$-closed graded ideals of $k[x_1, ..., x_n, x_{n+1}]$ obtained in the last section, we are about to determine the exact correspondence between the algebraic sets in $\mathbf{A}^n$ and those in $\mathbf{P}^n$.

Let $W \subset \mathbf{A}^n$ be an affine algebraic set, and let J be an ideal of $k[x_1, ..., x_n]$ such that $W = \mathbf{V}(J)$. If J^* is the homogenized ideal of J in $k[x_1, ..., x_n, x_{n+1}]$ with respect to x_{n+1}, then we have the projective algebraic set

$$\mathbf{V}(J^*) \subset \mathbf{P}^n.$$

Conversely consider a projective algebraic set $V \subset \mathbf{P}^n$ and let I be a graded ideal of $k[x_1, ..., x_n, x_{n+1}]$ such that $V = \mathbf{V}(I)$. If I_* is the dehomogenized ideal of I in $k[x_1, ..., x_n]$, with respect to x_{n+1}, then we have the affine algebraic set

$$\mathbf{V}(I_*) \subset \mathbf{A}^n.$$

6.1. Lemma (i) Let $W \subset \mathbf{A}^n$ be an affine algebraic set, and let J be *any* ideal of $k[x_1, ..., x_n]$ such that $W = \mathbf{V}(J)$. Then $\mathbf{V}(J^*) = \mathbf{V}\left(\left(\sqrt{J}\right)^*\right)$.
(ii) Let $V \subset \mathbf{P}^n$ be a projective algebraic set, and let I be *any* graded ideal of $k[x_1, ..., x_n, x_{n+1}]$ such that $V = \mathbf{V}(I)$. Then $\mathbf{V}(I_*) = \mathbf{V}\left(\left(\sqrt{I}\right)_*\right)$.

Proof Using Proposition 5.2, verification becomes straightforward. □

6.2. Lemma Let I be a graded ideal of $k[x_1, ..., x_n, x_{n+1}]$.
(i) If I is $(\phi_*)^*$-closed, then so is $\sqrt{I}$.
(ii) If I is $(\phi_*)^*$-closed, then $\sqrt{I_*} = \left(\sqrt{I}\right)_*$.

Proof (i) Using Theorem 5.7(ii), this becomes a straightforward verification.
(ii) Let $f \in \sqrt{I_*}$. Then $f^n \in I_*$ for some $n > 0$. Proposition 5.2 and the assumption entail $f^{*n} \in (I_*)^* = I$. It follows that $f^* \in \sqrt{I}$, and $f = (f^*)_* \in \left(\sqrt{I}\right)_*$. This shows that $\sqrt{I_*} \subset \left(\sqrt{I}\right)_*$. Conversely, since $\sqrt{I}$ is a graded ideal (CH.I §6 Exercise 1), let $F \in \sqrt{I}$ be a homogeneous element. Then $F^N \in I$ for some $N > 0$. It follows from Proposition 5.2 that $F_*^N \in I_*$, and hence $F_* \in \sqrt{I_*}$. This shows that $\left(\sqrt{I}\right)_* \subset \sqrt{I_*}$. Therefore, $\sqrt{I_*} = \left(\sqrt{I}\right)_*$, as desired. □

Now, for each affine algebraic set $W = \mathbf{V}(J) \subset \mathbf{A}^n$ defined by *some* ideal $J \subset k[x_1, ..., x_n]$, let us associate W to $\mathbf{V}(J^*) \subset \mathbf{P}^n$; Similarly, for each projective algebraic set $V = \mathbf{V}(I) \subset \mathbf{P}^n$ defined by *some* graded ideal $I \subset k[x_1, ..., x_n, x_{n+1}]$, let us associate V to $\mathbf{V}(I_*) \subset \mathbf{A}^n$.

Then the Nullstellensatz, Proposition 5.5(i), Lemma 6.1 and Lemma 6.2 together yield the following

6.3. Theorem If k is algebraically closed, there is an inclusion-preserving bijective correspondence:

$$\left\{\begin{array}{c}\text{projective algebraic sets } V = \mathbf{V}(I) \subset \mathbf{P}^n \\ \text{defined by some } (\phi_*)^* - \text{closed} \\ \text{graded ideal } I \text{ of } k[x_1, ..., x_n, x_{n+1}]\end{array}\right\}$$

$$\updownarrow$$

$$\left\{\begin{array}{c}\text{affine algebraic sets} \\ \\ \text{of } \mathbf{A}^n\end{array}\right\}$$

□

To deal with varieties, we need more preparation.

6.4. Proposition (i) Let J be an ideal in $k[x_1, ..., x_n]$. If J is a prime, then so is J^* in $k[x_1, ..., x_n, x_{n+1}]$.
(ii) Let I be a graded ideal in $k[x_1, ..., x_n, x_{n+1}]$. If I is $(\phi_*)^*$-closed, then the following are equivalent:
  (a) I is a prime ideal;
  (b) I_* is a prime ideal in $k[x_1, ..., x_n]$.

Proof Using Proposition 5.2 and the fact that I is prime if and only if it is graded prime (CH.I. §1), this is straightforward. □

6.5. Proposition Let I be a graded prime ideal of $k[x_1, ..., x_n, x_{n+1}]$ and $V = \mathbf{V}(I) \subset \mathbf{P}^n$ the projective variety defined by I. Then the following are equivalent:
  (a) I is $(\phi_*)^*$-closed;
  (b) $V \not\subset H_\infty^{n+1} = \mathbf{P}^n - U_{n+1} = \mathbf{V}(x_{n+1})$, i.e., V is not contained in the hyperplane at infinity.

Proof (i) ⇒ (ii) If we assume the opposite, then $V \subset H_\infty^{n+1}$ and consequently $x_{n+1} \in I$. Thus, $1 = \phi(x_{n+1}) \in I_*$, and hence $1 = \phi(x_{n+1})^* \in (I_*)^* = I$, a contradiction, because I is a prime ideal. So $V \not\subset H_\infty^{n+1}$.

(ii) ⇒ (i) That $V \not\subset H_\infty^{n+1}$ implies $x_{n+1}^N \notin I$ for any $N \geq 0$. It follows that x_{n+1} does not annihilate any nonzero element of $k[x_1, ..., x_{n+1}]/I$. By Theorem 5.7, I is $(\phi_*)^*$-closed, as desired. This finishes the proof of the proposition. □

So we arrive at the correspondence between affine and projective varieties.

6.6. Theorem Let k be an algebraically closed field. The correspondence exhibited in Theorem 6.3 defines an inclusion-preserving bijective correspondence:

$$\left\{\begin{array}{c}\text{projective varieties}\\ \text{of } \mathbf{P}^n\\ \text{not lying in } H_\infty^{n+1}\end{array}\right\} \longleftrightarrow \left\{\begin{array}{c}\text{affine varieties}\\ \\ \text{of } \mathbf{A}^n\end{array}\right\}$$

□

Taking the imbedding ϕ: $\mathbf{A}^n \to U_{n+1} \subset \mathbf{P}^n$ into account, for any affine algebraic set $V \subset \mathbf{A}^n$ and any defining ideal I of V, the natural question is

- What is the exact geometric relation between V, identified with $\phi(V)$, and the projective algebraic set $\mathbf{V}(I^*) \subset \mathbf{P}^n$?

The answer to this question is given by the next proposition.

6.7. Proposition With notation as before, let $I \subset k[x_1, ..., x_n]$ be *any* ideal such that $V = \mathbf{V}(I)$. Then
(i) $\phi(V) = \mathbf{V}(I^*) \cap U_{n+1}$. If $I = \langle f_1, ..., f_s \rangle$, then $\phi(V) = \mathbf{V}(f_1^*, ..., f_s^*) \cap U_{n+1}$.
(ii) $\mathbf{I}(V) = \mathbf{I}(\phi(V))_*$.
(iii) $\mathbf{V}(\mathbf{I}(V)^*) \subset \mathbf{V}\left(\sqrt{I^*}\right) \subset \mathbf{V}(I^*)$, and $\mathbf{V}(\mathbf{I}(V)^*)$ is the smallest projective algebraic set in $\mathbf{P}^n$ containing $\phi(V)$.
(iv) $\mathbf{I}(\mathbf{V}(\mathbf{I}(V)^*)) = \mathbf{I}(V)^*$.

Proof (i) For any $P = (a_1, ..., a_n) \in V$, $\phi(P) = (a_1, ..., a_n, 1) \in U_{n+1}$. For any nonzero $f \in I$, $f^*(\phi(P)) = f(P) = 0$. It follows that $\phi(P) \in \mathbf{V}(I^*) \cap U_{n+1}$, and hence $\phi(V) \subset \mathbf{V}(I^*) \cap U_{n+1}$. Conversely, for any $P = (a_1, ..., a_n, a_{n+1}) \in \mathbf{V}(I^*) \cap U_{n+1}$, we may assume that $a_{n+1} = 1$. Now for any nonzero $f \in I$, we have $0 = f^*(P) =$

$f(a_1, ..., a_n)$. Hence, $Q = (a_1, ..., a_n) \in V$ and $P = \phi(Q)$. This shows that $\mathbf{V}(I^*) \cap U_{n+1} \subset \phi$ (V). Therefore, $\phi(V) = \mathbf{V}(I^*) \cap U_{n+1}$.

(ii) From (i) we have $\phi(V) = \mathbf{V}(\mathbf{I}(V)^*) \cap U_{n+1}$. It follows that for any $f \in \mathbf{I}(V)$, $f^* \in \mathbf{I}(V)^* \subset \mathbf{I}(\mathbf{V}(\mathbf{I}(V)^*)) \subset \mathbf{I}(\phi(V))$. Hence, $f = (f^*)_* \in \mathbf{I}(\phi(V))_*$ by Proposition 5.2(iii), i.e., $\mathbf{I}(V) \subset \mathbf{I}(\phi(V))_*$. Conversely, if $F \in \mathbf{I}(\phi(V))$, then for any $P \in V$, $0 = F(\phi(P)) = F_*(P)$. It follows that $F_* \in \mathbf{I}(V)$. This proves $\mathbf{I}(\phi(V))_* \subset \mathbf{I}(V)$. Therefore, $\mathbf{I}(V) = \mathbf{I}(\phi(V))_*$.

(iii) Suppose that W is an algebraic set in $\mathbf{P}^n$ which contains $\phi(V)$. Then $\mathbf{I}(W) \subset \mathbf{I}(\phi(V))$. If $F \in \mathbf{I}(W)$ is any homogeneous element, then $F_* \in \mathbf{I}(\phi(V))_* = \mathbf{I}(V)$ by (ii). So $F = x_{n+1}^r(F_*)^* \in \mathbf{I}(V)^*$ for some $r \geq 0$ by Proposition 5.2(iii). Hence, $\mathbf{I}(W) \subset \mathbf{I}(V)^*$, so $W \supset \mathbf{V}(\mathbf{I}(V)^*)$. This finishes the proof of the Proposition.

(iv) We only have to prove $\mathbf{I}(\mathbf{V}(\mathbf{I}(V)^*)) \subset \mathbf{I}(V)^*$. Since $\mathbf{I}(\mathbf{V}(\mathbf{I}(V)^*))$ is a graded ideal, we only consider the homogeneous elements. Let $F \in \mathbf{I}(\mathbf{V}(\mathbf{I}(V)^*))$ be a homogeneous element. Then by (i) we have $0 = F(\phi(P)) = F_*(P)$ for all $P \in V$, and hence $F_* \in \mathbf{I}(V)$. It follows from Proposition 5.2(iii) that $F = x_{n+1}^r(F_*)^* \in \mathbf{I}(V)^*$, as desired. □

6.8. Definition Let V be an affine algebraic set in $\mathbf{A}^n$. Then the *projective closure* of V, denoted by $\overline{V}$, is defined to be the projective algebraic set $\mathbf{V}(\mathbf{I}(V)^*) \subset \mathbf{P}^n$.

Using the projective closure of an affine algebraic set as defined above, we may strengthen the correspondence in Theorem 6.3 and Theorem 6.6.

6.9. Proposition Let $V \subset \mathbf{P}^n$ be a projective algebraic set. Let I be *any* graded ideal of $k[x_1, ..., x_n, x_{n+1}]$ such that $V = \mathbf{V}(I)$. then

(i) $\phi(\mathbf{V}(I_*)) = V \cap U_{n+1}$.

(ii) $\phi(\mathbf{V}(I_*)) = \mathbf{V}((I_*)^*) \cap U_{n+1}$.

(iii) If k is algebraically closed, then $\mathbf{V}((I_*)^*) = \overline{\mathbf{V}(I_*)}$; if furthermore I is $(\phi_*)^*$-closed, then $V = \overline{\mathbf{V}(I_*)}$. Hence, any nonempty projective variety is the projective closure of some affine variety.

Proof (i) This is straightforward.

(ii) This follows from Proposition 6.7(i).

(iii) Since k is algebraically closed, we have

$$\begin{aligned}\overline{\mathbf{V}(I_*)} &= \mathbf{V}(\mathbf{I}(\mathbf{V}(I_*))^*) \text{ by Definition 6.8}\\ &= \mathbf{V}\left(\left(\sqrt{I_*}\right)^*\right) \text{ by the Nullstellensatz}\\ &= \mathbf{V}((I_*)^*) \text{ by Lemma 6.1(i)}\end{aligned}$$

The second conclusion is clear from the first and Proposition 6.5. □

If k is algebraically closed, then for an affine algebraic set $V = \mathbf{V}(I)$ with $I \subset k[x_1, ..., x_n]$, $\overline{V} = \mathbf{V}(I^*) \subset \mathbf{P}^n$. Unfortunately, this fails over fields that are not algebraically closed. Here is an example showing what can go wrong.

Example (i) Consider $I = \langle x^2 + y^4 \rangle \subset I\!R[x, y]$. Then $W = \mathbf{V}(I)$ consists of the single point $(0,0)$ in $\mathbf{A}^2_{I\!R}$, and, hence, the projective closure is the single point $\overline{W} = \{(0,0,1)\} \subset \mathbf{P}^2_{I\!R}$ (since this is obviously the smallest projective algebraic set containing W which is identified with $\phi(W) \subset \mathbf{P}^2_{I\!R} - \mathbf{V}(z)$).
On the other hand, $I^* = \langle x^2z^2 + y^4 \rangle \subset I\!R[x, y, z]$, it is easy to check that

$$\mathbf{V}(I^*) = \{(0,0,1), (1,0,0)\} \subset \mathbf{P}^2_{I\!R}.$$

This shows that $\mathbf{V}(I^*)$ is strictly larger than the projective closure of $W = \mathbf{V}(I)$.

(ii) Let $V = \mathbf{V}(I)$ be an affine algebraic set in $\mathbf{A}^n$ defined by an ideal I of $k[x_1, ..., x_n]$. By Lemma 6.3 we know that if k is algebraically closed then the projective closure V^* of V in $\mathbf{P}^n$ is defined by I^*. If furthermore we assume that $I = \langle f_1, ..., f_s \rangle$, then we might expect that I^* is generated by $\{f_1^*, ..., f_s^*\}$ in $k[x_1, ..., x_n, x_{n+1}]$. Unfortunately, that is not always the case. For instance, let $I = \langle f = y - x^2, g = z - x^3 \rangle \subset k[x, y, z]$. Then $zw - xy \in I^* \subset k[x, y, z, w]$, but $zw - xy \notin \langle f^*, g^* \rangle$, i.e., $I^* \neq \langle f^*, g^* \rangle$. Nevertheless, we might still expect $\overline{V} = \mathbf{V}(f_1^*, ..., f_s^*)$ because of Proposition 6.7(i). But this is unfortunately wrong again if we look at the twisted cubic $W = \mathbf{V}(f = y - x^2, g = z - x^3) \subset \mathbf{A}^3_{I\!R}$ once more. Note that for $V = \mathbf{V}(f^* = yw - x^2, g^* = zw^2 - x^3) \subset \mathbf{P}^3_{I\!R}$, $V = \phi(W) \cup (V \cap H^4_\infty)$ where $V \cap H^4_\infty = \mathbf{V}(f^*, g^*, w)$. One easily sees that $\mathbf{V}(f^*, g^*, w) = \mathbf{V}(w, x)$. The coordinates y and z are arbitrary here, so $V \cap H^4_\infty$ is the projective line $\mathbf{V}(x, w) \subset \mathbf{P}^3_{I\!R}$. Thus, we have $V = \phi(W) \cup \mathbf{V}(x, w)$.

But in CH.IV §5 we will see that $I^* = \langle x^2 - yw, xy - zw, xz - y^2\rangle$. In a similar way we derive that $\mathbf{V}(I^*) \cap H^4_\infty = \{(0,0,1,0)\}$, a single point. Since W is a curve in $\mathbf{A}^3_{\mathbb{R}}$, this fits our intuition that W should only have a finite number of points at infinity (see §1 Exercise 3). Moreover, this also establishes clearly that V is really "bigger" than $V(I^*)$.
In CH.IV §5, we will see that there is an algorithm which can be used to produce a generating set $\{g_1, ..., g_m\}$ of I^* from any given generating set $\{f_1, ..., f_s\}$ of I.

The next theorem deals with the relation between the coordinate ring resp. function field of an affine algebraic set resp. an affine variety V and those of $\overline{V}$. (Indeed it gives more.)

6.10. Theorem Let $V \subset \mathbf{A}^n$ be a *nonempty* affine algebraic set, and let I be *any* ideal of $k[x_1, ..., x_n]$ such that $V = \mathbf{V}(I)$.
(i) The map

$$\begin{array}{rccc} \alpha: & k[x_1, ..., x_{n+1}]/I^* & \longrightarrow & k[V] \\ & [F] & \mapsto & [F_*] \end{array}$$

is a ring epimorphism with $\mathrm{Ker}\alpha = \langle 1 - [x_{n+1}]\rangle$, where $[x_{n+1}]$ is the class of x_{n+1} in $k[x_1, ..., x_n, x_{n+1}]/I^*$.
(ii) If V is a variety, then α induces a field isomorphism

$$\begin{array}{rccc} F_\alpha: & k\left(\overline{V}\right) & \xrightarrow{\cong} & k(V) \\ & \dfrac{[F]}{[G]} & \mapsto & \dfrac{[F_*]}{[G_*]} \end{array}$$

where F and G are homogeneous elements of the same degree and $G \notin \mathbf{I}(V^*)$.
(iii) In the situation of (ii), for any $P \in V$, considering $Q = \phi(P) \in \overline{V}$, F_α induces an isomorphism of local rings:

$$L_\alpha: \qquad \mathcal{O}_Q \xrightarrow{\cong} \mathcal{O}_P$$

such that $L_\alpha(\mathcal{M}_Q) = \mathcal{M}_P$.

Proof (i) By proposition 5.2, α is well-defined. Every polynomial $F \in k[x_1, ..., x_n, x_{n+1}]$ may be expressed as $F = H + D$ where H is a homogeneous element and $D \in \langle 1 - x_{n+1} \rangle$ (§5 Exercise 1). Hence $\mathrm{Ker}\alpha = \langle 1 - [x_{n+1}] \rangle$.
(ii) This follows from Proposition 6.4(i), Proposition 6.7(iv) and a direct verification.
(iii) Direct verification. □

From Proposition 6.4, Lemma 6.5, Lemma 6.6 and Proposition 6.9 we immediately obtain the following theorem.

6.11. Theorem Let $V \subset \mathbf{P}^n$ be a projective algebraic set, and $I \subset k[x_1, ..., x_n, x_{n+1}]$ *any* graded ideal such that $V = \mathbf{V}(I)$.
(i) The map

$$\begin{array}{rccc} \alpha: & k[x_1, ..., x_{n+1}]/I & \longrightarrow & k[\mathbf{V}(I_*)] \\ & [F] & \mapsto & [F_*] \end{array}$$

is a ring epimorphism with $\mathrm{Ker}\alpha = \langle 1 - [x_{n+1}] \rangle$, where $[x_{n+1}]$ is the class of x_{n+1} in $k[x_1, ..., x_n, x_{n+1}]/I$.
(ii) If V is a variety which is not contained in H_{∞}^{n+1}, then α induces a field isomorphism

$$\begin{array}{rccc} F_\alpha: & k(V) & \xrightarrow{\cong} & k(\mathbf{V}(I_*)) \\ & \dfrac{[F]}{[G]} & \mapsto & \dfrac{[F_*]}{[G_*]} \end{array}$$

where F and G are homogeneous elements of the same degree and $G \notin \mathbf{I}(V^*)$.
(iii) In the situation of (ii), for any $P \in \mathbf{V}(I_*)$ and $Q = \phi(P) \in V$, F_α induces an isomorphism of local rings:

$$L_\alpha: \qquad \mathcal{O}_Q \xrightarrow{\cong} \mathcal{O}_P$$

such that $L_\alpha(\mathcal{M}_Q) = \mathcal{M}_P$.

□

Summary: Let us fix notation as follows: $U_{n+1} \subset \mathbf{P}^n$ and

$$\varphi: \quad U_{n+1} \to \mathbf{A}^n$$

$$\phi: \quad \mathbf{A}^n \to U_{n+1}$$

are as before such that $\varphi = \phi^{-1}$. Let $W = \mathbf{V}(J) \subset \mathbf{A}^n$ be an affine algebraic set and $V = \mathbf{V}(I) \subset \mathbf{P}^n$ a projective algebraic set, respectively. Then

- $\phi(W) = \mathbf{V}(J^*) \cap U_{n+1}$ by Proposition 6.7(i), and
- $\phi(\mathbf{V}(I_*)) = V \cap U_{n+1}$ by Proposition 6.9(i).

Moreover, suppose V is a projective variety, then $\mathbf{I}(V)$ is a graded prime ideal (Proposition 2.2). If $V \cap U_{n+1} \neq \emptyset$, then

- $V \not\subset \mathbf{V}(x_{n+1}) = H_\infty^{n+1} = \mathbf{P}^n - U_{n+1}$. It follows from Lemma 6.6 that $(\mathbf{I}(V)_*)^* = \mathbf{I}(V)$, i.e., $\mathbf{I}(V)$ is $(\phi_*)^*$-closed. Hence $\mathbf{V}(\mathbf{I}(V)_*)$ is irreducible by Lemma 6.5.

The above discussion leads to the following important result that is fundamental in modern algebraic geometry. We refer to CH.V, CH.VI for further applications.

6.12. Theorem With notation as before, the following properties hold with respect to the Zariski topology on an algebraic set.
(i) For $1 \leq i \leq n+1$, φ_i: $U_i \to \mathbf{A}^n$ is a homeomorphism with inverse ϕ_i: $\mathbf{A}^n \to U_i$.
(ii) If Y is a projective (respectively, quasi-projective) variety, then $Y = \cup_{i=1}^{n+1}(Y \cap U_i)$ is covered by the open sets $Y \cap U_i$, $i = 1, ..., n+1$, which are homeomorphic to affine (respectively, quasi-affine) varieties via the mapping φ_i. In particular, if $Y = \mathbf{V}_p(T)$ is a closed subset of $\mathbf{P}^n$ and $Y \cap U_i \neq \emptyset$, then $Y \cap U_i$ is homeomorphic to $\mathbf{V}(T_*)$ where the latter is defined via x_i; if $W = \mathbf{V}_a(S)$ is a closed subset of $\mathbf{A}^n$, then W is homeomorphic to $\mathbf{V}(S^*) \cap U_i$ where $\mathbf{V}(S^*)$ is defined via x_i.

□

Let V be a projective algebraic set in $\mathbf{P}^n$. Then

$$V = \bigcup_{i=1}^{n+1}(V \cap U_i).$$

If we form $\mathbf{V}(I_*)$ with respect to U_i (as with U_{n+1}), then the previous discussion allows to identify $\mathbf{V}(I_*)$ and $V \cap U_i$ via the identification $\mathbf{A}^n \overset{\phi_i}{\longleftrightarrow} U_i$; and if we restrict attention to varieties V, then the corresponding local rings are isomorphic. Indeed, if we restrict to a point $P \in V$, then up to an appropriate projective change of coordinates (§4), we may always assume that $P \in V \cap U_{n+1}$. In CH.VI we will see that questions about V near a point P may be reduced to questions about an affine variety V_* (at least if the question can be answered by looking at $\mathcal{O}_P$).

Example (iii) This example may be viewed as an illustration of Theorem 6.12.
Let k be any field, $\mathbf{P}^1 = \mathbf{P}^1_k = U_1 \cup U_2$, as before. Consider the variety $\mathbf{V}_p(x_1) = \{(0,1)\} = \mathbf{V}_p(x_1) \cap U_2$ consisting of the single point at infinity in x_1-direction. Then this variety corresponds to the affine variety

$$\mathbf{V}_a\left(\mathbf{I}_p(\mathbf{V}(x_1))_*\right) = \mathbf{V}_a(x_1) = \{0\} \subset \mathbf{A}^1 = \mathbf{A}^1_k$$

via the homeomorphism φ_2: $U_2 \to \mathbf{A}^1$. Similarly, the variety $\mathbf{V}_p(x_2) = \{(1,0)\} \subset \mathbf{P}^1$ (or the point at infinity in x_2-direction) corresponds to the affine variety

$$\mathbf{V}_a\left(\mathbf{I}_p(\mathbf{V}(x_2))_*\right) = \mathbf{V}_a(x_2) = \{0\} \subset \mathbf{A}^1 = \mathbf{A}^1_k$$

via the homeomorphism φ_1: $U_1 \to \mathbf{A}^1$. Viewing $X = \mathbf{A}^1 - \{0\}$ and $Y = \mathbf{A}^1 - \{0\}$ as open subsets in two copies of the affine line $\mathbf{A}^1$, respectively, we obtain the following homeomorphisms of open subsets:

$$\begin{array}{ccccccc} X & \xrightarrow{\phi_2} & U_1 \cap U_2 & \qquad & Y & \xrightarrow{\phi_1} & U_1 \cap U_2 \\ a & \mapsto & (a,1) & & b & \mapsto & (1,b) \end{array}$$

Moreover, note that

$$\begin{cases} \phi_2(0) = (0,1) \text{ point at infinity in } x_1 \text{ -direction} \\ U_2 = (U_1 \cap U_2) \bigcup \{(0,1)\} \end{cases}$$

$$\begin{cases} \phi_1(0) = (1,0) \text{ point at infinity in } x_2 \text{ -direction} \\ U_1 = (U_1 \cap U_2) \bigcup \{(1,0)\} \end{cases}$$

and the composite map

$$\begin{array}{ccccccc} \psi: & X & \xrightarrow{\phi_2} & U_1 \cap U_2 & \xrightarrow{\varphi_1} & Y \\ & a & \mapsto & (a,1) & \mapsto & \dfrac{1}{a} \end{array}$$

is bijective. We may express this by saying that $\mathbf{P}^1$ is obtained by *gluing* two copies of $\mathbf{A}^1$ via ψ which exchanges 0 and ∞.
In CH.VI §3 we will see that X and Y are indeed (isomorphic to) affine varieties in $\mathbf{A}^2$ and ψ is a birational map.

Remark The connection between projective algebraic geometry and (complex) analytic geometry may be sketched as follows. Since polynomials $f(z_1, ..., z_n) \in \mathbb{C}[z_1, ..., z_n]$ with complex coefficients are holomorphic functions of their variables $z_1, ..., z_n$, an algebraic set V in $\mathbf{A}^n_{\mathbb{C}}$ or $\mathbf{P}^n_{\mathbb{C}}$ will be in particular a complex analytic subvariety of these complex manifolds (i.e., a subset given locally as the zero locus of holomorphic functions). The famous Chow's Theorem claims that the converse of the above statement is also true in the projective case, i.e., if $X \subset \mathbf{P}^n_{\mathbb{C}}$ is any complex analytic subvariety then X is a projective algebraic set. Note that this may be false if we replace $\mathbf{P}^n_{\mathbb{C}}$ by $\mathbf{A}^n_{\mathbb{C}}$. For example, the subset $\mathbb{Z} \subset \mathbb{C} \cong \mathbf{A}^1_{\mathbb{C}}$ of integers is an analytic subvariety (e.g., see J.-P. Serre, *Géométrie algébrique et géométrie analytique*, Ann. Inst. Fourier, Grenoble, 6, 1956, 1–42).

Exercises for §6

1. If $I = \langle f \rangle$ is the ideal of an affine hypersurface, show that $I^* = \langle f^* \rangle$.
2. Suppose $V = \mathbf{V}(I)$ is a variety in $\mathbf{P}^n$ and $V \supset H^{n+1}_\infty = \mathbf{P}^n - U_{n+1}$. Show that $V = \mathbf{P}^n$ or $V = H^{n+1}_\infty$. If $V = \mathbf{P}^n$, then $\mathbf{V}(I_*) = \mathbf{A}^n$; and if $V = H^{n+1}_\infty$, then $\mathbf{V}(I_*) = \emptyset$.
3. If $V = \mathbf{V}(J)$ is a proper algebraic set in $\mathbf{A}^n$, then no component of $\mathbf{V}(J^*)$ lies in or contains $H^{n+1}_\infty = \mathbf{P}^n - U_{n+1}$.
4. Let $V = \mathbf{V}(I)$ be an affine (projective) algebraic set defined by an ideal of $k[x_1, ..., x_n]$ (graded ideal of $k[x_1, ..., x_n, x_{n+1}]$). Let $V = V_1 \cup \cdots \cup V_s$ be the decomposition of V by its irreducible components and $p_i = \mathbf{I}(V_i)$ the prime ideal of V_i, $i = 1, ..., s$. Show that $\mathrm{Min}(I) = \{p_i \mid p_i = \mathbf{I}(V_i),\ i = 1, ..., s\}$ is the set of

all minimal elements in the set { prime ideals containing I}, and $\sqrt{I} = p_1 \cap \cdots \cap p_s$.

5. Let $V = \mathbf{V}(I) \subset \mathbf{P}^n$ be a projective algebraic set defined by the graded ideal I, and $\sqrt{I} = p_1 \cap \cdots \cap p_s$ the decomposition of $\sqrt{I}$ by its minimal prime ideals given in Exercise 4. Show that I is $(\phi_*)^*$-closed if and only if $p_i \not\subset \langle x_{n+1}\rangle$ and $\langle x_{n+1}\rangle \not\subset p_i$ for all $i = 1, ..., s$.
6. Let k be an algebraically closed field. If $V = \mathbf{V}(I)$ is a proper algebraic set of $\mathbf{P}^n$, and no component of V lies in or contains H_∞^{n+1}, then $\mathbf{V}(I_*)$ is a proper algebraic set in $\mathbf{A}^n$ and $\overline{\mathbf{V}(I_*)} = V$.
7. if k is not algebraically closed, can you describe how the correspondence given in Theorem 6.4 will look like?
8. Prove that the element $[x_{n+1}]$ in Theorem 6.10 does not annihilate any nonzero element of $k[x_1, ..., x_n, x_{n+1}]/I^*$; and moreover the ideal $\langle 1 - [x_{n+1}]\rangle$ does not contain any homogeneous element of $k[x_1, ..., x_n, x_{n+1}]/I^*$.
9. Using the notation as in Example (iii), if we *glue* X and Y by the bijective map

$$\begin{array}{rccc} \alpha: & X & \longrightarrow & Y \\ & a & \mapsto & a \end{array}$$

can we still obtain $\mathbf{P}^1$?
Formulate a corresponding *gluing* description of $\mathbf{P}^n$.

§7. Multiprojective Space and Segre Product

In this section, we want to make the Cartesian product of two algebraic sets into an algebraic set. Let $\mathbf{A}^m = \mathbf{A}_k^m$ and $\mathbf{A}^n = \mathbf{A}_k^n$ be *affine spaces*. If $P = (x_1, ..., x_m)$, $Q = (y_1, ..., y_n)$, then the mapping

$$\begin{array}{rccc} \phi: & \mathbf{A}^m \times \mathbf{A}^n & \longrightarrow & \mathbf{A}^{m+n} \\ & (P, Q) & \mapsto & (x_1, ..., x_m, y_1, ..., y_n) \end{array}$$

is a one-to-one and onto mapping. If A is an algebraic set in A^m and W is an algebraic set in A^n then $\phi(V \times W)$ is clearly an algebraic set in A^{m+n}.

The product $\mathbf{P}^m \times \mathbf{P}^n$ requires some discussion, however. For $\mathbf{P}^m \times \mathbf{P}^n$, the attempt to set up a mapping analogous to the ϕ used in the affine case breaks down. To see why this is so, let $P = (x_1, ..., x_{m+1}) \in \mathbf{P}^m$ and $Q = (y_1, ..., y_{n+1}) \in \mathbf{P}^n$. We would try to map (P, Q) into the point $(x_1, ..., x_{m+1}, y_1, ..., y_{n+1})$ in $\mathbf{P}^{m+n+1}$. However, this mapping is not well-defined. Indeed, since the coordinate of P can be multiplied by any nonzero $\lambda \in k$ and so can those of Q, if $\phi(P, Q) = (x_1, ..., x_{m+1}, y_1, ..., y_{n+1}) \in \mathbf{A}^{n+m+2}$ was a point of $\mathbf{P}^{m+n+1}$, then for $\lambda, \mu \in k$, $\lambda \neq \mu$, we would have to have

$$(\lambda x_1, ..., \lambda x_{m+1}, \mu y_1, ..., \mu y_{n+1}) = (\nu x_1, ..., \nu x_{m+1}, \nu y_1, ..., \nu y_{n+1})$$

for some $\nu \neq 0$ in k. Since some $x_i \neq 0$, $y_j \neq 0$, this would entail that $\nu = \lambda = \mu$. To avoid this difficulty, we define the algebraic sets in $\mathbf{P}^m \times \mathbf{P}^n$ by considering the *bigraded* structure of the polynomial ring $k[\mathbf{x}, \mathbf{y}] = k[x_1, ..., x_m, y_1, ..., y_n]$.

A polynomial $F \in k[\mathbf{x}, \mathbf{y}]$ is called a *biform* of *bidegree* (p, q) if F is a form of degree p resp. q when considered as a polynomial in $x_1, ..., x_{m+1}$ resp. in $y_1, ..., y_{n+1}$ with coefficients in $k[y_1, ..., y_n]$ resp. in $k[x_1, ..., x_m]$. Every $F \in k[\mathbf{x}, \mathbf{y}]$ may be written uniquely as $F = \sum_{p,q} F_{p,q}$, where $F_{p,q}$ is a biform of bidegree (p, q).

If S is any set of biforms in $k[\mathbf{x}, \mathbf{y}]$ we let

$$V_b(S) = \left\{(\mathbf{x}, \mathbf{y}) \in \mathbf{P}^m \times \mathbf{P}^n \;\middle|\; F(\mathbf{x}, \mathbf{y}) = 0 \text{ for all } F \in S\right\}.$$

A subset V of $\mathbf{P}^m \times \mathbf{P}^n$ will be called an *algebraic set* if $V = V_b(S)$ for some S consisting of biforms of $k[\mathbf{x}, \mathbf{y}]$. Thus, we may define the *Zariski topology* on $\mathbf{P}^m \times \mathbf{P}^n$ as was did on $\mathbf{P}^n$. For any $V \subset \mathbf{P}^m \times \mathbf{P}^n$, define

$$\mathbf{I}_b(V) = \left\{F \in k[\mathbf{x}, \mathbf{y}] \;\middle|\; F(\mathbf{x}, \mathbf{y}) = 0 \text{ for all } (\mathbf{x}, \mathbf{y}) \in V\right\}.$$

We leave it to the reader to define a bihomogeneous ideal, show that $\mathbf{I}_b(V)$ is bihomogeneous, and likewise to carry out the entire development for algebraic sets and varieties in $\mathbf{P}^m \times \mathbf{P}^n$ as has been done for $\mathbf{P}^n$ in CH.III. If $V \subset \mathbf{P}^m \times \mathbf{P}^n$ is a variety (i.e., irreducible), then $k_b^g[V] = k[\mathbf{x},\mathbf{y}]/\mathbf{I}_b(V)$ is the *bihomogeneous coordinate ring*, $k_b(V)$ its quotient field, and

$$k(V) = \left\{ z \in k_b(V) \;\middle|\; \begin{array}{l} z = \dfrac{f}{g}, \\ f, g \text{ biforms of the same degree in } k_b^g[V] \end{array} \right\}$$

is the function field of V. The local rings $\mathcal{O}_{P,V}$ are defined as before. We leave it to the reader to develop the theory of multiprojective varieties in $\mathbf{P}^{n_1} \times \mathbf{P}^{n_2} \cdots \times \mathbf{P}^{n_r}$.

There is another way to make $\mathbf{P}^m \times \mathbf{P}^n$ into a projective variety by its Segre embedding which is constructed below.

For any point $(P, Q) = (x_1, ..., x_m, y_1, ..., y_n) \in \mathbf{P}^m \times \mathbf{P}^n$, Put

$$Z_{ij} = x_i y_j, \qquad i = 1, ..., m+1, \; j = 1, ..., n+1.$$

Since some x_i and some y_j are nonzero, it follows that some Z_{ij} is nonzero. Consequently we may view the Z_{ij} as homogeneous coordinate of a point in $\mathbf{P}_k^N$, where $N = (m+1)(n+1) - 1$. It is understood that these coordinates are written in some *fixed order*. For instance we could use lexicographical order which in the case $m = 1$, $n = 2$ would look as follows: $(Z_{00}, Z_{01}, Z_{02}, Z_{10}, Z_{11}, Z_{12})$. With this convention, the point with coordinates Z_{ij} is denoted by by (Z_{ij}).

7.1. Proposition The mapping

$$\begin{array}{rccc} \mathcal{S}: & \mathbf{P}^m \times \mathbf{P}^n & \longrightarrow & \mathbf{P}^N \\ & (P, Q) & \mapsto & (Z_{ij}) \end{array}$$

is injective.

Proof The mapping $\mathcal{S}$ is well-defined. If the x_i are replaced by λx_i and the y_j by μy_j, where λ, μ are nonzero elements of k, then $Z_{ij} = x_i y_j$ is replaced by $\lambda\mu x_i y_j = \lambda\mu Z_{ij}$, which defines the same point in $\mathbf{P}^N$.
We now show that $\mathcal{S}$ is one-to-one. Suppose (Z_{ij}) is in the image of $\mathcal{S}$. Then $Z_{ij} = x_i y_j$ and we have to show that the *points* $(x_1, ..., x_{m+1})$ and $(y_1, ..., y_{n+1})$ are uniquely determined. We know that some

y_h is nonzero. Suppose $y_h \neq 0$. Then as *points*, $(x_1, ..., x_{m+1}) = (x_1 y_h, ..., x_{m+1} y_h) = (Z_{1h}, ..., Z_{m+1,h})$. Similarly, some $x_l \neq 0$ and as *points*, $(y_1, ..., y_{n+1}) = (x_l y_1, ..., x_l y_{n+1}) = (Z_{l1}, ..., Z_{l,n+1})$. Thus for a given point in the image, the element (P, Q) from which it comes is uniquely reconstructible, hence S is one-to-one. □

The mapping S is not surjective. If (Z_{ij}) is an image point, it is immediately verified by taking $Z_{ij} = x_i y_j$, that it satisfies the system of $\binom{m+1}{2} \cdot \binom{n+1}{2}$ equations:

$$(*) \qquad Z_{ih} Z_{jl} = Z_{il} Z_{jh}, \quad i \neq j, \ h \neq l.$$

It is trivial to produce values of the Z_{ij}, naturally not of the form $x_i y_j$, such that these equations are not satisfied. Thus there are points of $\mathbf{P}^N$ not in the image of ϕ. The equations $(*)$ determine a proper algebraic subset $V_{m,n}$ in $\mathbf{P}^N$. The ideal generated by all $Z_{ih} Z_{jl} - Z_{il} Z_{jh}$ is a graded prime ideal, hence $V_{m,n}$ is a variety in $\mathbf{P}^N$ (cf. Exercises).

7.2. Proposition The mapping S: $\mathbf{P}^m \times \mathbf{P}^n \to V_{m,n}$ is surjective.

Proof Given a point (c_{ij}) in $V_{m,n}$, we have to produce a point $(a_1, ..., a_{m+1}) \in \mathbf{P}^m$ and a point $(b_1, ..., b_{n+1}) \in \mathbf{P}^n$ such that $c_{ij} = a_i b_j$.
First let c_{st} be a nonzero coordinate of (c_{ij}). We first suppose that the required a_i and b_j can be found. Then $c_{sj} = a_s b_j$ for $j = 1, ..., n+1$. Take $a_s = 1$ (note that $c_{st} = a_s b_t \neq 0$). Then $b_j = c_{sj}$. On the other hand, $c_{it} = a_i b_t$ for $i = 1, ..., m+1$. Now $b_t = c_{st} \neq 0$ and so $a_i = b_t^{-1} c_{it} = c_{st}^{-1} c_{it}$. (Note that this is consistent with the previous choice $a_s = 1$.) We take $a_i = c_{st}^{-1} c_{it}$ and $b_j = c_{sj}$. We have to show that $c_{ij} = a_i b_j$ for all i, j. Now $a_i b_j = c_{st}^{-1} c_{it} c_{sj}$. If $i = s$ or if $t = j$ then $a_i b_j = c_{ij}$. So we may assume $i \neq s$ and $t \neq j$. Using the equations $(*)$, it follows that $c_{it} c_{sj} = c_{ij} c_{st}$ and so

$$a_i b_j = c_{st}^{-1} c_{it} c_{sj} = c_{st}^{-1} c_{ij} c_{st} = c_{ij}.$$

This completes the proof of the proposition. □

7.3. Definition (i) The mapping S used above is called the *Segre imbedding* of $\mathbf{P}^m \times \mathbf{P}^n$ in $\mathbf{P}^{n+m+mn}$.

(ii) The algebraic set $V_{m,n}$ given by the equations $(*)$ is called the *Segre variety* associated with the product $\mathbf{P}^m \times \mathbf{P}^n$.

In the particular case where $m = n$, the product $\mathbf{P}^m \times \mathbf{P}^m$ allows the *diagonal subset* $D = \{(P, P) \mid P \in \mathbf{P}^m\}$. The mapping $\mathcal{S}$ maps D into a certain subset of the Segre variety $V_{m,m}$. If $P = (x_1, ..., x_{m+1})$ then $\mathcal{S}$: $(P, P) \mapsto (Z_{ij} = x_i x_j)$. Thus the points of $\mathcal{S}(D)$ satisfy the equations

$$(*) \qquad \begin{cases} Z_{ik}Z_{jl} = Z_{il}Z_{jk} \text{ for } i \neq j \text{ and } k \neq l, \\ \\ Z_{ij} = Z_{ji} \text{ for } i, j = 1, ..., m+1. \end{cases}$$

These equations determine an algebraic subset W_m of $V_{m,m}$ which is irreducible (exercise 3), and this algebraic subset is called the *Veronese variety* associated with the product $\mathbf{P}^m \times \mathbf{P}^m$.

7.4. Proposition The mapping $\mathcal{S}$: $D \to W_m$ is surjective.

Proof Since $W_m \subset V_{m,m}$ it follows that every point (c_{ij}) of W_m has the form $c_{ij} = a_i b_j$ where $(a_1, ..., a_{m+1})$ and $(b_1, ..., b_{m+1})$ are in $\mathbf{P}^m$. Now (c_{ij}) must satisfy the second of the equations $(*)$ and so $a_i b_j = a_j b_i$ for all i, j. Some $b_t \neq 0$ and so $a_i = a_t b_t^{-1} b_i$ for all i. Set $\lambda = a_t b_t^{-1}$. Then $\lambda \neq 0$ since some $a_i \neq 0$, and so $a_i = \lambda b_i$ for all i. Thus $(a_1, ..., a_{m+1}) = (b_1, ..., b_{m+1})$ as points of $\mathbf{P}^m$ and the restriction of $\mathcal{S}$ to D is surjective. □

Exercises for §7

1. Let $k[T_{ij}]$ be the polynomial ring in variables T_{ij}, $i = 1, ..., m+1$, $j = 1, ..., n+1$. The Segre embedding $\mathcal{S}$ induces a ring homomorphism α: $k[T_{ij}] \to k[x_1, ..., x_{m+1}, y_1, ..., y_{n+1}]$ with $\alpha(T_{ij}) = x_i y_j$. Establish the following:
   a. α is surjective and $\operatorname{Ker}\alpha = \langle T_{ij}T_{kl} - T_{kj}T_{il} \rangle$.
   b. $\operatorname{Ker}\alpha$ is a prime ideal. (This illustrates why $V_{m,n}$ is a variety.)
2. If V is an algebraic set in $\mathbf{P}^m$ and W is an algebraic set in $\mathbf{P}^n$, show that $\mathcal{S}(V \times W)$ is an algebraic set in $V_{m,n}$. Conversely, show that if W is an algebraic set of $\mathbf{P}^N$, then $\mathcal{S}^{-1}(W)$ is an algebraic subset of $\mathbf{P}^n \times \mathbf{P}^m$ in the sense of §1.
3. Show that the algebraic set W_m is irreducible.

CHAPTER IV
Groebner Basis

In this chapter we introduce the notion of Groebner basis for an ideal I in $k[x_1, ..., x_n]$ through a nice combination of the "ordered" structure of ideals in $k[x_1, ..., x_n]$ and a division algorithm in $k[x_1, ..., x_n]$. We give some applications of Groebner bases to some basic questions in algebraic geometry posed earlier in CH.I §§5–6, CH.II §4 and CH.III §6.

For a general theory concerning Groebner bases we refer to [CLO'] and [BW].

Throughout this chapter, k is a field with $\text{char}k = 0$. We say that k is *computable* if the operations on k can be performed on a computer.

§1. Monomial Orderings

If we check in detail the division algorithm in $k[x]$ and the row-reduction (Gaussian elimination) algorithm for systems of linear equations (or matrices), we see that the notion of *ordering of terms* in polynomials is a key ingredient of both (though this is not often stressed). For example, in dividing $f(x) = x^5 - 3x^2 + 1$ by $g(x) = x^2 - 4x + 7$ using the standard method, we follow the following procedure.

- Write the terms in the polynomials in decreasing order with respect to the degree in x.
- At the first step, the leading term (the term of highest degree) in f is $x^5 = x^3 \cdot x^2 = x^3 \cdot$(leading term in g). Thus, we would subtract $x^3 \cdot g(x)$ from f to cancel the leading term, leaving $4x^4 - 7x^3 - 3x^2 + 1$.
- Then, we would repeat the same process on $f(x) - x^3 \cdot g(x)$, etc., until we obtain a polynomial of degree less than 2.

For the division algorithm on polynomials in one variable, we consider the degree ordering on the one-variable monomials:

$$\cdots > x^{m+1} > x^m > \cdots > x^2 > x > 1. \tag{1}$$

The success of the algorithm depends on working systematically with the leading terms in f and g, and not removing terms "at random" from f using arbitrary terms from g.

Similarly, when applying the row-reduction algorithm on matices, in any given row, we systematically work with entries on the left, i.e., first-leading entries are those nonzero entries farthest to the left in the row. On the level of linear equations, this is expressed by ordering the variables $x_1, ..., x_n$ as follows:

$$x_1 > x_2 > \cdots > x_n. \tag{2}$$

We write the terms in our equations in decreasing order. Furthermore, in an echelon form system, the equations are listed with their leading terms in decreasing order. (In fact, the precise definition of an echelon form system could be given in terms of this ordering, we leave this as an exercise.)

The above observation suggests that the main step of an extension of division and row-reduction to arbitrary polynomials in several variables will be the consideration of an ordering on the terms in polynomials in $k[x_1, ..., x_n]$. In this section, we will discuss the desirable properties such an ordering should have, and we will construct several examples that satisfy these requirements. The reader may look up in [CLO'] or [BW] that each of these orderings will be useful in different contexts.

First, recall from CH.I. §1 (Notation 1.2) that, writing $k[\mathbf{x}] = k[x_1, ..., x_n]$,

$$\mathbb{Z}^n_{\geq 0} = \left\{(\alpha_1, ..., \alpha_n) \,\middle|\, \alpha_i \in \mathbb{Z}_{\geq 0},\ i = 1, ..., n\right\}$$

$$\mathrm{M}(k[\mathbf{x}]) = \left\{x^\alpha = x_1^{\alpha_1} \cdots x_n^{\alpha_n} \,\middle|\, \alpha = (\alpha_1, ..., \alpha_n) \in \mathbb{Z}^n_{\geq 0}\right\}.$$

This observation establishes a one-to-one correspondence between the monomials in $k[x_1, ..., x_n]$ and elements in $\mathbb{Z}^n_{\geq 0}$. Furthermore, it is clear that any ordering $>$ on $\mathbb{Z}^n_{\geq 0}$ will provide an ordering (also denoted $>$) on $\mathrm{M}(k[\mathbf{x}])$:

- $x^\alpha > x^\beta$ if and only if $\alpha > \beta$.

There are many different ways to define orderings on $\mathbb{Z}^n_{\geq 0}$. For our purposes, most of these orderings will not be useful, because we want the orderings to be "*compatible*" with the algebraic structure of polynomial rings.

First, since a polynomial is a sum of monomials, *we would like to be able to arrange the terms in a polynomial unambiguously in descending (or ascending) order*. To do this, we must be able to compare every pair of monomials to establish their proper relative positions. Thus, we require the ordering to be a *linear* or a *total* ordering. This means that for every pair of monomials x^α and x^β, exactly one of the following three statements should hold:

$$x^\alpha > x^\beta,\ x^\alpha = x^\beta,\ x^\beta > x^\alpha.$$

Next, *we must take into account the effect of the sum and product operations on polynomials*. When we add polynomials, after combining similar monomial terms, we may simply rearrange the terms present into the appropriate order; so sums present no difficulties. Products are more subtle, however. Since multiplication in a polynomial ring distributes over addition, it suffices to consider what happens when we multiply a monomial by a polynomial. If this process would entail a change in the relative ordering of terms, significant problems result in any process similar to the division algorithm in $k[x]$, in particular when identification of the "leading" terms in polynomials is necessary. The reason is that the leading term in the product could be different from the product of the monomial and the leading term of the original polynomial.

Hence, we will require that all term orderings have the following additional property. If $x^\alpha > x^\beta$ and x^γ is any monomial, then we require that $x^\alpha x^\gamma > x^\beta x^\gamma$. In terms of the exponent vectors, this property means that if $\alpha > \beta$ in our ordering on $\mathbb{Z}^n_{\geq 0}$, then, for all $\gamma \in \mathbb{Z}^n_{\geq 0}$, $\alpha + \gamma > \beta + \gamma$.
With these considerations in mind, we propose the following definition.

1.1. Definition A *monomial ordering* on $k[\mathbf{x}]$ is any relation $>$ on the set $\mathbb{Z}^n_{\geq 0}$, or equivalently, any relation on $M(k[\mathbf{x}])$, satisfying:
(i) $>$ is a total (or linear) ordering on $\mathbb{Z}^n_{\geq 0}$.
(ii) if $\alpha > \beta$ and $\gamma \in \mathbb{Z}^n_{\geq 0}$, then $\alpha + \gamma > \beta + \gamma$.
(iii) $>$ is a well-ordering on $\mathbb{Z}^n_{\geq 0}$. This means that every nonempty subset of $\mathbb{Z}^n_{\geq 0}$ has a smallest element under $>$.

The following lemma will help us understand why the well-ordering condition in part (iii) of the definition is important.

1.2. Lemma An order relation $>$ on $\mathbb{Z}^n_{\geq 0}$ is a well-ordering if and only if every strictly decreasing sequence in $\mathbb{Z}^n_{\geq 0}$

$$\alpha(1) > \alpha(2) > \alpha(3) > \cdots$$

eventually terminates.

Proof Observe that $>$ is not a well-ordering if and only if there is an infinite strictly decreasing sequence in $\mathbb{Z}^n_{\geq 0}$.
If $>$ is not a well-ordering, then some nonempty subset $S \subset \mathbb{Z}^n_{\geq 0}$ has no least element. Now pick $\alpha(1) \in S$. Since $\alpha(1)$ is not the least element, we can find $\alpha(1) > \alpha(2)$ in S. Then $\alpha(2)$ is also not the least element, so that there is $\alpha(2) > \alpha(3)$ in S. By repetition of this procedure we obtain an infinite strictly decreasing sequence

$$\alpha(1) > \alpha(2) > \alpha(3) > \cdots.$$

Conversely, given such an infinite sequence, then $\{\alpha(1), \alpha(2), \alpha(3), \cdots\}$ is a nonempty subset of $\mathbb{Z}^n_{\geq 0}$ without least element, thus, $>$ cannot be a well-ordering. □

The importance of this lemma will become evident in the proof Theorem 2.1 of §2. In computational algebra theory (e.g., see [CLO']) it

is used to show that various algorithms must terminate because some term strictly decreases (with respect to a fixed monomial order) at each step of the algorithm.
In §3. we will see that given parts (i) and (ii) in Definition 1.1., the well-ordering condition of part (iii) is equivalent to $\alpha \geq 0$ for all $\alpha \in \mathbb{Z}^n_{\geq 0}$.

For a simple example of a monomial order, note that the usual numerical order

$$\cdots > m+1 > m > \cdots > 3 > 2 > 1 > 0$$

on the elements of $\mathbb{Z}_{\geq 0}$ satisfies the three conditions of Definition 1.1. Hence, the degree ordering (1) on $\mathbf{M}(k[x])$ is a monomial ordering.

The first example of an ordering on $\mathbb{Z}^n_{\geq 0}$ will be lexicographic order (or *lex* order, for short).

1.3. Definition (Lexicographic Order) Let $\alpha = (\alpha_1, ..., \alpha_n)$, and $\beta = (\beta_1, ..., \beta_n) \in \mathbb{Z}^n_{\geq 0}$. We say $\alpha >_{lex} \beta$ if, in the vector difference $\alpha - \beta \in \mathbb{Z}^n$, the left-most nonzero entry is positive. We will write $x^\alpha >_{lex} x^\beta$ if $\alpha >_{lex} \beta$.

Here are some examples:

a. $(1, 2, 0) >_{lex} (0, 3, 4)$.
b. $(3, 2, 4) >_{lex} (3, 2, 1)$.
c. The variables $x_1, ..., x_n$ are ordered in the usual way (see (2)) by the lex ordering:

$$(1, 0, ..., 0) >_{lex} (0, 1, 0, ..., 0) >_{lex} \cdots >_{lex} (0, ..., 0, 1),$$

so $x_1 >_{lex} x_2 >_{lex} \cdots >_{lex} x_n$.

In practice, when we work with polynomials in two or three variables, we will call the variables x, y, z rather than x_1, x_2, x_3. we will also assume that the alphabetical order $x > y > z$ on the variables is used to define the lexicographic ordering unless we explicitly say otherwise.
Lex order is analogous to the ordering of words used in dictionaries (hence the name). We may view the entries of an n-tuple $\alpha \in \mathbb{Z}^n_{\geq 0}$

as analogues of the letters in a word. The letters are ordered alphabetically:

$$a > b > \cdots > y > z.$$

Then, for instance,

$$\text{arrow} >_{lex} \text{arson}$$

since the third letter of "arson" comes after the third letter of "arrow" in alphabetical order, whereas the first two letters are the same in both. Since all elements $\alpha \in \mathbb{Z}^n_{\geq 0}$ have length n, this analogy only applies to words with a fixed number of letters.
For completeness, we must check that the lexicographic order satisfies the three conditions of Definition 1.1.

1.4. Proposition The lex ordering on $\mathbb{Z}^n_{\geq 0}$ is a monomial ordering.

Proof (i) That $>_{lex}$ is a total ordering follows directly from the definition and the fact that the usual numerical order on $\mathbb{Z}_{\geq 0}$ is a total ordering.
(ii) If $\alpha >_{lex} \beta$, then we have that the left-most nonzero entry in $\alpha - \beta$, say $\alpha_k - \beta_k$, is positive. But $x^\alpha \cdot x^\gamma = x^{\alpha+\gamma}$ and $x^\beta \cdot x^\gamma = x^{\beta+\gamma}$. Then in $(\alpha + \gamma) - (\beta + \gamma) = \alpha - \beta$, the left-most nonzero entry is again $\alpha_k - \beta_k > 0$.
(iii) Suppose that $>_{lex}$ were not a well-ordering. Then by Lemma 1.2, there would be an infinite strictly decreasing sequence

$$\alpha(1) >_{lex} \alpha(2) >_{lex} \alpha(3) >_{lex} \cdots$$

of elements of $\mathbb{Z}^n_{\geq 0}$. We will show that this leads to a contradiction. Consider the first entries of the vectors $\alpha(i) \in \mathbb{Z}^n_{\geq 0}$. By the definition of the lex order, these first entries form a nonincreasing sequence of non-negative integers. Since $\mathbb{Z}_{\geq 0}$ is well-ordered, the first entries of the $\alpha(i)$ must "stabilize". That is, there exists a k such that all the first components of the $\alpha(i)$ with $i \geq k$ are equal.
Begining at $\alpha(k)$, the second and subsequent entries come into play in determining the lex order. The second entries of $\alpha(k), \alpha(k+1), \ldots$ for a nonincreasing sequence. By the same reasoning as before, finally the second entries "stabilize" as well. Continuing in the same way, we see that for some l, the $\alpha(l), \alpha(l+1), \ldots$ are all equal. This contradicts the fact that $\alpha(l) >_{lex} \alpha(l+1)$. □

One should note that there are many lex orders, corresponding to how the variables are ordered. So far, we have used lex order with $x_1 > x_2 > \cdots > x_n$. But given *any* ordering of the variables $x_1, ..., x_n$, there is a corresponding lex order. For example, if the variables are x and y, then we get one lex order with $x > y$ and a second with $y > x$. In the general case of n variables, there are $n!$ lex orders. In what follows, the phrase "lex order" will refer to the one with $x_1 > \cdots > x_n$ unless otherwise stated.

In lex order, a variable dominates *any* monomial involving only smaller variables, neglecting its total degree. Thus, for the lex order with $x > y > z$, we have $x >_{lex} y^5 z^3$. For the purpose of making a nice connection with the graded structure of the polynomial ring (which will be powerful in applying Groebner basis theory to algebraic geometry, see §5 later and CH.V §3), it is *necessary* to take the total degrees of the monomials into account and order monomials of bigger degree first. One way to do this is by using the graded lexicographic order (or *grlex* order).

1.5. Definition (Graded Lex Order) Let $\alpha, \beta \in \mathbb{Z}^n_{\geq 0}$. We say $\alpha >_{grlex} \beta$ if

$$|\alpha| = \sum_{i=1}^{n} \alpha_i > |\beta| = \sum_{i=1}^{n} \beta_i, \quad \text{or } |\alpha| = |\beta| \text{ and } \alpha >_{lex} \beta.$$

We see that grlex orders by total degree first, then "breaks ties" using lex order. Here are some examples:

a. $(1,2,3) >_{grlex} (3,2,0)$.
b. $(1,2,4) >_{grlex} (1,1,5)$.
c. The variables are ordered according to the lex order, i.e., $x_1 >_{grlex} \cdots >_{grlex} x_n$.

It is an exercise to show that the grlex ordering satisfies the three conditions of Definition 1.1. As in the case of lex order, there are $n!$ grlex orders on n variables, depending on how the variables are ordered.

Another (somewhat less intuitive) order on monomials is the graded reverse lexicographical order (or *grevlex* order). Even though this ordering "takes some getting used to", it has recently been shown that for certain operations, the grevlex ordering is most efficient for computations.

1.6. Definition (Graded Reverse Lex Order) Let $\alpha, \beta \in \mathbb{Z}^n_{\geq 0}$. We say $\alpha >_{grevlex} \beta$ if

$$|\alpha| = \sum_{i=1}^{n} \alpha_i > |\beta| = \sum_{i=1}^{n} \beta_i, \quad \text{or } |\alpha| = |\beta|$$

and, in $\alpha - \beta \in \mathbb{Z}^n$, the right-most nonzero entry is negative.

Like grlex, grevlex orders by total degree, but it "breaks ties" in a different way. For example:

a. $(4,7,1) >_{grevlex} (4,2,3)$.
b. $(1,5,2) >_{grevlex} (4,1,3)$.

One verifies that the grevlex ordering defines a monomial ordering.

Note also that lex and grevlex define the same ordering on the variables. That is,

$$(1,0,...,0) >_{grevlex} (0,1,0,...,0) >_{grevlex} \cdots >_{grevlex} (0,...,0,1)$$

or

$$x_1 >_{grevlex} x_2 >_{grevlex} \cdots >_{grevlex} x_n.$$

Thus, grevlex is really different from the grlex order with the variables rearranged (as one might be tempted to believe from the name). To explain the relation between grlex and grevlex, note that both use total degree in the same way. To break a tie, grlex uses lex order, so that it looks at the left-most (or largest) variable and takes the *larger* power. In contrast, when grevlex finds the same total degree, it looks at the right-most (or smallest) variable and takes the *smaller* power. For example, we have $x^5yz >_{grlex} x^4yz^2$ and $x^5yz >_{grevlex} x^4yz^2$ but for different reasons.

As with lex and grlex, there are $n!$ grevlex orderings corresponding to how the n variables are ordered.

There are many other monomial orders besides the ones considered here. Some of these are explored in [CLO'] pp.72–74. Most computer algebra systems implement lex order, and some (such as *Macaulay* and REDUCE) also allow other orders, such as grlex and grevlex. Once such an order is chosen these systems allow the user to specify any of the $n!$ orderings of the variables.

We end this section with a discussion of how a monomial ordering may be applied to polynomials.
If $f = \sum_\alpha a_\alpha x^\alpha$ is a polynomial in $k[\mathbf{x}]$ and we have selected a monomial ordering $>$, then we can order the monomials of f in an unambiguous way with respect to $>$. For example, let $f = 4xy^2z + 4z^2 - 5x^3 + 7x^2z^2 \in k[x, y, z]$. Then:

a. with respect to the lex order, we reorder the terms of f in decreasing order as

$$f = -5x^3 + 7x^2z^2 + 4xy^2z + 4z^2.$$

b. With respect to the grlex order, we have

$$f = 7x^2z^2 + 4xy^2z - 5x^3 + 4z^2.$$

c. with respect to the grevlex order, we have

$$f = 4xy^2z + 7x^2z^2 - 5x^3 + 4z^2.$$

1.7. Definition Let $f = \sum_\alpha a_\alpha x^\alpha$ be a nonzero polynomial in $k[\mathbf{x}]$ and let $>$ be a monomial order.
(i) The *multidegree* of f is

$$\mathrm{md}(f) = \max\left\{\alpha \in \mathbb{Z}^n_{\geq 0} \;\middle|\; a_\alpha \neq 0\right\}$$

(the maximum is taken with respect to $>$).
(ii) The *leading coefficient* of f is

$$\mathrm{LC}(f) = a_{\mathrm{md}(f)} \in k.$$

(iii) The *leading monomial* of f is

$$\mathrm{LM}(f) = x^{\mathrm{md}(f)}$$

(with coefficient 1).
(iv) The *leading term* of f is

$$\mathrm{LT}(f) = \mathrm{LC}(f) \cdot \mathrm{LM}(f).$$

To illustrate, let $f = 4xy^2z + 4z^2 - 5x^3 + 7x^2z^2$ as before and let $>$ denote the lex order. Then

$$\begin{aligned}\text{md}(f) &= (3,0,0),\\ \text{LC}(f) &= -5,\\ \text{LM}(f) &= x^3,\\ \text{LT}(f) &= -5x^3.\end{aligned}$$

As in the one variable case, the multidegree has the following useful properties (we leave the proof as an exercise).

1.8. Lemma Let $f, g \in k[\mathbf{x}]$ be nonzero polynomials. Then:
(i) $\text{md}(fg) = \text{md}(f) + \text{md}(g)$.
(ii) If $f + g \neq 0$, then $\text{md}(f + g) \leq \max\{\text{md}(f), \text{md}(g)\}$. If, in addition, $\text{md}(f) \neq \text{md}(g)$, then equality occurs.

□

Convention From now on in this chapter, unless otherwise stated, We assume that *one particular monomial order has been selected*, and that leading terms, etc., will always be computed relative to that order.

Exercises for §1

1. Write a precise definition of what it means for a system of linear equations to be in echelon form, using the ordering given in equation (2).
2. Show that grlex ordering satisfies the three conditions of Definition 1.1.
3. Prove Lemma 1.8.

§2. A Division Algorithm in $k[x_1, ..., x_n]$

Any monomial ordering on $\mathbb{Z}^n_{\geq 0}$ in the sense of the foregoing section makes it possible to express a polynomial $f \in k[x_1, ..., x_n]$ by *ordered* monomials in a unique way. The aim of this section is to extend the division algorithm in $k[x]$ to $k[\mathbf{x}] = k[x_1, ..., x_n]$ with $n > 1$ variables.

One of the main motivations of doing this is the ideal membership problem for polynomials in $k[\mathbf{x}]$. If I is a nonzero ideal of $k[\mathbf{x}]$, then by Hilbert basis theorem (See CH.I §1 or §4 below) I has a finite generating set $\{f_1, ..., f_s\} \subset I$ such that $I = \langle f_1, ..., f_s \rangle$. Hence, if $g \in k[\mathbf{x}]$, then $g \in I$ if and only if

$$g = \sum_{i=1}^{s} a_i f_i, \quad a_i \in k[\mathbf{x}].$$

But in practice, how do we know whether $g \in I$ or not? When $n = 1$ we have $I = \langle f \rangle$ for some $f \in I$, and the Eucledian division algorithm yields

$$g = qf + r, \quad q, r \in k[x], \ \deg(r) < \deg(g),$$

hence, $g \in I$ if and only if $r = 0$ since r is unique. In the general case, the goal is to divide g by $f_1, ..., f_s$. As we will see, this means expressing g in the form

$$g = a_1 f_1 + \cdots + a_s f_s + r,$$

where the "quotients" $a_1, ..., a_s$ and remainder r lie in $k[\mathbf{x}]$.

The *basic* idea of the algorithm is the same as in the one-variable case. Based on Lemma 1.8 of §1, we may say that

- for $x^\alpha, x^\beta \in \mathrm{M}(k[\mathbf{x}])$, x^β is *divisible* by x^α if $x^\beta = x^\gamma x^\alpha$ for some $x^\gamma \in \mathrm{M}(k[\mathbf{x}])$, or equivalently, if $\beta = \alpha + \gamma$ for some $\gamma \in \mathbb{Z}^n_{\geq 0}$.

In this sense of divisibility, we want to cancel the leading term of a nonzero polynomial g (with respect to a fixed monomial order) by multiplying some f_i by an appropriate monomial and subtracting. Then the latter monomial becomes a term in the corresponding a_i.

Rather than stating the algorithm in general, let us first work through some examples to see what is involved.

Example (i) We will first divide $f = xy^2 + 1$ by $f_1 = xy + 1$ and $f_2 = y + 1$ in $k[x, y]$, using lex order with $x > y$. We want to employ the same procedure as for division of one-variable polynomials, the difference being that there are now several divisors and quotients.

Listing the divisors f_1, f_2 and the quotients a_1, a_2 *vertically*, we have the following set up:

$$\begin{array}{rl} a_1: & \\ a_2: & \\ xy+1 & \big) \; \overline{xy^2+1} \\ y+1 & \end{array}$$

The leading terms $\mathrm{LT}(f_1) = xy$ and $\mathrm{LT}(f_2) = y$ both divide the leading term $\mathrm{LT}(f) = xy^2$. Since f_1 is listed first, we will use it. Thus, we divide xy into xy^2, leaving y, and then subtract $y \cdot f_1$ from f:

$$\begin{array}{rl} a_1: & \quad y \\ a_2: & \\ xy+1 & \big) \; \overline{xy^2+1} \\ y+1 & \quad \underline{xy^2+y} \\ & \quad -y+1 \end{array}$$

Now we repeat the same process on $-y+1$. This time we must use f_2 since $\mathrm{LT}(f_1) = xy$ does not divide $\mathrm{LT}(-y+1) = -y$. We obtain

$$\begin{array}{rl} a_1: & \quad y \\ a_2: & \quad -1 \\ xy+1 & \big) \; \overline{xy^2+1} \\ y+1 & \quad \underline{xy^2+y} \\ & \quad -y+1 \\ & \quad \underline{-y-1} \\ & \quad 2 \end{array}$$

Since $\mathrm{LT}(f_1)$ and $\mathrm{LT}(f_2)$ do not divide 2, the remainder is $r = 2$ and we are done. Thus, we have written $f = xy^2 + 1$ in the form

$$xy^2 + 1 = y \cdot (xy + 1) + (-1) \cdot (y + 1) + 2.$$

(ii) In this example, we will encounter an unexpected subtlety that can occur when dealing with polynomials of more than one variable. Let us divide $f = x^2y + xy^2 + y^2$ by $f_1 = xy - 1$ and $f_2 = y^2 - 1$. As in the previous example, we will use lex order with $x > y$. The first two steps of the algorithm are as usual, providing with the following partially completed division (remember that when both leading terms divide, we use f_1):

$$
\begin{array}{rl}
a_1 : & x + y \\
a_2 : & \\
xy - 1 & \sqrt{x^2y + xy^2 + y^2} \\
y^2 - 1 & \quad x^2y - x \\
\hline
 & \quad xy^2 + x + y^2 \\
 & \quad xy^2 - y \\
\hline
 & \quad x + y^2 + y
\end{array}
$$

Note that neither $\mathrm{LT}(f_1) = xy$ nor $\mathrm{LT}(f_2) = y^2$ divides $\mathrm{LT}(x + y^2 + y) = x$. However, $x+y^2+y$ is *not* the remainder since $\mathrm{LT}(f_2)$ divides y^2. Thus, if we move x to the remainder, we can continue dividing. (This is something that never happens in the one-variable case: Once the leading term of the divisor no longer divides the leading term of what is left under the radical, the algorithm terminates.)

To implement this idea, we create a remainder column r, to the right of the radical, where we put the terms belonging to the remainder. Also we call the polynomial under the radical the *intermediate dividend*. Then we continue dividing until no term of the intermediate dividend is divisible by some $\mathrm{LT}(f_i)$. Here is the next step, where we move x to the remainder column (as indicated by the arrow):

$$
\begin{array}{rlcl}
a_1 : & x + y & & r \\
a_2 : & & & -- \\
xy - 1 & \sqrt{x^2y + xy^2 + y^2} & & \\
y^2 - 1 & \quad x^2y - x & & \\
\hline
 & \quad xy^2 + x + y^2 & & \\
 & \quad xy^2 - y & & \\
\hline
 & \quad x + y^2 + y & & \\
\hline
 & \quad y^2 + y & \longrightarrow & x
\end{array}
$$

Now we continue dividing. If we can divide by $\mathrm{LT}(f_1)$ or $\mathrm{LT}(f_2)$, we proceed as usual, and if none of these divides, we move the leading term of the intermediate dividend to the remainder column. Here is the rest of the division:

$$
\begin{array}{rl|l}
a_1: & x+y & r \\
a_2: & 1 & ---- \\
\hline
xy-1 & \sqrt{x^2y+xy^2+y^2} & \\
y^2-1 & x^2y-x & \\
\cline{2-2}
 & xy^2+x+y^2 & \\
 & xy^2-y & \\
\cline{2-2}
 & x+y^2+y & \\
\cline{2-2}
 & y^2+y & \longrightarrow\ x \\
 & y^2-1 & \\
\cline{2-2}
 & y+1 & \longrightarrow\ x+y+1
\end{array}
$$

Thus, the remainder is $x+y+1$, and we obtain

(1) $x^2y+xy^2+y^2=(x+y)\cdot(xy-1)+1\cdot(y^2-1)+x+y+1.$

Note that the remainder is a sum of monomials, none of which is divisible by the leading terms $\mathrm{LT}(f_1)$ or $\mathrm{LT}(f_2)$.

The above example is a fairly complete illustration of how the division algorithm works. It also makes it clear what property we want the remainder to have:

- none of its terms should be divisible by the leading terms of the polynomials by which we are dividing.

We can now state the general form of the division algorithm (we refer to [CLO'] for the algorithm written with pseudo-codes).

2.1. Theorem (Division Algorithm in $k[x_1,...,x_n]$) Fix a monomial order $>$ on $\mathbb{Z}^n_{\geq 0}$, and let $F=(f_1,...,f_s)$ be an ordered s-tuple of polymonials in $k[\mathbf{x}]$. Then every $f\in k[\mathbf{x}]$ can be written as

$$f=a_1f_1+\cdots+a_sf_s+r,$$

where $a_i, r\in k[\mathbf{x}]$, and either $r=0$ or r is a k-linear combination of monomials, none of which is divisible by any of $\mathrm{LT}(f_1),...,\mathrm{LT}(f_s)$. We will call r a *remainder* of f for division by F. Furthermore, if $a_if_i\neq 0$, then we have

$$\mathrm{md}(f)\geq \mathrm{md}(a_if_i).$$

Proof (We refer the reader to [CLO'] p.63 for another proof where the algorithm is given by pseudocode) Suppose that $\mathrm{LM}(f)$ is divisible by some $\mathrm{LM}(f_{i1}) \in \mathrm{LM}(F)$. Then $f = b_{i1}f_{i1} + p_1$ with $\mathrm{md}(p_1) < \mathrm{md}(f)$, where

$$b_{i1} = \frac{\mathrm{LC}(f)}{\mathrm{LC}(f_{i1})} x^{\mathrm{md}(f) - \mathrm{md}(f_{i1})}.$$

If $\mathrm{LM}(p_1)$ is divisible by some $\mathrm{LM}(f_{i2}) \in \mathrm{LM}(F)$, we also have $p_1 = b_{i2}f_{i2} + p_2$, and hence $f = b_{i1}f_{i1} + b_{i2}f_{i2} + p_2$ with $\mathrm{md}(p_2) < \mathrm{md}(p_1) < \mathrm{md}(f)$. Since $\mathrm{md}(f) < \infty$, this procedure should stop after a finite number of steps, say t, and it follows that

$$f = b_{i1}f_{i1} + \cdots + b_{it}f_{it} + p_t,$$

where $p_t = 0$ or $\mathrm{LM}(p_t)$ is not divisible by any $\mathrm{LM}(f_i)$. Suppose $p_t \neq 0$. If we put $r_1 = \mathrm{LT}(p_t)$, $g_1 = p_t - \mathrm{LT}(p_t)$, then $f = b_{i1}f_{i1} + \cdots + b_{it}f_{it} + g_1 + r_1$ with $\mathrm{md}(g_1) < \mathrm{md}(p_t) < \mathrm{md}(f)$, and $\dfrac{1}{\mathrm{LC}(p_t)} r_1$ is not divisible by any $\mathrm{LM}(f_i)$. Repeating the same procedure for g_1 as before we obtain

$$g_1 = c_{j1}f_{j1} + \cdots + c_{jm}f_{jm} + g_2 + r_2$$

with $\mathrm{md}(g_2) < \mathrm{md}(g_1)$, and $\frac{1}{c} r_2$ is a monomial (for some $0 \neq c \in k$) which is not divisible by any $\mathrm{LM}(f_i)$. Thus, $f = b_{i1}f_{i1} + \cdots + b_{it}f_{it} + c_{j1}f_{j1} + \cdots + c_{jm}f_{jm} + g_2 + r_1 + r_2$ with $\mathrm{md}(g_2) < \mathrm{md}(g_1) < \mathrm{md}(f)$, and every monomial of $r_1 + r_2$ is not divisible by any $\mathrm{LM}(f_i)$. Finally, a finite number, say q, of repetations of the above procedure will yield:

$$f = a_1f_1 + a_2f_2 + \cdots + a_sf_s + r_1 + r_2 + \cdots + r_q.$$

Put $r = r_1 + \cdots + r_q$. Then r has the requred property. Moreover, from the proof it is clear that if $a_if_i \neq 0$ then $\mathrm{md}(a_if_i) \leq \mathrm{md}(f)$. □

We conclude this section by checking whether the division algorithm has the same nice properties as the one-variable version. Unfortunately, the examples given below will show that the division algorithm is far from perfect. In fact, the algorithm achieves its full potential only when coupled with the Groebner bases studied later.

A first important property of the division algorithm in $k[x]$ is that the remainder is uniquely determined. To see how this can fail when there is more than one variable, consider the following example.

Example (iii) Let us divide $f = x^2y + xy^2 + y^2$ by $f_1 = y^2 - 1$ and $f_2 = xy - 1$. We will use lex order with $x > y$. This is the same as Example (ii) above, except that we have changed the order of the divisors. For practice, we suggest that the reader should do the division. You should get the following answer:

$$
\begin{array}{rl l}
a_1: & x+1 & r \\
a_2: & x & ---- \\
y^2-1 & \overline{x^2y+xy^2+y^2} & \\
xy-1 & x^2y - x & \\
 & \overline{xy^2+x+y^2} & \\
 & xy^2 - x & \\
 & \overline{2x+y^2} & \\
 & \overline{y^2} & \longrightarrow \ 2x \\
 & y^2-1 & \\
 & 1 & \longrightarrow \ 2x+1
\end{array}
$$

This shows that

$$x^2y + xy^2 + y^2 = (x+1)\cdot(y^2-1) + x\cdot(xy-1) + 2x + 1.$$

If we compare this with equation (1), we see that the remainder is different from what we obtained in Example (ii).

This shows that the remainder r is not uniquely characterized by the requirement that none of its terms be divisible by $\mathrm{LT}(f_1), \dots, \mathrm{LT}(f_s)$. The situation is not completely bad: If we follow the algorithm presisely as stated (most importantly testing $\mathrm{LT}(p)$ for divisibility by $\mathrm{LT}(f_1), \mathrm{LT}(f_2), \dots$ in that order), then $a_1, \dots, a_s$ and r are uniquely determined. However, the foregoing examples show that the *ordering* of the s-tuple of polynomials $(f_1, \dots, f_s)$ definitely matters, both in the number of steps the algorithm will take to complete the calculation and in the results. The a_i and r can change if we simply rearrange the f_i. (The a_i and r may also change if we change the monomial ordering, but that is another story.)

Returning to the ideal membership problem, we know that one nice feature of the division algorithm in $k[x]$ is the way it solves the ideal membership problem. Do we get something similar for several variables? One implication is an easy corollary of Theorem 2.1. If after division of f by $F = (f_1, ..., f_s)$ we obtain a remainder $r = 0$, then $f \in \langle f_1, ..., f_s\rangle$. Thus $r = 0$ is a *sufficient* condition for ideal membership. However, as the following example shows, $r = 0$ is not a *necessary* condition.

Example (iv) Let $f_1 = xy + 1$, $f_2 = y^2 - 1 \in k[x, y]$ with the lex order. Dividing $f = xy^2 - x$ by $F = (f_1, f_2)$, the result is

$$xy^2 - x = y \cdot (xy + 1) + 0 \cdot (y^2 - 1) + (-x - y).$$

With $F = (f_2, f_1)$, however, we have

$$xy^2 - x = x \cdot (y^2 - 1) + 0 \cdot (xy + 1) + 0.$$

The second calculation shows that $f \in \langle f_1, f_2\rangle$. Then the first calculation shows that even if $f \in \langle f_1, f_2\rangle$, it is still possible to obtain a nonzero remainder for division by $F = (f_1, f_2)$.

Thus, we must conclude that the division algorithm given in Theorem 2.1. is an imperfect generalization of its one-variable counterpart. To remedy this situation, the ideal structure theory allows the possibility of passing from $f_1, ..., f_s$ to a different generating set for I. The problem may thus be rephrased as follows.

- Is there a "good" generating set for I? For such a set, we want the remainder r for division by the "good" generators to be uniquely determined and the condition $r = 0$ should be *equivalent* to membership in the ideal.

In the next two sections, we establish that such a "good" generating set does exist for I.

Exercises for §2

1. Compute the remainder for division of the given polynomial f by the ordered set F (by hand). Use the grlex order, then the lex order in each case.
   a. $f = x^7y^2 + x^3y^2 - y + 1$, $F = (xy^2 - x, x - y^3)$.

b. Repeat part a with the order of the pair F reversed.

2. Compute the remainder for divisionof f by F:
 a. $f = xy^2z^2 + xy - yz$, $F = (x - y^2, y - z^3, z^2 - 1)$.
 b. Repeat part a but applying a cyclic permutation to the order of the set F.
3. Using a computer algebra system (e.g. *Macaulay*, MAPLE, MATHEMATIK, MODULA-2), check your work for Exercises 1 and 2.
4. Let $V = \mathbf{V}(y - x^2, z - x^3)$ be the twisted cubic curve in $\mathbf{A}^3_{\mathbb{R}}$.
 a. Using the division algorithm, prove that every polynomial $f \in \mathbb{R}[x, y, z]$ can be written as

$$f = h_1(y - x^2) + h_2(z - x^3) + r,$$

 where r is a polynomial in x alone. (Hint: Carefully specify the monomial ordering to be used.)
 b. Use the parametrization of the twisted cubic to show that $z^2 - x^4y$ vanishes at every point of the twisted cubic.
 c. Find the explicit representation

$$z^2 - x^4y = h_1(y - x^2) + h_2(z - x^3)$$

 using the division algorithm.
5. Let $V \subset \mathbf{A}^3_{\mathbb{R}}$ be the curve parametrized by (t, t^m, t^n), $n, m \geq 2$.
 a. Show that V is an affine algebraic set.
 b. Adapt the ideas in Exercise 4 to determine $\mathbf{I}(V)$.
6. Show that the operation of computing remainders for division by $F = (f_1, ..., f_s)$ is k-linear. That is, if the remainder for division of g_i by F is r_i, $i = 1, 2$, then, for any $c_1, c_2 \in k$, the remainder for division of $c_1g_1 + c_2g_2$ is $c_1r_1 + c_2r_2$.

§3. Monomial Ideals and Dickson's Lemma

In this section we first prove that any monomial ideal of $k[\mathbf{x}] = k[x_1, ..., x_n]$ has a "good" generating set (as described at the end of the foregoing section). This is realized by Dickson's Lemma which is a special version of Hilbert basis theorem (Ch.I §1) for monomial ideals.

3.1. Definition An ideal $I \subset k[\mathbf{x}]$ is a *monomial ideal* if there is a subset $A \subset \mathbb{Z}^n_{\geq 0}$ (possibly infinite) such that $I = \langle x^\alpha \mid \alpha \in A \rangle$.

We first need to characterize *all* monomials that lie in a given monomial ideal.

3.2. Lemma Let $I = \langle x^\alpha \mid \alpha \in A \rangle$ be a monomial ideal. Then a monomial x^β lies in I if and only if x^β is divisible by x^α for some $\alpha \in A$.

Proof If x^β is a multiple of x^α for some $\alpha \in A$, then it is obvious that $x^\beta \in I$. Conversely, if $x^\beta \in I$, then $x^\beta = \sum_{i=1}^{s} h_i x^{\alpha(i)}$, where $h_i \in k[x_1, ..., x_n]$ and $\alpha(i) \in A$. If we expand each h_i as a linear combination of monomials, we see that every term on the right side of the equation is divisible by some $x^{\alpha(i)}$. Hence, the left side x^β must have the same property.

□

Checking whether a given polynomial f lies in a monomial ideal can be done by looking at the monomials of f. This is a consequence of the following lemma.

3.3. Lemma Let I be a monomial ideal, and let $f \in k[\mathbf{x}]$. Then the following are equivalent:
(i) $f \in I$.
(ii) Every term of f lies in I.
(iii) f is a k-linear combination of the monomials in I.

Proof The implications (iii) $\Rightarrow$ (ii) $\Rightarrow$ (i) are trivial. The proof of (i) $\Rightarrow$ (iii) is similar to the proof for Lemma 3.2 and is left as an exercise. □

An immediate consequence of part (iii) of the lemma is that a monomial ideal is uniquely determined by its monomials.

3.4. Corollary Two monomial ideals are the same if and only if they contain the same monomials.

□

Now we are ready to announce the main result of this section.

3.5. Lemma (Dickson's Lemma) A monomial ideal $I = \langle x^\alpha, \mid \alpha \in A\rangle \subset k[x]$ can be written in the form $I = \langle x^{\alpha(1)}, ..., x^{\alpha(s)}\rangle$, where $\alpha(1), ..., \alpha(s) \in A$. In particular, I has a finite basis.

Proof (By induction on n, the number of variables.) If $n = 1$, then I is generated by the monomials x^α, where $\alpha \in A \subset \mathbb{Z}_{\geq 0}$. Let β be the smallest element of $A \subset \mathbb{Z}_{\geq 0}$. Then x^β divides all the other generators, and, thus, $I = \langle x^\beta\rangle$.
Now assume that $n > 1$ and that the theorem is true for $n - 1$. We will write the variables as $x_1, ..., x_{n-1}, y$, so that monomials in $k[x_1, ..., x_{n-1}, y]$ may be written as $x^\alpha y^m$, where $\alpha = (\alpha_1, ..., \alpha_{n-1}) \in \mathbb{Z}^{n-1}_{\geq 0}$ and $m \in \mathbb{Z}_{\geq 0}$.
Suppose that $I \subset k[x_1, ..., x_{n-1}, y]$ is a monomial ideal. To find generators for I, let J be the ideal in $k[x_1, ..., x_{n-1}]$ generated by the monomials x^α for which $x^\alpha y^m \in I$ for some $m \geq 0$. Since J is a monomial ideal in $k[x_1, ..., x_{n-1}]$, the induction hypothesis implies that finitely many of the x^α's generate J, say $J = \langle x^{\alpha(1)}, ..., x^{\alpha(s)}\rangle$. The ideal J may be viewed as the "projection" of I to $k[x_1, ..., x_{n-1}]$. For each i between 1 and s, the definition of J entails that $x^{\alpha(i)} y^{m_i} \in I$ for some $m_i \geq 0$. Let m be the largest of the m_i. Then for each k between 0 and $m-1$, consider the ideal $J_k \subset k[x_1, ..., x_{n-1}]$ generated by the monomials x^β such that $x^\beta y^k \in I$. One may think of J_k as the "slice" of I generated by monomials containing y exactly to the kth power. Using the induction hypothesis again, J_k has a finite generating set of monomials, say $J_k = \langle x^{\alpha_k(1)}, ..., x^{\alpha_k(s_k)}\rangle$.
We claim that I is generated by the monomials in the following list:

$$\begin{aligned}
\text{from } J &: \; x^{\alpha(1)} y^m, ..., x^{\alpha(s)} y^m, \\
\text{from } J_0 &: \; x^{\alpha_0(1)}, ..., x^{\alpha_0(s_0)}, \\
\text{from } J_1 &: \; x^{\alpha_1(1)} y, ..., x^{\alpha_1(s_1)} y, \\
&\;\;\vdots \\
\text{from } J_{m-1} &: \; x^{\alpha_{m-1}(1)} y^{m-1}, ..., x^{\alpha_{m-1}(s_{m-1})} y^{m-1}.
\end{aligned}$$

First note that every monomial in I is divisible by one on the list. Indeed, let $x^\alpha y^p \in I$. If $p \geq m$, then $x^\alpha y^p$ is divisible by some $x^{\alpha(i)} y^m$ by construction of J. On the other hand, if $p \leq m - 1$, then $x^\alpha y^p$ is divisible by some $x^{\alpha_p(j)} y^p$ by construction of J_p. It follows from Lemma 3.2 that the above monomials generate an ideal having the same monomials as I. In view of Corollary 3.4, this forces the ideals to be the same, and the claim is proved.

To complete the proof of the theorem, we need to show that the finite set of generators can be chosen from a given set of generators for the ideal. If we switch to writing the variables as $x_1, ..., x_n$, then the monomial ideal is $I = \langle x^\alpha \mid \alpha \in A\rangle \subset k[x_1, ..., x_n]$. We have to show that I is generated by finitely many of the x^α's, where $\alpha \in A$. By the previous paragraph, we know that $I = \langle x^{\beta(1)}, ..., x^{\beta(s)}\rangle$ for some monomials $x^{\beta(i)}$ in I. Since $x^{\beta(i)} \in I = \langle x^\alpha \mid \alpha \in A\rangle$, Lemma 3.2 entails that each $x^{\beta(i)}$ is divisible by $x^{\alpha(i)}$ for some $\alpha(i) \in A$. From this it follows that $I = \langle x^{\alpha(1)}, ..., x^{\alpha(s)}\rangle$ (see Exercise 1 for the details). This completes the proof. □

To provide more insight in the method of proof for theorem 3.5, let us apply it to the ideal $I = \langle x^4y^2, x^3y^4, x^2y^5\rangle$. It is easily seen that the "projection" is $J = \langle x^2\rangle \subset k[x]$. Since $x^2y^5 \in I$, we have $m = 5$. Then we get the "slices" J_k, $0 \le k \le 4 = m - 1$, generated by monomials containing y^k:

$$\begin{aligned} J_0 = J_1 &= \langle 0\rangle \\ J_2 = J_3 &= \langle x^4\rangle \\ J_4 &= \langle x^3\rangle. \end{aligned}$$

Then the proof of Theorem 5 yields $I = \langle x^2y^5, x^4y^2, x^4y^3, x^3y^4\rangle$.

Let $I = \langle x^\alpha \mid \alpha \in A\rangle$ be a monomial ideal. Then by Dickson's lemma I must have a finite generating set $\{x^{\alpha(1)}, ..., x^{\alpha(s)}\} \subset \{x^\alpha \mid \alpha \in A\}$. The first application of this finiteness property is the following important fact about monomial orderings on $\mathrm{M}(k[\mathbf{x}])$.

3.6. Corollary Let $>$ be a relation on $\mathbb{Z}^n_{\ge 0}$ satisfying:
(i) $>$ is a total ordering on $\mathbb{Z}^n_{\ge 0}$,
(ii) if $\alpha > \beta$ and $\gamma \in \mathbb{Z}^n_{\ge 0}$, then $\alpha + \gamma > \beta + \gamma$.
Then $>$ is a well-ordering if and only if $\alpha \ge 0$ for all $\alpha \in \mathbb{Z}^n_{\ge 0}$.

Proof ⇒: Assuming $>$ is a well-ordering, let α_0 be the smallest element of $\mathbb{Z}^n_{\ge 0}$. Suppose that $0 > \alpha_0$. Adding $n\alpha_0$ to both sides, hypothesis (ii) implies that $n\alpha_0 > (n+1)\alpha_0$. Thus

$$0 > \alpha_0 > 2\alpha_0 > \cdots > n\alpha_0 > (n+1)\alpha_0 > \cdots.$$

This infinite descending sequence contradicts the assumption that $>$ was a well-ordering.
$\Leftarrow$: Assuming that $\alpha \geq 0$ for all $\alpha \in \mathbb{Z}^n_{\geq 0}$, let $A \subset \mathbb{Z}^n_{\geq 0}$ be nonempty. We have to show that A has a smallest element. Since $I = \langle x^\alpha, \quad \alpha \in A \rangle$ is a monomial ideal, $I = \langle x^{\alpha(1)}, ..., x^{\alpha(s)} \rangle$ for some $\alpha(1), ..., \alpha(s) \in A$. Relabeling if necessary, we may assume that $\alpha(1) < \alpha(2) < \cdots < \alpha(s)$. We claim that $\alpha(1)$ is the smallest element of A. To prove this, take $\alpha \in A$. Then $x^\alpha \in I = \langle x^{\alpha(1)}, ..., x^{\alpha(s)} \rangle$, so that by Lemma 3.2., x^α is divisible by some $x^{\alpha(i)}$. This means that $\alpha = \alpha(i) + \gamma$ for some $\gamma \in \mathbb{Z}^n_{\geq 0}$. Then $\gamma \geq 0$ and the hypothesis (ii) imply that

$$\alpha = \alpha(i) + \gamma \geq \alpha(i) + 0 = \alpha(i) \geq \alpha(1).$$

Thus, $\alpha(1)$ is the least element of A. □

As a result of this corollary, the definition of monomial ordering given in Definition 1.1 of §1 can be simplified. Conditions (i) and (ii) in the definition remain unchanged, but we replace (iii) by the simpler condition that $\alpha \geq 0$ for all $\alpha \in \mathbb{Z}^n_{\geq 0}$. This makes it *much* easier to verify that a given ordering is actually a monomial ordering. We refer to [CLO'] p.71 for some examples.

Another application of the finiteness property is that it allows us to solve the ideal membership problem for monomial ideals.

3.7. Corollary Let $I = \langle x^{\alpha(1)}, ..., x^{\alpha(s)} \rangle$ be a monomial ideal, and let $f \in k[\mathbf{x}]$. Then $f \in I$ if and only if the remainder of f on division by $x^{\alpha(1)}, ..., x^{\alpha(s)}$ is zero.

Proof $\Leftarrow$: Trivial.
$\Rightarrow$: Suppose $f \in I$. After dividing f by $x^{\alpha(1)}, ..., x^{\alpha(s)}$, we obtain $f = \sum a_i x^{\alpha(i)} + r$, where r is the remainder. Suppose $r \neq 0$. Then $r = f - \sum a_i x^{\alpha(i)} \in I$. By Lemma 3.3. we have $\mathrm{LT}(r) \in I$, and hence, $\mathrm{LT}(r)$ is divisible by some $x^{\alpha(i)}$ by Lemma 3.2. This contradicts the definition of r. Thus, r must be zero, as desired. □

Exercises for §3

1. Let $I \subset k[x_1, ..., x_n]$ be an ideal with the property that for every $f = \sum_\alpha c_\alpha x^\alpha \in I$, every monomial x^α appearing in f is also in I. Show that I is a monomial ideal.

2. Let $I = \langle x^\alpha \mid \alpha \in A \rangle$ be a monomial ideal, and assume that we have a finite basis $I = \langle x^{\beta(1)}, \ldots, x^{\beta(s)} \rangle$. In the proof of Dickson's lemma, we observed that each $x^{\beta(i)}$ is divisible by $x^{\alpha(i)}$ for some $\alpha(i) \in A$. Prove that $I = \langle x^{\alpha(1)}, \ldots, x^{\alpha(s)} \rangle$.
3. Prove that Dickson's Lemma is equivalent to the following statement: Given a subset $A \subset \mathbb{Z}^n_{\geq 0}$, there are finitely many elements $\alpha(1), \ldots, \alpha(s) \in A$ such that for every $\alpha \in A$, there exists some i and some $\gamma \in \mathbb{Z}^n_{\geq 0}$ such that $\alpha = \alpha(i) + \gamma$.
4. A generating set $\{x^{\alpha_1}, \ldots, x^{\alpha_s}\}$ for a monomial ideal I is said to be *minimal* if no x^{α_i} in the generating set divides another x^{α_j} for $i \neq j$.
   a. Prove that every monomial ideal has a minimal generating set.
   b. Show that every monomial ideal has a unique minimal generating set.

§4. Hilbert Basis Theorem and Groebner Basis

In this section, we study a general ideal $I \subset k[\mathbf{x}] = k[x_1, \ldots, x_n]$ by passing to a "nice" monomial ideal. Indeed, since the leading terms play an important role in the division algorithm, the most natural monomial ideal which could be connected to a given ideal is the ideal generated by all leading terms of elements of I.

4.1. Definition Let I be a nonzero ideal in $k[\mathbf{x}]$.
(i) We denote by $\mathrm{LT}(I)$ the set of leading terms of elements of I. Thus,

$$\mathrm{LT}(I) = \left\{ c_\alpha x^\alpha \;\middle|\; \text{there exists } f \in I \text{ with } \mathrm{LT}(f) = c_\alpha x^\alpha \right\}.$$

(ii) We denote by $\langle \mathrm{LT}(I) \rangle$ the ideal generated by $\mathrm{LT}(I)$.

4.2. Lemma Let $I \subset k[\mathbf{x}]$ be an ideal.
(i) $\langle \mathrm{LT}(I) \rangle$ is a monomial ideal, and there are $g_1, \ldots, g_t \in I$ such that $\langle \mathrm{LT}(I) \rangle = \langle \mathrm{LT}(g_1), \ldots, \mathrm{LT}(g_t) \rangle$.

(ii) If I is a monomial ideal, say $I = \langle m_1, ..., m_s \rangle$ with m_i being monomials, then $\mathrm{LT}(m_i) = m_i$ for $i = 1, ..., s$ and hence $\langle \mathrm{LT}(I) \rangle = \langle m_1, ..., m_s \rangle = I$.

Proof (i) Obviously, $\langle \mathrm{LT}(I) \rangle = \langle \mathrm{LM}(g) \mid g \in I \rangle$, i.e., $\langle \mathrm{LT}(I) \rangle$ is a monomial ideal. Since $\langle \mathrm{LT}(I) \rangle$ is generated by the monomials $\mathrm{LM}(g)$ for $g \in I - \{0\}$, Dickson's lemma yields that $\langle \mathrm{LT}(I) \rangle = \langle \mathrm{LM}(g_1), ..., \mathrm{LM}(g_t) \rangle$ for finitely many $g_1, ..., g_t \in I$. Since $\mathrm{LM}(g_i)$ differs from $\mathrm{LT}(g_i)$ by a nonzero constant, it follows that $\langle \mathrm{LT}(I) \rangle = \langle \mathrm{LT}(g_1), ..., \mathrm{LT}(g_t) \rangle$.
(ii) Easy. □

Generally, if $I = \langle f_1, ..., f_s \rangle$, one has to be careful with the ideals $\langle \mathrm{LT}(I) \rangle$ and $\langle \mathrm{LT}(f_1), ..., \mathrm{LT}(f_s) \rangle$, in fact, these may be *different* ideals. It is true that $\mathrm{LT}(f_i) \in \mathrm{LT}(I) \subset \langle \mathrm{LT}(I) \rangle$ by definition, which implies $\langle \mathrm{LT}(f_1), ..., \mathrm{LT}(f_s) \rangle \subset \langle \mathrm{LT}(I) \rangle$. However, $\langle \mathrm{LT}(I) \rangle$ can be strictly larger. For example, let $I = \langle f_1 = x^3 - 2xy, f_2 = x^2y - 2y^2 + x \rangle$. Using the grlex ordering on $M(k[x, y])$, then

$$x \cdot (x^2y - 2y^2 + x) - y \cdot (x^3 - 2xy) = x^2,$$

so that $x^2 \in I$. Thus, $x^2 = \mathrm{LT}(x^2) \in \langle \mathrm{LT}(I) \rangle$. However, $x^2 \notin \langle \mathrm{LT}(f_1) = x^3, \mathrm{LT}(f_2) = x^2y \rangle$ because x^2 is not divisible by x^3 or by x^2y.
Hence, although both $\langle \mathrm{LT}(I) \rangle$ and $\langle \mathrm{LT}(f_1), ..., \mathrm{LT}(f_s) \rangle$ are monomial ideals, that does not help us to solve the ideal membership problem, even if we only consider monomials.
But on the other hand, we have seen from Corollary 3.7.that $\langle \mathrm{LT}(I) \rangle$ is good enough to solve the membership problem. This motivates us to explore the relation between I and $\langle \mathrm{LT}(I) \rangle$. Using Lemma 4.2 and the division algorithm, we can now prove the existence of a finite generating set for *every* polynomial ideal. As always, we have selected one particular monomial order to use in the division algorithm and in computing leading terms.

4.3. Theorem (Hilbert basis theorem) Let I be a nonzero ideal of $k[\mathbf{x}]$. Then there are $g_1, ..., g_t \in I$ such that every element $f \in I$ may be written as

$$f = a_1g_1 + \cdots + a_tg_t, \quad a_i \in k[\mathbf{x}],$$

with $\mathrm{md}(f) \geq \mathrm{md}(a_i g_i)$ whenever $a_i g_i \neq 0$.
Consequently $I = \langle g_1, ..., g_t \rangle$.

Proof By Lemma 4.2, $\langle \mathrm{LT}(I) \rangle = \langle \mathrm{LT}(g_1), ..., \mathrm{LT}(g_t) \rangle$ for a finite number $g_1, ..., g_t \in I$. We claim that $I = \langle g_1, ..., g_t \rangle$.
It is clear that $\langle g_1, ..., g_t \rangle \subset I$. Conversely, let $f \in I$ be any polynomial. If we apply the division algorithm from §2 to divide f by $g_1, ..., g_t$, then we get an expression of the form

$$f = a_1 g_1 + \cdots + a_t g_t + r,$$

where r is the remainder and $\mathrm{md}(f) \geq \mathrm{md}(a_i g_i)$ whenever $a_i g_i \neq 0$. We claim that $r = 0$. To see this, note that $r = f - \sum a_i g_i \in I$. If $r \neq 0$, then $\mathrm{LT}(r) \in \langle \mathrm{LT}(I) \rangle = \langle \mathrm{LT}(g_1), ..., \mathrm{LT}(g_t) \rangle$, and by Lemma 3.2., $\mathrm{LT}(r)$ must be divisible by some $\mathrm{LT}(g_i)$. This contradicts the definition of a remainder. Consequently, r must be zero. Thus $f = a_1 g_1 + \cdots + a_t g_t \in \langle g_1, ..., g_t \rangle$. Since f is arbitrary, this shows that $I \subset \langle g_1, ..., g_t \rangle$, finishing the proof. □

From the above theorem and its proof we retain that for any nonzero ideal I of $k[\mathbf{x}]$, nice generating set may be selected from the monomial ideal $\langle \mathrm{LT}(I) \rangle$, and using this generating set, the membership problem can be solved.

4.4. Definition Let I be an ideal of $k[\mathbf{x}]$ and $G = \{g_1, ..., g_t\}$ a finite subset of I.
(i) Let $f \in I$. If f can be expressed as

$$f = \sum_{i=1}^{t} a_i g_i, \quad a_i \in k[\mathbf{x}],$$

with $\mathrm{md}(f) \geq \mathrm{md}(a_i g_i)$ whenever $a_i g_i \neq 0$, then we say that f has a *standard representation* with respect to G.
(ii) If every $f \in I$ has a standard representation with respect to G, then we call G a *Groebner basis* (or *standard basis*).

4.5. Theorem Let $>$ be a fixed ordering on $\mathrm{M}(k[\mathbf{x}])$.
(i) Every nonzero ideal of $k[\mathbf{x}]$ has a Groebner basis.
(ii) If I is a nonzero ideal of $k[\mathbf{x}]$ and $G = \{g_1, ..., g_t\}$ is a finite subset of I, then G is a Groebner basis of I if and only if

$$\langle \mathrm{LT}(g_1), ..., \mathrm{LT}(g_t) \rangle = \langle \mathrm{LT}(I) \rangle .$$

Proof (i) This follows from Theorem 4.3.
(ii) Suppose that G is a Groebner basis of I. If $f \in I$, then

$$f = \sum_{i=1}^{t} a_i g_i, \quad a_i \in k[\mathbf{x}],$$

with $\mathrm{md}(f) \geq \mathrm{md}(a_i g_i)$ whenever $a_i g_i \neq 0$. Since $\mathrm{LT}(a_i g_i) = \mathrm{LT}(a_i)\mathrm{LT}(g_i)$, it follows that $\mathrm{LT}(f) = \mathrm{LT}(a_j)\mathrm{LT}(g_j)$ for some j. Hence, $\mathrm{LT}(f) \in \langle \mathrm{LT}(g_1), ..., \mathrm{LT}(g_t)\rangle$, i.e., $\langle \mathrm{LT}(I)\rangle = \langle \mathrm{LT}(g_1), ..., \mathrm{LT}(g_t)\rangle$.
Conversely, if $\langle \mathrm{LT}(g_1), ..., \mathrm{LT}(g_t)\rangle = \langle \mathrm{LT}(I)\rangle$, then a similar argument as in the proof of Theorem 4.3 will finish the proof. □

4.6. Corollary Let $G = \{g_1, ..., g_t\}$ be a Groebner basis for an ideal $I \subset k[\mathbf{x}]$ and let $f \in k[\mathbf{x}]$. Then there is a unique $r \in k[\mathbf{x}]$ with the following two properties:
(i) No term of r is divisible by one of $\mathrm{LT}(g_1), ..., \mathrm{LT}(g_t)$.
(ii) There is $g \in I$ such that $f = g + r$.
In particular, r is the remainder on division of f by G no matter how the elements of G are listed when used in the division algorithm.

Proof The division algorithm yields

$$f = \sum_{i=1}^{t} a_i g_i + r, \quad a_i \in k[\mathbf{x}],$$

where r satisfies (i). By setting $g = \sum_{i=1}^{t} a_i g_i \in I$, (ii) is also satisfied. This proves the existence of r. The uniqueness of r may be proved by a similar argument as in the proof of Theorem 10. □

Here we should point out that, although the remainder r is unique, even for a Groebner basis, the "quotients" a_i produced by the division algorithm $f = \sum a_i g_i + r$ can change if we list the generators in a different order (see Exercise 1).

As a corollary, let us explicitly mention the following criterion for deciding when a polynomial belongs to an ideal.

4.7. Corollary Let $G = \{g_1, ..., g_t\}$ be a Groebner basis for an ideal $I \subset k[\mathbf{x}]$ and let $f \in k[\mathbf{x}]$. Then $f \in I$ if and only if the remainder on division of f by G is zero.

□

Using Corollary 4.7., we arrive at an algorithm for solving the ideal membership problem *provided* that we know a Groebner basis G for the ideal in question–we only need to compute a remainder with respect to G to determine whether $f \in I$.

In the sequel, we use the following notation.

4.8. Definition We will denote the remainder for division of f by the ordered s-tuple $F = (f_1, ..., f_s)$ by $\overline{f}^F$. If F is a Groebner basis for $\langle f_1, ..., f_s\rangle$, then we may view F as a set (without considering any specific order) by Corollary 3.11 and write $F = \{f_1, ..., f_s\}$.

Example (i) By the foregoing calculation and Theorem 4.5, $\{f_1 = x^3 - 2xy, f_2 = x^2y - 2y^2 + x\}$ is not a Groebner basis for the ideal $I\langle f_1, f_2\rangle$.

(ii) Consider the ideal $J = \langle g_1 = x + z, g_2 = y - z\rangle \subset \mathbb{R}[x, y, z]$. We claim that $G = \{g_1, g_2\}$ is a Groebner basis for J with respect to the lex ordering in $\mathbb{R}[x, y, z]$. Thus, we must show that $\langle \mathrm{LT}(J)\rangle = \langle \mathrm{LT}(g_1), \mathrm{LT}(g_2)\rangle = \langle x, y\rangle$. Let $f = Ag_1 + Bg_2$ be any nonzero element of J. By Lemma 3.2 we have to show that $\mathrm{LT}(f)$ is divisible by x or by y. Suppose on the contrary that $\mathrm{LT}(f)$ is divisible by neither x nor y. Then by the definition of lex order, f must be a polynomial in z alone. However, f vanishes on the linear subspace $L = \mathbf{V}(x+z, y-z) \subset \mathbf{A}^3_{\mathbb{R}}$ since $f \in J$. It is easy to check that $(x, y, z) = (-t, t, t) \in L$ for any real number t. The only polynomial in z alone that vanishes at all of these points is the zero polynomial, which is a contradiction. It follows that G is a Groebner basis for J.

We refer the reader to [CLO'] for

- a more systematic way to detect when a generating set is a Groebner basis by using the *S-polynomial*,
- the famous Buchberger's Algorithm which produces a Groebner basis containing a given finite generating set of an ideal,

and

- a computation of Groebner bases for the sum, product, and the intersection of two ideals in $k[\mathbf{x}]$.

Exercises for §4

1. Above we established that $G = \{x+z, y-z\}$ is a Groebner basis with respect to the lex order. Let us use this basis to study the uniqueness of the division algorithm.
   a. Divide xy by $x+z$, $y-z$.
   b. Now reverse the order and divide xy by $y-z$, $x+z$.
   You should get the same remainder but the "quotients" should be different for the two divisions. This shows that the uniqueness of the remainder is the best one can hope for.
2. Let $A = (a_{ij})$ be an $m \times n$ matrix with real entries in row echelon form and let $J \subset I\!R[x_1, ..., x_n]$ be an ideal generated by the linear polynomials $\sum_{j=1}^{n} a_{ij}x_j$ for $1 \le i \le m$. Show that the given generators form a Groebner basis for J with respect to a suitable lexicographic order. (Hint: Order the variables corresponding to the leading one before the other variables.)
3. Let $I \subset k[x_1, ..., x_n]$ be a *principal ideal*. Show that any finite subset of I containing a generator for I is a Groebner basis for I.
4. Let $f \in k[x_1, ..., x_n]$. If $f \notin \langle x_1, ..., x_n \rangle$, then show that $\langle x_1, ..., x_n, f\rangle = k[x_1, ..., x_n]$.
5. Consider the algebraic set $V = \mathbf{V}(x^2 - y, y + x^2 - 4) \subset \mathbb{C}^2$. Note that $V = \mathbf{V}(I)$, where $I = \langle x^2 - y, y + x^2 - 4\rangle$.
   a. Prove that $I = \langle x^2 - y, x^2 - 2\rangle$.
   b. Using the basis in part a. prove that
   $$\mathbf{V}(I) = \left\{(\sqrt{2}, 2), (-\sqrt{2}, 2)\right\}.$$
   One reason why the second basis made V easier to understand was that $x^2 - 2$ could be *factored*. This implied that V "split" into two pieces. The next exercise is a general version of this.
6. When an ideal has a basis containing some elements that can be factored, we may use the factorization in the description of the algebraic set.

a. Show that if $g \in k[x_1, ..., x_n]$ factors as $g = g_1 g_2$, then for any f, $V(f, g) = \mathbf{V}(f, g_1) \cup \mathbf{V}(f, g_2)$.
b. Show that in $\mathbb{R}^3$, $\mathbf{V}(y - x^2, xz - y^2) = \mathbf{V}(y - x^2, xz - x^4)$.
c. Use part a. to describe and/or sketch the algebraic set in part b.

§5. Applications to Previous Chapters

As before, we put $k[\mathbf{x}] = k[x_1, ..., x_n]$.

• Consistency of Nullstellensatz
Let k be an algebraically closed field, or let $k = \mathbb{C}$. Recall that this problem asks whether a system

$$\begin{cases} f_1 = 0 \\ f_2 = 0 \\ \vdots \\ f_s = 0 \end{cases}$$

of polynomial equations has a common solution in k. By the weak Nullstellensatz, $\mathbf{V}(f_1, ..., f_s) = \emptyset$ if and only if $1 \in \langle f_1, ..., f_s \rangle$. Thus, to solve the consistency problem, we must be able to determine whether 1 belongs to an ideal.

5.1. Proposition Let I be an ideal of $k[\mathbf{x}]$ and $G = \{g_1, ..., g_s\}$ a Groebner basis of I with respect to a monomial ordering. The following are equivalent:
(i) $1 \in I$;
(ii) $1 \in G$.

Proof (i) $\Rightarrow$ (ii) if $1 \in I$, then 1 has a standard presentation by G:

$$1 = \sum_{i=1}^{s} h_i g_i, \qquad h_i \in k[\mathbf{x}],$$

where $0 = \text{md}(1) \geq \text{md}(h_i g_i)$ whenever $h_i g_i \neq 0$. But by Corollary 3.5 we claim that $\text{md}(g_i) = 0$ for some i and hence g_i is a constant. Thus we may say that $g_i = 1$.

(ii) ⇒ (i) This is obvious. □

Summarizing, we have obtained the following *consistency algorithm*: If we have polynomials $f_1, ..., f_s \in k[\mathbf{x}]$, we compute a Groebner basis of the ideal they generate with respect to any ordering. If the basis contains 1, the polynomials have no common zero in k; if the basis does not contain 1, they must have a common zero.
If we are working over a field k which is not algebraically closed, then the consistency algorithm still works in one direction: if 1 is contained in a Groebner basis of $\langle f_1, ..., f_s \rangle$, then $\mathbf{V}(f_1, ..., f_s) = \emptyset$. The converse is not true, for instance, consider in $\mathbb{R}[x, y]$ the polynomial $1 + x^2 + y^2$.

For systematic reasons, we postpone the algorithmic solution to the "finiteness of zeros" problem posed in CH.I §5 untill CH.V Theorem 2.4 and Corollary 2.5.

• The Radical membership problem
Let I be an ideal in $k[\mathbf{x}]$ and $\sqrt{I}$ the radical of I. For $f \in k[\mathbf{x}]$, to test whether $f \in \sqrt{I}$, we could use the ideal membership algorithm to check whether $f^m \in I$ for all integers $m > 0$. This is not satisfactory because we might have to go to very large powers of m, and it will never tells us if $f \notin \sqrt{I}$ (at least, not untill we work out a priori bounds on m). Fortunately, we can adapt the proof of Hilbert's Nullstellensatz to give an algorithm for determining whether $f \in \sqrt{I}$.

5.2. Proposition Let k be an arbitrary field and let $I = \langle f_1, ..., f_s \rangle$ be an ideal of $k[\mathbf{x}]$. Then for $f \in k[\mathbf{x}]$, $f \in \sqrt{I}$ if and only if the constant polynomial 1 belongs to the ideal $\tilde{I} = \langle f_1, ..., f_s, 1 - yf \rangle \subset k[\mathbf{x}, y]$ (in which case, $\tilde{I} = k[\mathbf{x}, y]$).

Proof If $1 \in \tilde{I}$, then a similar argument as in the proof of Hilbert's Nullstellensatz (CH.I) shows that $1 \in \tilde{I}$ implies $f^m \in I$ for some m, which, in turn, implies $f \in \sqrt{I}$. Conversely, suppose that $f \in \sqrt{I}$. Then $f^m \in I \subset \tilde{I}$ for some m. But we also have $1 - yf \in \tilde{I}$, and, consequently,

$$\begin{aligned} 1 &= y^m f^m + (1 - y^m f^m) \\ &= y^m f^m + (1 - yf)\left(1 + yf + \cdots + y^{m-1} f^{m-1}\right) \\ &\in \widetilde{I}, \end{aligned}$$

as desired. □

Proposition 5.2, together with the discussion of the consistency problem above, immediately leads to the *radical membership algorithm*. That is, to determine if $f \in \sqrt{\langle f_1, ..., f_s \rangle} \subset k[\mathbf{x}]$, we compute a Groebner basis G of the ideal $\langle f_1, ..., f_s, 1 - yf \rangle \subset k[\mathbf{x}, y]$ with respect to some ordering. If $1 \in G$, then $f \in \sqrt{I}$. Otherwise, $f \notin \sqrt{I}$.

Example (i) Consider the ideal $I = \langle xy^2 + 2y^2, x^4 - 2x^2 + 1 \rangle$ in $k[x, y]$. Let us test if $f = y - x^2 + 1$ lies in $\sqrt{I}$. Using lex order on $k[x, y, z]$, one checks that the ideal

$$\widetilde{I} = \left\langle xy^2 + 2y^2, x^4 - 2x^2 + 1, 1 - z(y - x^2 + 1) \right\rangle \subset k[x, y, z]$$

has a Groebner basis $\{1\}$. It follows that $y - x^2 + 1 \in \sqrt{I}$ by Proposition 5.2.

• **Recognition of the invertibility of polynomial mappings**
Let $V \subset \mathbf{A}_k^n$, $W \subset \mathbf{A}_k^m$ be affine algebraic sets and ϕ: $W \to V$ a polynomial mapping with coordinate functions $f_1, ..., f_n \in k[y_1, ..., y_m]$. The algorithm asked for in the second question posed in CH.II §4 now follows.

A. The case $V = W = \mathbf{A}_k^n$

Recall from [CLO'] that for a polynomial ideal I with a Groebner basis G, G is called *reduced* if

a. $\mathrm{LC}(g) = 1$ for all $g \in G$; and
b. for all $g \in G$, no monomial of g lies in $\langle \mathrm{LT}(G - \{g\}) \rangle$.

For any nonzero polynomial ideal I, I has a *unique* reduced Groebner basis with respect to a given monomial ordering.

5.3. Proposition Let $y_1, ..., y_n$ be new indeterminates and $I = \langle y_1 - f_1, ..., y_n - f_n \rangle \subset k[x_1, ..., x_n, y_1, ..., y_n]$. Furthermore, let $>$

be a monomial order on $k[x_1, ..., x_n, y_1, ..., y_n]$ satisfying $y_j < x_i$ for $1 \leq i, j \leq n$. Then the following are equivalent:

a. ϕ has a polynomial inverse mapping;

b. The reduced Groebner basis G of I with respect to $>$ is of the form

$$G = \{x_1 - g_1, ..., x_n - g_n\} \text{ with } g_1, ..., g_n \in k[y_1, ..., y_n].$$

Moreover, if b holds, then $g_i(f_1, ..., f_n) = x_i$ for $1 \leq i \leq n$.

Proof We refer to (A. Van den Essen, "A criterion to decide if a polynomial map is invertible and to compute the inverse", *Comm. Alg.* 10(18)(1990)) or ([BW] P.331) for a detailed proof. □

B. The general case

Put

$I_1 = \mathbf{I}(V), \quad I_2 = \mathbf{I}(W),$
$J = \langle I_2, x_1 - f_1, , ..., x_n - f_n\rangle \subset k[x_1, ..., x_n, y_1, ..., y_m],$
G a reduced Groebner basis for J,
$G_{\mathbf{x}} = G \cap k[x_1, ..., x_n],$
$G_M = G - G_{\mathbf{x}},$
ϕ_*: $k[V] \to k[W]$ the coordinate ring homomorphism induced by ϕ.

Then we have

5.4. Proposition With notation as above, ϕ_* is an isomorphism, or equivalently, ϕ is a polynomial isomorphism if and only if $G_{\mathbf{x}}$ is a Groebner basis of I_1 and $G_M = \{y_1 - h_1, ..., y_m - h_m\}$ with $h_i \in k[y_1, ..., y_m]$.

Proof We refer to (M. Kwiecinski, "A Groebner basis criterion for isomorphism of algebraic varieties," *J. Pure and Applied Algebra,* 74(1991)) for a detailed proof.

• Determination of the defining equations of the projective closure

Let I be an ideal of $k[\mathbf{x}] = k[x_1, ..., x_n]$ and $V = \mathbf{V}(I)$ the algebraic set defined by I in $\mathbf{A}^n$. By CH.III Lemma 6.1(i) we know that $\mathbf{V}(I^*) = \mathbf{V}((\sqrt{I})^*)$, where I^* resp. $(\sqrt{I})^*$ is the homogenization of I resp. $(\sqrt{I})^*$ in $k[\mathbf{x}][x_{n+1}] = k[x_1, ..., x_n, x_{n+1}]$ with respect to x_{n+1}.

It follows from the Nullstellensatz that, if k is algebraically closed, $\mathbf{V}(I^*)$ is equal to the projective closure of $\mathbf{V}(I)$ in $\mathbf{P}^n$. Furthermore, let $I = \langle f_1, ..., f_s \rangle$. We now proceed to produce a generating set for I^* from the one for I, as promised at the end of CH.III §6.

Recall from §1 that a *graded* monomial order in $k[\mathbf{x}]$ is defined first by total degree:

$$x^\alpha > x^\beta$$

whenever $|\alpha| > |\beta|$. Note that grlex and grevlex are graded orders, whereas lex is not.
Let $>$ be a graded order on monomials in $k[\mathbf{x}]$. We can extend $>$ to a monomial order $>_h$ in $k[\mathbf{x}][x_{n+1}]$ as follows:

$$x^\alpha x_{n+1}^d >_h x^\beta x^e \Longleftrightarrow x^\alpha > x^\beta \text{ or } x^\alpha = x^\beta \text{ and } d > e.$$

It is not hard to check that $>_h$ is a monomial order on monomials in $k[\mathbf{x}][x_{n+1}]$, we leave it as an exercise to the reader. Note that with this ordering, we have $x_i >_h x_{n+1}$ for all $1 \leq i \leq n$.

5.5. Lemma Taking the homogenization f^* of f with respect to x_{n+1}, if $f \in k[\mathbf{x}]$ and $>$ is a graded order on $k[\mathbf{x}]$, then

$$\mathrm{LM}_{>_h}(f^*) = \mathrm{LM}_>(f),$$

where $\mathrm{LM}_>(f)$ resp. $\mathrm{LM}_{>_h}(f^*)$ denotes the leading monomial of f with respect to $>$ resp. the leading monomial of f^* with respect to $>_h$.

Proof Since $>$ is a graded order, for any $f \in k[\mathbf{x}]$, $\mathrm{LM}_>(f)$ is one of the monomials x^α appearing in the homogeneous component of f of *maximal* total degree. Under homogenization this term is unchanged. If $x^\beta x_{n+1}^e$ is any one of the other monomials appearing in f^*, then $\alpha > \beta$. By definition of $>_h$, it follows that $x^\alpha >_h x^\beta x_{n+1}^e$. Hence, $x^\alpha = \mathrm{LM}_{>_h}(f^*)$, and the lemma is proved. □

5.6. Proposition Let I be an ideal in $k[\mathbf{x}]$ and let $G = \{g_1, ..., g_s\}$ be a Groebner basis for I with respect to a graded monomial order $>$ in $k[\mathbf{x}]$. Taking the homogenization I^* of I in $k[\mathbf{x}][x_{n+1}]$ with respect to x_{n+1}, then $G^* = \{g_1^*, ..., g_s^*\}$ is a Groebner basis for I^* with respect to $>_h$.

Proof Each $g_i^* \in I^*$ by definition. Thus, it suffices to show that the ideal of leading terms $\langle \mathrm{LT}_{>_h}(I)\rangle$ is generated by $\mathrm{LT}_{>_h}(G^*) = \{g_1^*, ..., g_s^*\}$. To prove this, consider $F \in I^*$. Since I^* is a graded ideal, each homogeneous component of F is in I^* and, hence, we may assume that F is homogeneous. Because $F \in I^*$, we have

$$(1) \qquad F = \sum_j H_j f_j^*, \quad H_j \in k[\mathbf{x}][x_{n+1}] \text{ and } f_j \in I.$$

Then by taking the dehomogenization of F with respect to x_{n+1} and applying CH.III Proposition 5.2 we derive from (1) that:

$$F_* = \sum_j H_{j*}(f_j^*)_* = \sum_j H_{j*} f_j.$$

Consequently, $F_* \in I \subset k[\mathbf{x}]$. Again by CH.III Proposition 5.2 we obtain $F = x_{n+1}^e (F_*)^*$ for some $e \geq 0$. Thus,

$$(2) \qquad \mathrm{LM}_{>_h}(F) = x_{n+1}^e \cdot \mathrm{LM}_{>_h}((F_*)^*) = x_{n+1}^e \cdot \mathrm{LM}_{>}(F_*),$$

where the last equality follows from Lemma 4.3. Since G is a Groebner basis fir I, we know that $\mathrm{LM}_>(F_*)$ is divisible by some $\mathrm{LM}_>(g_i) = \mathrm{LM}_{>_h}(g_i^*)$ (using Lemma 4.3 again). Then (2) shows that $\mathrm{LM}_{>_h}(F)$ is divisible by $\mathrm{LM}_{>_h}(g_i^*)$, as desired. This completes the proof of the proposition. □

To illustrate the proposition, consider the ideal $I = \langle y - x^2, z - x^3\rangle \subset \mathbb{R}[x, y, z]$ of the affine twisted cubic $W \subset \mathbf{A}^3_{\mathbb{R}}$ once again (see the the final remark of CH.III §6). Computing a Groebner basis for I with respect to grlex order, we find

$$G = \left\{x^2 - y, xy - z, xz - y^2\right\}.$$

(The reader may verify this example either directly by Buchberger's algorithm (see [CLO'] or [BW]) or by using a computer algebra system that produces a Groebner basis for a polynomial ideal (e.g. Macaulay).) By Proposition 5.6, the homogenizations of these polynomials generate I^*. Thus

$$I^* = \left\langle x^2 - yw, xy - zw, xz - y^2\right\rangle \subset \mathbb{R}[x, y, z, w].$$

Note that this ideal defines the projective algebraic set $V' = \mathbf{V}(I^*) \subset \mathbf{P}^3_R$ which is the projective closure of the affine twisted cubic. If we compare the algebraic sets $\mathbf{V}(yw - x^2, zw^2 - x^3)$ and V', then as we have seen in CH.III §6 first one is really "bigger" than the second one.

• Recognition of the $(\phi_*)^*$-closed ideals

Let $I \subset k[x_1, ..., x_{n+1}]$ be a graded ideal with respect to the natural gradation on $k[x_1, ..., x_{n+1}]$ (see CH.I §1), and let ϕ: $\mathbf{A}^n \to U_{n+1} \subset \mathbf{P}^n$ be the $n+1$st imbedding of $\mathbf{A}^n$ into $\mathbf{P}^n$. From CH.III Proposition 6.9 we retain if I is $(\phi_*)^*$-closed, then $\mathbf{V}(I) = \overline{\mathbf{V}(I_*)}$ under the assumption that k is algebraically closed. It follows from CH.III Theorem 5.7 that I is $(\phi_*)^*$-closed if and only if

$$(*) \qquad x_{n+1}k[x_1, ..., x_{n+1}] \cap I = x_{n+1}I.$$

To obtain an algorithmic criterion for the equality $(*)$ above, we start from the *algorithm for computing intersections of ideals* (e.g. see [CLO']):

If $I = \langle f_1, ..., f_r \rangle$ and $J = \langle g_1, ..., g_s \rangle$ are ideals in $k[x_1, ..., x_n]$, we consider the ideal

$$\langle tf_1, ..., tf_r, (1-t)g_1, ..., (1-t)g_s \rangle \subset k[x_1, ..., x_n, t]$$

and compute a Groebner basis with respect to an appropriate order in which t is greater than the x_i. Then the elements of this basis which do not contain the variable t will form a basis (in fact, a Groebner basis) of $I \cap J$.

With this algorithm in hand, we have two ways for checking $(*)$:

(i) Compute the Groebner bases for the ideals on both sides of $(*)$, and then check the divisibility of one Groebner basis by another.

(ii) Using the reduced Groebner basis for a polynomial ideal (mentioned before), we have an *ideal equality algorithm* for checking when $(*)$ holds:

- Simply fix a monomial order and compute a reduced Groebner basis for

$$x_{n+1}k[x_1, ..., x_{n+1}] \cap I \text{ and } x_{n+1}I,$$

respectively. Then the ideals are equal if and only if their Groebner bases are the same.

CHAPTER V
Dimension of Algebraic Sets

The notion of dimension is at the basis of human thinking, not in the least because our observations of reality around us necessitate an interpretation in terms of "dimensions". This is absolutely obvious when we think about visual observations. Nevertheless if the human race would consist of blind beings, the notion of dimension would most probably be discovered and developed in the same way but perhaps based on "feeling" or motions of body parts. Anyway the mathematical notion would be the same. The dimension of a vector space is well-understood on an intuitive level but intuition may deceive us if we apply it to algebraic varieties that are not linear subspaces of $\mathbf{A}_k^n$. In fact, an in-depth study of the theory we have developed so far quickly leads to the discovery that really we are dealing with objects having an algebraic, geometric as well as a topological side. Each of these aspects of the theory of varieties will have a specific notion of "dimension" associated to it . As H. Weyl once stated, the angel of Topology and the devil of abstract Algebra seem to be fighting for the soul of every discipline in Mathematics! No statement could be more to the point here; indeed, the notion of dimension in geometry is standing firmly at the crossroads

of Algebra and Topology.
In algebraic geometry, there are many ways to formulate the concept of dimension, each of them has its own advantage. However, a computational approach to dimension theory will provide us with a more natural, effective and elegant way to understand other interesting approaches (see later §6 – §9).

Trying to introduce "dimension" in an abstract or axiomatized way we should keep in mind what kind of properties we want a notion of dimension to have. Let us list some of the most obvious characteristics we would require.

a. For both affine and projective algebraic sets, it should have a natural geometric meaning.
b. It can be characterized by an appropriate "algebraic invariant" of the the corresponding algebraic object $k[V]$ (or $k(V)$ when V is a variety).
c. It is computable whenever the ground field is computable.
d. If V and W are isomorphic algebraic sets (or birationally equivalent varieties, see CH.VI), then V and W have the same dimension.

Before obtaining a good definition, we better get rid of some intuitive illusions by looking at the following examples.

(i) Let $\mathbf{A}^n_k = k^n$ be the affine n-space over k. Then, as a k-linear space, $\dim_k \mathbf{A}^n = n$.

(ii) Let $V = \mathbf{V}(y - f(x))$ be a curve in $\mathbf{A}^2$. Since there is an isomorphism

$$\begin{array}{ccccc} k & \longrightarrow & V & \longrightarrow & k \\ x & \mapsto & (x, f(x)) & \mapsto & x \end{array}$$

we may say that V has dimension 1. From this point of view we may say that curves in $\mathbf{A}^2$ have dimension 1.

(iii) Let $V = \mathbf{V}(z - f(x,y))$ be a surface in $\mathbf{A}^3$. Since there is an isomorphism

$$\begin{array}{ccccc} \mathbf{A}^2 & \longrightarrow & V & \longrightarrow & \mathbf{A}^2 \\ (x,y) & \mapsto & (x, y, f(x,y)) & \mapsto & (x,y) \end{array}$$

we may say that V has dimension 2. From this point of view we may say that surfaces in $\mathbf{A}^3$ has dimension 2.

(iv) More generally, let $V = \mathbf{V}(f(x_1, ..., x_n))$ be a hypersurface in $\mathbf{A}^n$. Then we would say that V has dimension $n-1$.

(v) Now consider the twisted cubic $V = \mathbf{V}(y - x^2, z - x^3)$ in $\mathbf{A}^3$. Then V is a curve. And since there is an isomorphism

$$\begin{array}{ccccc} k & \longrightarrow & V & \longrightarrow & k \\ x & \mapsto & (x, x^2, x^3) & \mapsto & x \end{array}$$

we may say that V has dimension 1, i.e., the dimension drops by 2.

Note that since each equation imposes an extra constraint, intuition suggests that each equation drops the dimension by one. Thus, if we started in $\mathbb{R}^4$, one expects that an affine algebraic set defined by two equations is a surface (i.e., dimension 2). Now, consider the algebraic set $\mathbf{V}(xz, yz)$ in $\mathbf{A}^3$. One easily checks that the equations $xz = yz = 0$ define the union of the (x, y)-plane and the z-axis. Hence, this algebraic set consists of two pieces which have different dimensions, and one of the pieces (the plane) has the "wrong" dimension according to the above intuition.
Next, consider a system of m linear equations in $n(\geq m)$ variables $x_1, ..., x_n$ with coefficients in k:

$$\begin{cases} a_{11}x_1 + \cdots + a_{1n}x_n = 0 \\ \qquad\vdots \\ a_{m1}x_1 + \cdots + a_{mn}x_n = 0 \end{cases}$$

The solutions of these equations form an algebraic set in $\mathbf{A}^n$ (note that this example includes lines and planes in $\mathbf{A}^2$ and $\mathbf{A}^3$), denoted V. Again, the dimension of V need not be $n - m$ even thuogh V is defined by m equations. Instead, when V is nonempty, linear algebra tells us that V has dimension $n-r$, where r is the rank of the matrix (a_{ij}). So the dimension of V is determined by the number of *independent* equations. This intuition applies to more general affine algebraic sets, except that the notion of "independent" becomes more subtle.

Adopting the viewpoints and methods of [CLO'] and [BW], we are now just three steps from a definition of dimension! These steps are described as follows:

(i) Starting with an algebraic set consisting of linear subspaces defined by the simplest "coordinate system", define and compute dim$W$ for an algebraic set W given by a monomial ideal;
(ii) For a general affine algebraic set $V = \mathbf{V}(I)$, define and compute dim$V$ by finding the algebraic alternative for dim$\langle \mathrm{LT}(I)\rangle$ (which defines an algebraic set consisting of linear subspaces) from the coordinate ring $k[V]$ of V;
(iii) For a general projective algebraic set V, define and compute dim$V$ in a similar way.

Throughout this chapter k is a field with char$k$=0.

§1. The Algebraic Set of a Monomial Ideal

Since the "simplest" ideals in $k[\mathbf{x}] = k[x_1, ..., x_n]$ are ideals consisting of monomials, we begin our study of dimension by considering algebraic sets defined by monomial ideals.
Suppose, for example, we have the ideal $I = \langle x^2y, x^3\rangle$ in $k[x, y]$. Letting H_x denote the line in $\mathbf{A}^2$ defined by $x = 0$ (so $H_x = \mathbf{V}(x)$) and H_y the line $y = 0$, we have

$$\mathbf{V}(I) = \mathbf{V}\left(x^2y\right) \cap \mathbf{V}\left(x^3\right) = H_x.$$

Thus, $\mathbf{V}(I)$ is the y-axis H_x. Since H_x has dimension 1 as a vector subspace of $\mathbf{A}^2$, it is reasonable to say that it also has dimension 1 as an algebraic set.
As a second example, consider the ideal

$$I = \left\langle yz^3, x^5z^4, x^2yz^2\right\rangle \subset k[x, y, z].$$

Let H_x be the plane defined by $x = 0$ and define H_y and H_z similarly. Also let H_{xy} be the line $x = y = 0$. Then we have

$$\mathbf{V}(I) = \mathbf{V}\left(y^2z^3\right) \cap \mathbf{V}\left(x^5z^4\right) \cap \mathbf{V}\left(x^2yz^2\right) = H_z \cup H_{xy}.$$

Thus, $\mathbf{V}(I)$ is the union of the (x, y)-plane H_z and the z-axis H_{xy}. We will say that *the dimension of a union of finitely many vector subspaces of* $\mathbf{A}^n$ *is the maximum of the dimensions of the subspaces*, and so the dimension of $\mathbf{V}(I)$ is 2 in this example.
To an algebraic set given by a monomial ideal we may assign a dimension in the same way. But first we have to describe what an algebraic set of a general monomial ideal looks like.

1.1. Definition In $\mathbf{A}^n$, a vector subspace defined by setting some subset of the variables $x_1, \dots, x_n$ equal to zero is called a *coordinate subspace.*

1.2. Proposition The algebraic set of a monomial ideal in $k[\mathbf{x}]$ is a finite union of coordinate subspaces of $\mathbf{A}^n$.

Proof First note that if $x_{i_1}^{\alpha_1} \cdots x_{i_r}^{\alpha_r}$ is a monomial in $k[\mathbf{x}]$ with $\alpha_j \geq 1$ for $1 \leq j \leq r$, then

$$\mathbf{V}\left(x_{i_1}^{\alpha_1} \cdots x_{i_r}^{\alpha_r}\right) = H_{x_{i_1}} \cup \cdots \cup H_{x_{i_r}},$$

where $H_{x_h} = \mathbf{V}(x_h)$. Thus, the algebraic set defined by a monomial is a union of coordinate hyperplanes. Note also that there are only n such hyperplanes.
Since a monomial ideal is generated by a finite collection of monomials, the algebraic set corresponding to a monomial ideal is a finite intersection of unions of coordinate hyperplanes. By the distributive property of intersections over unions, any finite intersection of unions of coordinate hyperplanes can be rewritten as a finite union of intersections of coordinate hyperplanes. But the intersection of any collection of coordinate hyperplanes is a coordinate subspace.□

When we write the algebraic set of a monomial ideal I as a union of finitely many coordinate subspaces, we may omit a subspace if it is contained in another one appearing in the union. Thus we may write $\mathbf{V}(I)$ as a union of coordinate subspaces

$$\mathbf{V}(I) = V_1 \cup \cdots \cup \mathbf{V}_p,$$

where $V_i \not\subset V_j$ for $i \neq j$. In fact, such a decomposition is unique, as one can show easily.

Let us make the following temporary definition.

1.3. Definition Let V be an algebraic set which is the union of a finite number of linear subspaces of $\mathbf{A}^n$. Then the *dimension* of V, denoted $\dim V$, is the largest of the dimensions of the subspaces.

Thus, the dimension of the union of two planes and a line is 2, and the dimension of a union of three lines is 1. To compute the dimension of the algebraic set corresponding to a monomial ideal, we only need to find the maximum of the dimensions of the coordinate subspaces contained in $\mathbf{V}(I)$.
Although this is easy to do for any given example, it is worth systematizing the computation. Let $I = \langle m_1, ..., m_t \rangle$ be a *proper ideal* generated by the monomials m_j. In order to compute $\dim \mathbf{V}(I)$, we need to pick out the component of

$$\mathbf{V}(I) = \bigcap_{j=1}^{t} \mathbf{V}(m_j)$$

of largest dimension. If we can find a collection of variables $x_{i_1}, ..., x_{i_r}$ such that at least one of these variables appears in each m_j, then the coordinate subspace defined by the equations $x_{i_1} = \cdots = x_{i_r} = 0$ is contained in $\mathbf{V}(I)$. This means we should look for variables which occur in as many of the different m_j as possible. More precisely, put

$$M_j = \left\{h \in \{1, ..., n\} \;\middle|\; x_h \text{ divides the monomial } m_j\right\}, \quad 1 \le j \le t,$$

$$\mathcal{M} = \left\{J \subset \{1, ..., n\} \;\middle|\; J \cap M_j \neq \emptyset \text{ for all } 1 \le j \le t\right\}.$$

Note that M_j is nonempty by the assumption $I \neq k[\mathbf{x}]$, and $\mathcal{M}$ is not empty because $\{1, ..., n\} \in \mathcal{M}$.

1.4. Proposition With notation as above,

$$\dim \mathbf{V}(I) = n - \min\left\{|J| \;\middle|\; J \in \mathcal{M}\right\}.$$

Proof Let J be an element of $\mathcal{M}$ such that $|J| = r$ is minimal in $\mathcal{M}$. Since each monomial m_j contains some power of some x_{i_h}, $1 \le h \le r$, the coordinate subspace $W = \mathbf{V}(x_{i_1}, ..., x_{i_r})$ is contained

in $\mathbf{V}(I)$. The dimension of W is $n - r = n - |J|$, and, hence, by Definition 1.3., the $\dim\mathbf{V}(I) \geq n - |J|$.
If $\dim\mathbf{V}(I) > n-r$, then for some $s < r$ there would be a coordinate subspace $W' = \mathbf{V}(x_{h_1}, ..., x_{h_s})$ contained in $\mathbf{V}(I)$. Each monomial m_j would vanish on W' and, in particular, it would vanish at the point $p \in W'$ whose h_i-th coordinate is 0 for $1 \leq i \leq s$ and whose other coordinates are equal to 1. Hence, at least one of the x_{h_i} must divide m_j (CH.IV Lemma 3.2), and it would follow that $J' = \{h_1, ..., h_s\} \in \mathcal{M}$. Since $|J'| = s < r$, this would contradict the minimality of r. Thus, the dimension of $\mathbf{V}(I)$ must be as claimed.□

Example (i) Let us check the proposition on the second example given above. To match the notation of the proposition, we relabel the variables x, y, z as x_1, x_2, x_3, respectively. Then

$$I = \left\langle x_2^2x_3^3, x_1^5x_3^4, x_1^2x_2x_3^2 \right\rangle = \langle m_1, m_2, m_3\rangle,$$

where $m_1 = x_2^2x_3^3$, $m_2 = x_1^5x_3^4$, $m_3 = x_1^2x_2x_3^2$. Using the notation of the discussion preceding Proposition 1.4.,

$$M_1 = \{2,3\}, \quad M_2 = \{1,3\}, \quad M_3 = \{1,2,3\},$$

so that

$$\mathcal{M} = \{\{1,2,3\}, \{1,2\}, \{1,3\}, \{2,3\}, \{3\}\}.$$

Then $\min\{|J| \mid J \in \mathcal{M}\} = 1$, which implies that

$$\dim\mathbf{V}(I) = 3 - \min_{J\in\mathcal{M}} |J| = 3 - 1 = 2.$$

In the extreme case, note that if some variable, say x_i, appears in every monomial in a set of generators for a proper monomial ideal I, then it will be true that $\dim\mathbf{V}(I) = n - 1$ since $J = \{i\} \in \mathcal{M}$. For a converse, see Exercise 3.

Remark (i) It is natural to compare a monomial ideal to its radical $\sqrt{I}$. Indeed, one can easily show that $\sqrt{I}$ is a monomial ideal when I is. We also know that $\mathbf{V}(I) = \mathbf{V}(\sqrt{I})$ for any ideal I. It follows from Definition 1.3. that $\mathbf{V}(I)$ and $\mathbf{V}(\sqrt{I})$ have the same dimension (since we defined dimension in terms of the underlying algebraic set).

(ii) At the first glance, Definition 1.3 and Proposition 1.4 do not immediately tell us that if V and W are isomorphic algebraic sets defined by monomial ideals then $\dim V = \dim W$. But this is definitely true as we will see in §6 Corollary 6.3.

Exercises for §1

1. For each of the following monomial ideals I, write $\mathbf{V}(I)$ as a union of coordinate subspaces.
   a. $I = \langle z^5, x^4yz^2, x^3z \rangle \subset k[x, y, z]$.
   b. $I = \langle wx^2y, xyz^3, wz^5 \rangle \subset k[w, x, y, z]$.
   c. $I = \langle x_1x_2, x_3 \cdots x_n \rangle \subset k[x_1, ..., x_n]$.
2. Find $\dim \mathbf{V}(I)$ for each of the following monomial ideals.
   a. $I = \langle xy, yz, xz \rangle \subset k[x, y, z]$.
   b. $I = \langle wx^2z, w^3y, wxyz, x^5z^6 \rangle \subset k[w, x, y, z]$.
   c. $I = \langle u^2vwyz, wx^3y^3, uxy^7z, y^3z, uwx^3y^3z^2 \rangle \subset k[u, v, w, x, y, z]$.
3. Suppose that $I \subset k[x_1, ..., x_n]$ is a monomial ideal such that $\dim \mathbf{V}(I) = n - 1$.
   a. Show that the monomials in any generating set for I have a nonconstant common factor.
   b. Write $\mathbf{V}(I) = V_1 \cup \cdots \cup V_p$, where V_i is a coordinate subspace and $V_i \not\subset V_j$ for $i \neq j$. Suppose, in addition, that exactly one of the V_i has dimension $n - 1$. What is the maximum that p (the number of components) can be? Give an example in which this maximum is achieved.
4. Let I be a monomial ideal in $k[x_1, ..., x_n]$ such that $\dim \mathbf{V}(I) = 0$.
   a. What is $\mathbf{V}(I)$ in this case?
   b. Show that $\dim \mathbf{V}(I) = 0$ if and only if for each $1 \leq i \leq n$, $x_i^{l_i} \in I$ for some $l_i \geq 1$. (Hint: In Proposition 1.4, when will it be true that $\mathcal{M}$ contains only $J = \{1, ..., n\}$?
5. Let $\langle m_1, ..., m_r \rangle \subset k[x_1, ..., x_n]$ be a monomial ideal generatd by $r \leq n$ monomials. Show that $\dim \mathbf{V}(m_1, ..., m_r) \geq n - r$.
6. Show that a coordinate subspace is an irreducible algebraic set when the field k is infinite.
7. Let $I \subset k[x_1, ..., x_n]$ be a monomial ideal, and assume that k is an infinite field.
   a. Show that $\mathbf{I}(\mathbf{V}(I))$ is a monomial ideal.

b. Show that $\mathbf{I}(\mathbf{V}(I)) = \sqrt{I}$.

8. Let $\mathbf{F}_2$ be the field with two elements, and let $I = \langle x \rangle \subset \mathbf{F}_2[x, y]$. Show that $\mathbf{I}(\mathbf{V}(I)) = \langle x, y^2 - y \rangle$. This is bigger than $\sqrt{I}$ and is not a monomial ideal.

§2. The Vector Space $k[x_1, ..., x_n]/I$

In §1, we have defined and computed $\dim V$ for the affine algebraic set of a monomial ideal from a geometric viewpoint. As for the irreducibility of an algebraic set, it is natural to ask

- Is there an algebraic property of $k[V]$ that may be used to characterize $\dim V$?

In order to answer this question, we first further the study of the k-vector space $k[x_1, ..., x_n]/\mathbf{I}(V) \cong k[V]$. In this section, we consider the more general k-vector space $k[x_1, ..., x_n]/I$, where I is an *arbitrary* ideal. We will use the division algorithm given in CH.IV to produce a simple basis for the k-vector space $k[x_1, ..., x_n]/I$; as an added dividend, we derive an easy to check criterion to determine when a system of polynomial equations over $\mathbb{C}$ has only finitely many solutions.

The *basic idea* used in the discussion is (a direct consequence of) the fact that the remainder for division of a polynomial f by a Groebner basis G for an ideal I is uniquely determined by the polynomial f. If no confusion is possible, we will write $k[\mathbf{x}]$ for the polynomial ring $k[x_1, ..., x_n]$ as before.

2.1. Proposition Fix a monomial order on $\mathrm{M}(k[\mathbf{x}])$ and let $I \subset k[\mathbf{x}]$ be an ideal. As before, $\langle \mathrm{LT}(I) \rangle$ will denote the ideal generated by the leading terms of elements of I.
(i) For every $f \in k[\mathbf{x}]$, there is a unique r, which is a k-linear combination of the monomials in $\mathrm{M}(k[\mathbf{x}]) - \langle \mathrm{LT}(I) \rangle$, such that $f - r \in I$.
(ii) The elements of $x^\alpha \in \mathrm{M}(k[\mathbf{x}]) - \langle \mathrm{LT}(I) \rangle$ are "linearly independent (modulo I)" in the sense that if

$$\sum_\alpha c_\alpha x^\alpha \in I \text{ with } x^\alpha \in \mathrm{M}(k[\mathbf{x}]) - \langle \mathrm{LT} \rangle,\ c_\alpha \in k,$$

then $c_\alpha = 0$ for all α.

Proof (i) Let G be a Groebner basis for I and let $f \in k[\mathbf{x}]$. By the division algorithm, the remainder $r = \overline{f}^G$ satisfies $f = q + r$, where $q \in I$. Hence $f - r = q \in I$, so $f \equiv r \bmod I$. The division algorithm also tells us that r is a k-linear combination of the monomials $x^\alpha \notin \langle \mathrm{LT}(I) \rangle = \langle \mathrm{LT}(G) \rangle$. The uniqueness of r follows from CH.IV §2.
(ii) The argument needed to establish this part of the proposition is essentially the same as the proof of the uniqueness of the remainder in CH.IV §2. □

2.2. Proposition Let $I \subset k[\mathbf{x}]$ be an ideal. Then $k[\mathbf{x}]/I$ is isomorphic as a k-vector space to

$$S = k\text{-Span}\left(x^\alpha \;\middle|\; x^\alpha \in \mathrm{M}(k[\mathbf{x}]) - \langle \mathrm{LT}(I) \rangle\right).$$

Proof By Proposition 2.1 we may define a map

$$\begin{array}{rccc} \phi: & k[\mathbf{x}]/I & \longrightarrow & S \\ & [f] & \mapsto & \overline{f}^G \end{array}$$

It is easy to see that ϕ is bijective. It remains to show that ϕ preserves the vector space operations. But from CH.IV §2 and Proposition 2.1 it is easy to verify that $\overline{f+g}^G = \overline{f}^G + \overline{g}^G$, where $f, g \in k[\mathbf{x}]$ and G is a Groebner basis for I. Hence the claim is established. □

The above discussion provides us with a completely algorithmic way to handle computations in $k[\mathbf{x}]/I$. To summarize, we phrase the following result.

2.3. Proposition Let I be an ideal in $k[\mathbf{x}]$ and let G be a Groebner basis for I with respect to any monomial order. For each $[f] \in k[\mathbf{x}]/I$, we get the *standard representative* $\overline{f} = \overline{f}^G$ in $S = k\text{-Span}(x^\alpha \mid x^\alpha \in \mathrm{M}(k[\mathbf{x}]) - \langle \mathrm{LT}(I) \rangle)$. Then:
(i) $[f] + [g] = \overline{f} + \overline{g}$.
(ii) $[f] \cdot [g] = \overline{\overline{f} \cdot \overline{g}}^G \in S$.

□

We conclude this section by providing an algorithmic answer to the question posed in CH.I §5, i.e., to algorithmically determine when an algebraic set in $\mathbb{C}^n$ contains only a finite number of points or, equivalently, to determine when a system of polynomial equations has only a finite number of solutions in $\mathbb{C}$. (As in CH.I, we must work over an algebraically closed field to ensure that we are not "missing" any solutions of the equations with coordinates in a larger field.)

2.4. Theorem Let $V = \mathbf{V}(I)$ be an affine algebraic set in $\mathbf{A}^n_{\mathbb{C}s}$ and fix a monomial order in $\mathbb{C}[\mathbf{x}]$. Then the following statements are equivalent:
(i) V is a finite set.
(ii) For each i, $1 \le i \le n$, there is some $m_i \ge 0$ such that $x_i^{m_i} \in \langle \mathrm{LT}(I) \rangle$.
(iii) Let G be a Groebner basis for I. Then for each i, $1 \le i \le n$, there is some $m_i \ge 0$ such that $x_i^{m_i} = \mathrm{LM}(g)$ for some $g \in G$.
(iv) The $\mathbb{C}$-vector space $S = k\text{-Span}(x^\alpha \mid x^\alpha \in \mathrm{M}(k[\mathbf{x}]) - \langle \mathrm{LT}(I) \rangle)$ is finite dimensional.
(v) The $\mathbb{C}$-vector space $\mathbb{C}[\mathbf{x}]/I$ is finite dimensional.

Proof (i) $\Rightarrow$ (ii) If $V = \emptyset$, then $1 \in I$ by the weak Nullstellensatz. In this case, we can take $m_i = 0$ for all i. If V is nonempty, then for a fixed i, let a_j, $j = 1, ..., h$, be the distinct complex numbers appearing as i-th coordinates of points in V. Form the one-variable polynomial

$$f(x_i) = \prod_{j=1}^{h} (x_i - a_j).$$

By construction, f vanishes at every point in V, so $f \in \mathbf{I}(V)$. By the Nullstellensatz, there is some $m \ge 1$ such that $f^m \in I$. But this means that the leading monomial of f^m is in $\langle \mathrm{LT}(I) \rangle$. Examining the expression for f, we see that $x_i^{hm} \in \langle \mathrm{LT}(I) \rangle$.
(ii) $\Leftrightarrow$ (iii) Let $x_i^{m_i} \in \langle \mathrm{LT}(I) \rangle$. Since G is a Groebner basis of I, $\langle \mathrm{LT}(I) \rangle = \langle \mathrm{LT}(g) \mid g \in G \rangle$. By Lemma 3.2 of CH. IV. §3, there is some $g \in G$, such that $\mathrm{LT}(g)$ divides $x_i^{m_i}$. But this implies that $\mathrm{LT}(g)$ is a power of x_i, as claimed. The opposite implication follows directly from the definition of $\langle \mathrm{LT}(I) \rangle$.
(ii) $\Rightarrow$ (iv) If some power $x_i^{m_i} \in \langle \mathrm{LT}(I) \rangle$ for each i, then the mono-

mials $x_1^{\alpha_1} \cdots x_n^{\alpha_n}$ for which some $\alpha_i \geq m_i$ are all in $\langle \mathrm{LT}(I)\rangle$. The monomials in $\mathrm{M}(\mathbb{C}[\mathbf{x}]) - \langle \mathrm{LT}(I)\rangle$ must have $\alpha_i \leq m_i - 1$ for each i. As a result, the number of monomials in the $\mathrm{M}(\mathbb{C}[\mathbf{x}]) - \langle \mathrm{LT}(I)\rangle$ can be at most $m_1 \cdot m_2 \cdots m_n$.
(iv) $\Leftrightarrow$ (v) Follows from Proposition 2.2.
(v) $\Rightarrow$ (i) To show that V is finite, it suffices to show that for each i there can be only finitely many distinct i-th coordinates for the points of V. Fix i and consider the classes $[x_i^j] = x_i^j + I$ in $\mathbb{C}[\mathbf{x}]/I$, where $j = 0, 1, 2, \ldots$. Since $\mathbb{C}[\mathbf{x}]/I$ is finite dimensional, the $[x_i^j]$ must be linearly dependent in $\mathbb{C}[\mathbf{x}]/I$. That is, there exist constants c_j (not all zero) and some m such that

$$\sum_{j=0}^{m} c_j \left[x_i^j\right] = \left[\sum_{j=0}^{m} c_j x_i^j\right] = [0].$$

However, this implies that $\sum_{j=0}^{m} c_j x_i^j \in I$. Since a nonzero polynomial in one variable can have only finitely many roots in $\mathbb{C}$, this shows that the points of V have only finitely many different i-th coordinates.
We note that the hypothesis $k = \mathbb{C}$ was used only in showing that (i) $\Rightarrow$ (ii). The other implications are true even if k is not algebraically closed. □

In §5, we will see that the above theorem also provides an algorithmic criterion for detecting 0-dimensional algebraic sets. Moreover, close examination of the proof of the theorem also yields the following quantitative estimation of the number of solutions of a system of equations when that number is finite.

2.5. Corollary Let $I \subset \mathbb{C}[\mathbf{x}]$ be an ideal such that for each i, some power $x_i^{m_i} \in \langle \mathrm{LT}(I)\rangle$. Then the number of points of $\mathbf{V}(I)$ is at most $m_1 \cdot m_2 \cdots m_n \leq \dim_{\mathbb{C}}(\mathbb{C}[\mathbf{x}]/I)$.

Proof Exercise. □

In CH.VII §2, we will see that (over $\mathbb{C}$) if two curves $V(f)$, $V(g) \subset \mathbf{A}^2_{\mathbb{C}}$ have no common components, then $\dim_{\mathbb{C}}(\mathbb{C}[x,y]/\langle f, g\rangle)$ actually counts the "number of points" of $V(f) \cap V(g)$ by adding up the intersection number at every $P \in V(f) \cap V(g)$.

Exercises for §2

1. Let $I = \langle y + x^2 - 1, xy - 2y^2 + 2y\rangle \subset \mathbb{R}[x, y]$.
   a. Use a computer algebra system to check that $G = \{x^2 + y - 1, xy - 2y^2 + 2y, y^3 - \frac{7}{4}y^2 + \frac{3}{4}y\}$ is a Groebner basis for the ideal I with respect to the lex order with $x > y$.
   b. Construct a vector space isomorphism $\mathbb{R}[x, y]/I \cong \mathbb{R}^4$; and
   c. Compute a "multiplication table" for the elements $\{[1], [x], [y], [y^2]\}$ in $\mathbb{R}[x, y]/I$. (Express each product as a linear combination of these four classes.
   d. Is $\mathbb{R}[x, y]/I$ a field? Why or why not?
2. Let I be any ideal in $k[x_1, ..., x_n]$.
   a. Suppose that $S = k\text{-Span}(x^\alpha \mid x^\alpha \notin \langle \text{LT}(I)\rangle)$ is a k-vector space of finite dimension d for some choice of monomial order. Show that the dimension of $k[x_1, ..., x_n]/I$ as a k-vector space is equal to d.
   b. Deduce from part a. that the number of monomials in the complement of $\langle \text{LT}(I)\rangle$ is independent of the choice of the monomial order, when the number is finite.
3. Suppose that $I \subset \mathbb{C}[x_1, ..., x_n]$ is an ideal such that for each i, $x_i^{m_i} \in \langle \text{LT}(I)\rangle$. State and prove a criterion that can be used to determine when $\mathbf{V}(I)$ contains *exactly* $m_1 \cdot m_2 \cdots m_n$ points in $\mathbb{C}^n$. Does the criterion take the multiplicity of the roots into account?
4. Complete the proof of Corollary 2.5.

§3. Hilbert Function and dim$V$

In this section, we continue the study of the k-vector space

$$k[x_1, ..., x_n]/I$$

for an *arbitrary* ideal $I \subset k[x_1, ..., x_n]$, in order to answer the question posed in the begining of §2.

We know that, as a vector space over k, the polynomial ring $k[\mathbf{x}] = k[x_1, ..., x_n]$ has infinite dimension, and the same is true for any nonzero ideal (CH.I §1 Exercise 3). Furthermore, from the last section we know that $k[\mathbf{x}]/I$ may also be infinite dimensional. However, recall from CH.I §1 that if we look at the standard k-basis of $k[\mathbf{x}]$ resp, $k[\mathbf{x}]/I$ by considering the total degree of polynomials, we see that for each $m \geq 0$, the subspace

$$k[\mathbf{x}]_{\leq m} = \left\{ f \in k[\mathbf{x}] \;\middle|\; f \text{ has total degree } \leq m \right\} \text{ resp.}$$

$$\frac{k[\mathbf{x}]_{\leq m} + I}{I} \cong \frac{k[\mathbf{x}]_{\leq m}}{I \cap k[\mathbf{x}]_{\leq m}}$$

is finite dimensional because $\dim_k k[\mathbf{x}]_{\leq m} = \binom{n+m}{m}$ (CH.I. Lemma 1.6). It is clear that

$$k[\mathbf{x}]_{\leq m} \subset k[\mathbf{x}]_{\leq m+1} \text{ for } m \geq 0; \qquad k[\mathbf{x}] = \bigcup_{m \geq 0} k[\mathbf{x}]_{\leq m}, \text{ and}$$

$$\frac{k[\mathbf{x}]_{\leq m} + I}{I} \subseteq \frac{k[\mathbf{x}]_{\leq m+1} + I}{I} \text{ for } m \geq 0; \; k[\mathbf{x}]/I = \bigcup_{m \geq 0} \frac{k[\mathbf{x}]_{\leq m} + I}{I}$$

This may be rephrased in a vulgarizing way as stating that the k-dimension of $k[\mathbf{x}]/I$ is being built-up step by step from finite dimensional subspaces.

The task we set for the present section is:

- first we establish the following facts:
  a. the "growth" of the k-vector space $k[\mathbf{x}]/I$ can be measured by a polynomial $h(x) \in \mathbb{Q}[x]$, i.e., for $m \gg 0$, $\dim_k(k[\mathbf{x}]_{\leq m} + I/I) = h(m)$; and
  b. $\deg h(x) = \dim \mathbf{V}(\langle \mathrm{LT}(I) \rangle)$;
- motivated by a. and b. before, we generalize the notion of "dimension" defined in §1 to an arbitrary affine algebraic set, and point out an algorithm to compute the dimension.

Put

$$I_{\leq m} = I \cap k[\mathbf{x}]_{\leq m}.$$

Then $I_{\leq m}$ is a vector subspace of $k[\mathbf{x}]_{\leq m}$. We are now ready to define the affine Hilbert function of I.

3.1. Definition Let I be an ideal in $k[\mathbf{x}]$. The *affine Hilbert function* of I is the function on the non-negative integers defined by

$$\begin{aligned} {}^{a}HF_I(m) &= \dim_k \frac{k[\mathbf{x}]_{\le m}}{I_{\le m}} \\ &= \dim_k k[\mathbf{x}]_{\le m} - \dim_k I_{\le m} \quad \text{for } m \in I\!N. \end{aligned}$$

From §2 we retain that the structure of the k-vector space $k[\mathbf{x}]/I$ is completely determined by the monomial ideal $\langle \mathrm{LT}(I)\rangle$. Next we will see that

- a "graded monomial ordering" may combine the "algorithmic method depending on monomials" and the "algebraic structure depending on gradings" together to reduce the discussion of the Hilbert function of an *arbitrary* ideal I to the discussion of the Hilbert function of the monomial ideal $\langle \mathrm{LT}(I)\rangle$, and this reduction makes ${}^{a}HF_I$ always *computable*.

3.2. Proposition (Macaulay, see [CLO'] CH.9 §3, Proposition 4) Let $k[\mathbf{x}]$ be equipped with a it graded monomial order $>$, and let I be an ideal of $k[\mathbf{x}]$ and $\langle \mathrm{LT}(I)\rangle$ the monomial ideal generated by $\mathrm{LT}(I)$, the set of leading terms of elements of I. Then

$$ {}^{a}HF_I(m) = {}^{a}HF_{\langle \mathrm{LT}(I)\rangle}(m), \quad m \ge 0. $$

Proof Put $\mathrm{M}(k[\mathbf{x}])_{\le m} = k[\mathbf{x}]_{\le m} \cap \mathrm{M}(k[\mathbf{x}])$. We will prove the proposition by showing that there are k-space isomorphisms:

$$ \frac{k[\mathbf{x}]_{\le m}}{I_{\le m}} \xrightarrow{\cong} k\text{-Span}\left(x^{\alpha} \;\middle|\; x^{\alpha} \in \mathrm{M}(k[\mathbf{x}])_{\le m} - \langle \mathrm{LT}(I)\rangle\right) \xleftarrow{\cong} \frac{k[\mathbf{x}]_{\le m}}{\langle \mathrm{LT}(I)\rangle_{\le m}} $$

Taking a Groebner basis G for I, note that since $\langle \mathrm{LT}(I)\rangle$ is a monomial ideal, the second isomorphism follows from CH.IV Lemma 3.3. The first isomorphism will be obtained by showing that the k-spaces $I_{\le m}$ and $\langle \mathrm{LT}(I)\rangle_{\le m}$ have the same dimension.
First note that

$$ (*) \qquad \left\{\mathrm{LM}(f) \;\middle|\; f \in I_{\le m}\right\} = \{\mathrm{LM}(f_1), \ldots, \mathrm{LM}(f_s)\} $$

for a finite number of $f_1, ..., f_s \in I_{\leq m}$. By rearranging and deleting duplicates, we may assume that $\mathrm{LM}(f_1) > \mathrm{LM}(f_2) > \cdots > \mathrm{LM}(f_s)$. We claim that $f_1, ..., f_s$ are a basis of $I_{\leq m}$ as a k-space. To prove this, consider a nontrivial linear combination $c_1 f_1 + \cdots + c_s f_s$ and choose the smallest i such that $c_i \neq 0$. Given how we ordered the leading monomials, there is nothing to cancel $c_i \mathrm{LT}(f_i)$, so the linear combination is nonzero. Hence, $f_1, ..., f_s$ are linearly independent. Next, let $W = k\text{-Span}\{f_1, ..., f_s\} \subset I_{\leq m}$. If $W \neq I_{\leq m}$, pick $f \in I_{\leq m} - W$ with $\mathrm{LM}(f)$ minimal (note that $>$ is a well-ordering). By $(*)$, $\mathrm{LM}(f) = \mathrm{LM}(f_i)$ for some i, and hence, $\mathrm{LT}(f) = \lambda \mathrm{LT}(f_i)$ for some $\lambda \in k$. Then $f - \lambda f_i \in I_{\leq m}$ has a smaller leading monomial, so that $f - \lambda f_i \in W$ by the minimality of $\mathrm{LM}(f)$. This implies $f \in W$, a contradiction. It follows that $W = I_{\leq m}$, and we may conclude that $f_1, ..., f_s$ is a basis.
A similar argumentation as above shows that $\mathrm{LM}(f_1), ..., \mathrm{LM}(f_s)$ are k-linearly independent. Note that we are using the graded monomial order $>$. This means that for any nonzero $f \in A$, if $\mathrm{LM}(f) \in I_{\leq m}$, then so does f. It follows immediately from CH.IV Lemma 3.3 and the above $(*)$ that

$$\langle \mathrm{LT}(I) \rangle_{\leq m} = k\text{-Span}\{\mathrm{LM}(f_1), ..., \mathrm{LM}(f_s)\}.$$

So we are done. □

Starting from the foregoing proposition, we now proceed to prove that there exists a polynomial $h(x)$ in $\mathbb{Q}[x]$ with positive leading coefficient such that for $m \gg 0$, ${}^aHF_I(m) = h(m)$. Roughly speaking, the main idea is:

- to classify $S = \mathrm{M}(k[\mathbf{x}])_{\leq m} - \langle \mathrm{LT}(I) \rangle$ by a suitably defined equivalence relation on S, and then
- to compute the number of elements in each equivalence class.

To this end, we first have to introduce a few more technicalities.

Let $M \in \mathbb{N}$, $t \in \mathrm{M}(k[\mathbf{x}])$, and let $\deg_{x_i}(t)$ denote the degree in x_i of the polynomial t. Then we define

$$\mathrm{top}_M(t) = \left\{ i \in \{1, ..., n\} \;\middle|\; \deg_{x_i}(t) \geq M \right\},$$

i.e., $\mathrm{top}_M(t)$ is the set of indices where "t tops M." Furthermore, we set

$$\mathrm{sh}_M(t) = \prod_{i \in \mathrm{top}_M(t)} x_i^M \cdot \prod_{\substack{i=1 \\ i \notin \mathrm{top}_M(t)}}^{n} x_i^{\deg_{x_i}(t)},$$

i.e., $\mathrm{sh}_M(t)$ is "t shaved at M."

3.3. Lemma for $t \in \mathrm{M}(k[\mathbf{x}])$ and $M \in \mathbb{N}$, we have

$$\mathrm{sh}_M(\mathrm{sh}_M(t)) = \mathrm{sh}_M(t).$$

Proof Put $t' = \mathrm{sh}_M(t)$, then $\mathrm{top}_M(t') = \mathrm{top}_M(t)$. Hence the desired equality holds. □

3.4. Lemma Let $S \subset \mathrm{M}(k[\mathbf{x}])$ be a subset and $M \in \mathbb{N}$. The following hold:
(i) The relation $\sim_M$ defined by

$$s \sim_M t \qquad \text{iff } \mathrm{sh}_M(s) = \mathrm{sh}_M(t)$$

is an equivalence relation on S.
(ii) Let $[s]_{\sim_M}$ denote the equivalence class of $s \in S$. If S is such that $\mathrm{sh}_M(s) \in S$ for all $s \in S$, then the set

$$R_M = \left\{ t \in S \;\middle|\; \mathrm{sh}_M(t) = t \right\}$$

of "already shaved terms" is a system of unique representatives for the partition of S into equivalence classes with respect to $\sim_M$ (i.e., $[s]_{\sim_M} = [t]_{\sim_M}$ for a unique $t \in R_M$). The set R_M may also be described as

$$R_M = \left\{ t \in S \;\middle|\; \deg_{x_i}(t) \le M \text{ for } 1 \le i \le n \right\}.$$

(iii) Let S and R_M be as in (ii). For $t \in R_M$, we have

$$[t]_{\sim_M} = \left\{ w \cdot t \;\middle|\; \begin{array}{l} w \in \mathrm{M}(k[\mathbf{x}]), \\ \deg_{x_i}(w) = 0 \text{ for all } i \notin \mathrm{top}_M(t) \text{ and } w \cdot t \in S, \end{array} \right\}$$

i.e., the elements of $[t]_{\sim_M}$ are obtained by raising those exponents in t that equal M in such a way that the result remains in S.

Proof (i) is clear.
(ii) By the assumption we have $\mathrm{sh}_M(s) \in S$ for all $s \in S$. Hence $s \sim_M \mathrm{sh}_M(s)$ by Lemma 3.3, i.e., $[s]_{\sim_M} = [\mathrm{sh}_M(s)]_{\sim_M}$ with $\mathrm{sh}_M(s) \in R_M$. If $s \sim_M t$, then $\mathrm{sh}_M(s) = \mathrm{sh}_M(t)$. This means that the representative of $[s]_{\sim_M}$ in R_M is unique.
Furthermore, if $t \in R_M$, then

$$t = \mathrm{sh}_M(t) = \prod_{i\in \mathrm{top}_M(t)} x_i^M \cdot \prod_{\substack{i=1\\ i\notin \mathrm{top}_M(t)}}^{n} x_i^{\deg_{x_i}(t)},$$

and consequently $\deg_{x_i}(t) \leq M$ for $1 \leq i \leq n$. Conversely, if $t \in S$ with $\deg_{x_i}(t) \leq M$ for $1 \leq i \leq n$, then clearly $t = \mathrm{sh}_M(t)$, i.e., $t \in R_M$.
(iii) Let $t \in R_M$. If $s \sim_M t$, then $\mathrm{sh}_M(s) = \mathrm{sh}_M(t) = t$, i.e.,

$$\begin{aligned} t &= \mathrm{sh}_M(t) = \prod_{i\in \mathrm{top}_M(t)} x_i^M \cdot \prod_{\substack{i=1\\ i\notin \mathrm{top}_M(t)}}^{n} x_i^{\deg_{x_i}(t)} \\ &= \prod_{j\in \mathrm{top}_M(s)} x_j^M \cdot \prod_{\substack{i=1\\ j\notin \mathrm{top}_M(s)}}^{n} x_j^{\deg_{x_j}(s)} \\ &= \mathrm{sh}_M(s) \end{aligned}$$

Comparing the degrees of x_i, for any $i \notin \mathrm{top}_M(t)$ we have $\deg_{x_i}(s) = \deg_{x_i}(t) < M$; for $i \in \mathrm{top}_M(t)$, $\deg_{x_i}(s) \geq M = \deg_{x_i}(t)$. Hence there exists $w \in \mathrm{M}(k[\mathbf{x}])$ with $\deg_{x_i}(w) = 0$, $i \notin \mathrm{top}_M(t)$, such that $s = wt$.
Conversely, for any wt satisfying $wt \in S$ and $\deg_{x_i}(w) = 0$, $i \notin \mathrm{top}_M(t)$, it is clear that $t \sim_M wt$. □

3.5. Lemma Let I be a proper ideal of $k[\mathbf{x}]$ and G a Groebner basis of I with respect to the graded monomial order $>$ we fixed before. If we put

$$M = \max\left\{\beta_i \;\middle|\; \mathrm{LT}(g) = c_\beta x_1^{\beta_1} \cdots x_n^{\beta_n},\ g \in G,\ 1 \leq i \leq n\right\},$$

then for every $t \in \mathrm{M}(k[\mathbf{x}])$, whenever $i \in \mathrm{top}_M(t)$ and $\nu \in I\!N$, $t \in \mathrm{M}(k[\mathbf{x}]) - \langle \mathrm{LT}(I)\rangle$ if and only if $t \cdot x_i^\nu \in \mathrm{M}(k[\mathbf{x}]) - \langle \mathrm{LT}(I)\rangle$.

Proof If $tx_i^\nu \in \mathrm{M}(k[\mathbf{x}]) - \langle \mathrm{LT}(I)\rangle$, then $t \in \mathrm{M}(k[\mathbf{x}]) - \langle \mathrm{LT}(I)\rangle$. Conversely, let $t \in \mathrm{M}(k[\mathbf{x}]) - \langle \mathrm{LT}(I)\rangle$. Assume that $tx_i^\nu \in \langle \mathrm{LT}(I)\rangle = \langle \mathrm{LT}(g) \mid g \in G\rangle$. Then $\mathrm{LT}(g) | tx_i^\nu$ for some $g \in G$. On the other hand, since $\deg_{x_i}(t) \geq M$, the definition of M yields $\deg_{x_i}(t) \geq \deg_{x_i}(\mathrm{LT}(g))$ for every $g \in G$. It follows that $\mathrm{LT}(g)|t$, i.e., $t \in \mathrm{LT}(I)$, a contradiction. Hence $tx_i^\nu \in \mathrm{M}(k[\mathbf{x}]) - \langle \mathrm{LT}(I)\rangle$. □

Before proving the main result, let us recall a fact from Proposition 3.2 and its proof:

$$(\diamond) \quad \begin{cases} {}^aHF_I(m) &= {}^aHF_{\langle \mathrm{LT}(I)\rangle}(m) \\ &= \left|\mathrm{M}(k[\mathbf{x}])_{\leq m} - \langle \mathrm{LT}(I)\rangle\right|, \quad m \geq 0, \end{cases}$$

where $\mathrm{M}(k[\mathbf{x}])_{\leq m} = \mathrm{M}(k[\mathbf{x}]) \cap k[\mathbf{x}]_{\leq m}$ and $\langle \mathrm{LT}(I)\rangle$ is the ideal generated by all leading terms of elements of I.

Moreover, we need the following elementary observation. If $m \in \mathbb{N}$, we set

$$q = \frac{x(x-1)\cdots(x-m+1)}{m!}.$$

Then $q \in \mathbb{Q}[x]$ with $\deg(q) = m$, and for all $N \in \mathbb{N}$ with $N \geq m$, we obtain

$$q(N) = \binom{N}{m}.$$

It is in fact customary to write $q = \binom{x}{m}$ in this case.

3.6. Theorem Let I be an ideal of $k[\mathbf{x}]$ and G a Groebner basis of I with respect to a graded monomial order on $\mathrm{M}(k[\mathbf{x}])$. Let

$$M = \max\left\{\beta_i \;\middle|\; \mathrm{LT}(g) = c_\beta x_1^{\beta_1} \cdots x_n^{\beta_n},\ g \in G,\ 1 \leq i \leq n\right\}.$$

Then there exists a unique polynomial $h \in \mathbb{Q}[x]$ with positive leading coefficient such that

$${}^aHF_I(m) = h(m), \text{ for all } m \geq n \cdot M.$$

If the ground field is computable, then h and the number $n \cdot M$ can be computed from any given generating set of I.

Proof By the previous formula ($\diamond$), the desired polynomial $h(x)$ must satisfy

$$(1) \qquad h(m) = \left|\mathrm{M}(k[\mathbf{x}])_{\le m} - \langle \mathrm{LT}(I)\rangle\right|$$

for all $m \ge n{\cdot}M$. We will arrive at such a polynomial by counting the elements of $\mathrm{M}(k[\mathbf{x}])_{\le m} - \langle \mathrm{LT}(I)\rangle$. To this end, we let $m \in \mathbb{N}$ with $m \ge n{\cdot}M$. It follows from Lemma 3.5 that $S = \mathrm{M}(k[\mathbf{x}])_{\le m} - \langle \mathrm{LT}(I)\rangle$ satisfies $\mathrm{sh}_M(s) \in S$ for all $s \in S$. hence S is the disjoint union of the equivalence classes with respect to the equivalence relation $\sim_M$ of Lemma 3.4(i). Using the set R_M of Lemma 3.4(ii) as a system of the unique representatives, we have

$$(2) \qquad \left|\mathrm{M}(k[\mathbf{x}])_{\le m} - \langle \mathrm{LT}(I)\rangle\right| = \sum_{t \in R_M} |[t]_{\sim_M}|,$$

where $[t]_{\sim_M}$ is the equivalence class of $t \in R_M$.
Now from Lemma 3.4(iii) and Lemma 3.5 it is clear that for every $t \in R_M$, if $t = x_1^{\alpha_1} \cdots x_n^{\alpha_n}$, then

$$[t]_{\sim_M} = \left\{ w \cdot t \;\middle|\; \begin{array}{l} w = x_1^{\gamma_1} \cdots x_n^{\gamma_n}, \\ \gamma_i = 0 \text{ for } i \notin \mathrm{top}_M(t),\ |\gamma| \le m - |\alpha| \end{array} \right\}.$$

It follows from CH.I Lemma 1.6 that

$$|[t]_{\sim_M}| = \binom{m - |\alpha| + |\mathrm{top}_M(t)|}{|\mathrm{top}_M(t)|}$$

which is a polynomial of degree $|\mathrm{top}_M(t)|$ in $m - |\alpha|$. Combining this with (2), we have obtained a polynomial $h \in \mathbb{Q}[x]$ with positive leading coefficient that satisfies (1) and

$$\begin{aligned} \deg(h(x)) &= \max\left\{ |\mathrm{top}_M(t)| \;\middle|\; t \in R_M \right\} \\ &= \max\left\{ |\mathrm{top}_M(s)| \;\middle|\; s \in \mathrm{M}(k[\mathbf{x}])_{\le m} - \langle \mathrm{LT}(I)\rangle \right\}. \end{aligned}$$

Since we have $\langle \mathrm{LT}(I)\rangle = \langle \mathrm{LT}(G)\rangle$, the existence proof of the polynomial h that we have just given shows that the last statement of the theorem concerning computability is true. It is also clear that there can be only one $h(x) \in \mathbb{Q}[x]$ satisfying ${}^a HF_I(m) = h(m)$ for infinitely many $m \in \mathbb{N}$. □

3.7. Definition The polynomial h obtained in the above theorem is called the *affine Hilbert polynomial* of I and is denoted ${}^aHP_I$.

Example (i) Let $I = \langle x^4y^3, x^2y^5\rangle \subset k[x,y]$. Then it is obvious that $G = (x^4y^3, x^3y^5)$ is a Groebner basis of I. Thus

$$\mathrm{M}(k[x,y]) - \langle \mathrm{LT}(I)\rangle = \left\{ x^iy^j \;\middle|\; \begin{array}{l} x^4y^3 \text{ not divides } x^iy^j \\ \text{and } x^2y^5 \text{ not divides } x^iy^j \end{array} \right\}$$

Put $C = \{(i,j),\ x^iy^j \in \mathrm{M}(k[x,y]) - \langle \mathrm{LT}(I)\rangle\}$ Then $C = \{(i,j),\ 0 \le i \le 1\} \cup \{(i,j),\ 0 \le j \le 2\} \cup \{(2,3),(2,4),(3,3),(3,4)\}$. If $i,j \le 5$, then clearly $\mathrm{sh}_5(x^iy^j) = x^iy^j$. Hence

$$\begin{aligned} R_5 &= \left\{ t \in S \;\middle|\; \deg_{x_i}(t) \le 5,\ 1 \le i \le 2 \right\} \\ &= \left\{ x^iy^j \;\middle|\; 0 \le i \le 5,\ 0 \le j \le 5,\ (i,j) \in C \right\} \end{aligned}$$

and only when $t = y^5, xy^5, x^5, x^5y, x^5y^2$ $|\mathrm{top}_M(t)| = 1$, otherwise $|\mathrm{top}_M(t)| = 0$. Since $|R_5| = 28$, if $m \ge 2\cdot 5$ we have by the theorem that

$$\begin{aligned} h(m) &= \left|\mathrm{M}(k[x,y])_{\le m} - \langle \mathrm{LT}(I)\rangle\right| \\ &= \sum_{\substack{t\in R_5 \\ |top_5(t)|=1}} \binom{m-\deg(t)+1}{1} + \sum_{\substack{t\in R_5 \\ |top_5(t)|=0}} 1 \\ &= (m-4)+(m-5)+(m-4)+(m-5)+(m-6)+25 \\ &= 5m-1 \end{aligned}$$

Hence the Hilbert polynomial of I is $h_I(x) = 5x - 1$.

From Proposition 3.2 and Theorem 3.6 we retain that for an ideal I the equality ${}^aHP_I = {}^aHP_{\langle \mathrm{LT}(I)\rangle}$ holds. Next, we prove the equality $\dim \mathbf{V}(\langle \mathrm{LT}(I)\rangle) = \deg({}^aHP_I)$ by showing that the degree of ${}^aHP_I$ is closely related to the independence of subsets in $\{x_1, ..., x_n\}$ (modulo I) (in §6 we will further explore this independence in terms of the coordinate ring of $\mathbf{V}(I)$).

To see this, for $U = \{x_{i_1}, ..., x_{i_r}\} \subset \{x_1, ..., x_n\}$, we write $k[U]$ for the polynomial subring $k[x_{i_1}, ..., x_{i_r}]$ of $k[x_1, ..., x_n]$, and write $\mathrm{M}(U)$ for the set of all monomials of $k[U]$. For an ideal I of $k[\mathbf{x}]$, we say that U is *independent* (modulo I) if $\mathrm{M}(U) \cap \langle \mathrm{LT}(I)\rangle = \{0\}$. If we put

$$d = \max\left\{|U| \;\middle|\; U \subset \{x_1, ..., x_n\} \text{ is independent (modulo) } I\right\},$$

then we have

3.8. Proposition With notation as above, $\deg({}^aHP_I) = d$.

Proof Let G be a Groebner basis of I with respect to a graded monomial order on $k[\mathbf{x}]$. We first recall from the foregoing that for

$$M = \max\left\{\beta_i \;\middle|\; \mathrm{LT}(g) = c_\beta x_1^{\beta_1} \cdots x_n^{\beta_n},\ g \in G,\ 1 \le i \le n\right\},$$

$\deg({}^aHP_I) = \max\{|\mathrm{top}_M(t)| \mid t \in \mathrm{M}(k[\mathbf{x}])_{\le m} - \langle \mathrm{LT}(I)\rangle\}$, where $m \ge n \cdot M$.

To prove the inequality "$\le$", assume for a contradiction that there exists $t \in \mathrm{M}(k[\mathbf{x}])_{\le m} - \langle \mathrm{LT}(I)\rangle$, $t = x_1^{\alpha_1} \cdots x_n^{\alpha_n}$ with $\alpha_i \ge M$ for more than d many indices. Then there exists a subset $U = \{x_{i_1}, ..., x_{i_r}\} \subset \{x_1, ..., x_n\}$ with $i_1 > \cdots > i_r$, $r > d$, and a decomposition $t = t_1 \cdot t_2$ with $t_2 \in \mathrm{M}(U)$ and $t_1 \in M(U^c)$, where $U^c = \{x_1, ..., x_n\} - U$, such that

$$t_2 = x_{i_1}^{\alpha_{i_1}} \cdots x_{i_r}^{\alpha_{i_r}}, \quad \alpha_{i_j} \ge M,\ 1 \le j \le r.$$

We must have $t_2 \in \mathrm{M}(k[\mathbf{x}])_{\le m} - \langle \mathrm{LT}(I)\rangle$ because t_2 is a factor of t. On the other hand, since $r > d$, we have $\mathrm{M}(U) \cap \langle \mathrm{LT}(I)\rangle \ne \{0\}$, and so there exists $g \in G$ with $\mathrm{LT}(g) \in \mathrm{M}(U)$. Note that since for $\mathrm{LT}(g) = c_\beta x_1^{\beta_1} \cdots x_n^{\beta_n}$ we have $\beta_i \le M$, $1 \le i \le n$, it follows that t_2 is divisible by $\mathrm{LT}(g)$ and thus $t \in \langle \mathrm{LT}(g) \mid g \in G\rangle = \langle \mathrm{LT}(I)\rangle$, a contradiction.

For the inequality "$\ge$", let $U = \{x_{i_1}, ..., x_{i_d}\} \subset \{x_1, ..., x_n\}$ be such that $i_1 > \cdots > i_d$ and $\mathrm{M}(U) \cap \langle \mathrm{LT}(I)\rangle = \{0\}$. Then for $m \ge n \cdot M$, it is easy to see that $t = x_{i_1}^M \cdots x_{i_d}^M \in \mathrm{M}(k[\mathbf{x}])_{\le m} - \langle \mathrm{LT}(I)\rangle$ and $|\mathrm{top}_M(t)| = d$. □

3.9. Proposition If I is a proper monomial ideal, then

$$\dim \mathrm{V}(I) = d = \deg({}^aHP_I).$$

Proof Suppose $I = \langle m_1, ..., m_s \rangle$ where the m_j are monomials. Recall from §1 that

$$M_j = \{h \in \{1, ..., n\} \mid x_h \text{ divides the monomial } m_j\},\ 1 \le j \le s,$$

$$\mathcal{M} = \{J \subset \{1, ..., n\} \mid J \cap M_j \ne \emptyset \text{ for all } 1 \le j \le s\},$$

if we let $|J|$ denote the number of elements in a set J, then $\dim \mathbf{V}(I) = n - \min\{|J| \mid J \in \mathcal{M}\}$ (Proposition 1.4).
Now suppose $U = \{x_{i_1}, ..., x_{i_r}\} \subset \{x_1, ..., x_n\}$ with $i_1 > \cdots > i_r$ and $\mathrm{M}(U) \cap I = \{0\}$ (note that $\langle \mathrm{LT}(I) \rangle = I$). Then the set $J_U = \{1, ..., n\} - \{i_1, ..., i_r\}$ satisfies $J_U \cap M_j \ne \emptyset$, $1 \le j \le s$. Hence $J_U \in \mathcal{M}$, and consequently

$$r = n - |J_U| \le n - \min\Big\{|J| \ \Big|\ J \in \mathcal{M}\Big\}.$$

Conversely, for any $\{i_1, ..., i_t\} = J \in \mathcal{M}$, if we put $U = \{x_{j_1}, ..., x_{j_{n-t}}\}$, where $\{j_1, ..., j_{n-t}\} = \{1, ..., n\} - J$, then it is easy to see that $\mathrm{M}(U) \cap I = \{0\}$. Hence, $|U| = n - t \le d$. Thus we have proved the required equality. □

Proposition 3.9 highlights the following two facts:

- The degree of the affine Hilbert polynomial coincides with the intuitive notion of dimension for the algebraic set of a monomial ideal.
- For a general proper ideal I of $k[\mathbf{x}]$, the dimension of the affine algebraic set defined by $\langle \mathrm{LT}(I) \rangle$ is equal to the degree of the polynomial ${}^aHP_I$ which measures the growth of the k-dimension of the ring $k[\mathbf{x}]/I$ as a k-vector space (or how far $I_{\le m}$ is from being all of $k[\mathbf{x}]_{\le m}$).

This prompts the definition of dimension in terms of the degree of the affine Hilbert polynomial.

3.10. Definition The *dimension* of an affine algebraic set $V \subset \mathbf{A}_k^n$, denoted dim$V$, is the degree of the affine Hilbert polynomial of the corresponding ideal $I = \mathbf{I}(V) \subset k[\mathbf{x}]$.

Example (ii) Let $V = \mathbf{V}(y - x^2, z - x^3) \subset \mathbb{R}^3$ be the twisted cubic. It can be shown that $I = I(V) = \langle y - x^2, z - x^3 \rangle \subset \mathbb{R}[x, y, z]$. Using the grlex order, $G = \{y^3 - z^2, x^2 - y, xy - z, xz - y^2\}$ is a Groebner basis for I, so that $\langle \mathrm{LT}(I)\rangle = \langle y^3, x^2, xy, xz\rangle$. Then

$$\begin{aligned}\dim V &= \deg({}^a HP_I)\\ &= \deg\left({}^a HP_{\langle \mathrm{LT}(I)\rangle}\right)\\ &= \left\{\begin{array}{l}\text{maximum dimension}\\ \text{of a coordinate subspace in } \mathbf{V}(\langle \mathrm{LT}(I)\rangle)\end{array}\right.\end{aligned}$$

Since

$$\mathbf{V}(\langle \mathrm{LT}(I)\rangle) = \mathbf{V}\left(y^3, x^2, xy, xz\right) = \mathbf{V}(x,y) \subset \mathbb{R}^3,$$

we conclude that $\dim V = 1$. This agrees with our intuition that the twisted cubic should be 1-dimensional since it is a curve in $\mathbb{R}^3$.

(iii) Let I be a monomial ideal in $k[x_1, ..., x_n]$. Then since $\mathbf{I}(\mathbf{V}(I)) = \sqrt{I}$ (§1 Exercise 7), we have

$$\dim \mathbf{V}(I) = \deg\left({}^a HP_{\mathbf{I}(\mathbf{V}(I))}\right) = \deg\left({}^a HP_{\sqrt{I}}\right) = \deg({}^a HP_I),$$

and it follows from Proposition 3.9 that $\dim \mathbf{V}(I)$ is the maximum dimension of a coordinate subspace contained in $\mathbf{V}(I)$. This agrees with the temporary definition of dimension given in §1.

An interesting exceptional case is the empty variety. Note that $1 \in \mathbf{I}(V)$ if and only if $k[x_1, ..., x_n]_{\leq m} = \mathbf{I}(V)_{\leq m}$ for all m. Hence

$$V = \emptyset \iff {}^a HP_{\mathbf{I}(V)} = 0.$$

Since the zero polynomial does not have a degree, we do not assign a dimension to the empty variety.

By Definition 3.10, in order to compute the dimension of an algebraic set V, we need to know $\mathbf{I}(V)$, which, in general, is difficult to compute. It would be much nicer if $\dim V$ could be the degree of ${}^a HP_I$, where I is an arbitrary ideal defining V. Unfortunately, this is not true in general. For example, if $I = \langle x^2 + y^2\rangle \subset \mathbb{R}[x,y]$, it is easy to check that ${}^a HP_I(m)$ has degree 1, yet $V = \mathbf{V}(I) = \{(0,0)\} \subset \mathbb{R}^2$

is easily seen to have dimension 0. Thus, $\dim \mathbf{V}(I) \neq \deg({}^aHP_I)$ in this case.
To overcome these difficulties, we first need to compare the degree of the affine Hilbert polynomials for I and $\sqrt{I}$.

3.11. Proposition If $I \subset k[\mathbf{x}]$ is an ideal, then the affine Hilbert polynomials of I and $\sqrt{I}$ have the same degree.

Proof For a monomial ideal I, we know that $\dim \mathbf{V}(I) = \dim \mathbf{V}\left(\sqrt{I}\right)$ (see §1 Remark). Hence, $\deg({}^aHP_I) = \deg\left({}^aHP_{\sqrt{I}}\right)$ by Proposition 3.9.
Now let I be an arbitrary ideal in $k[\mathbf{x}]$ and pick any graded monomial order on $\mathrm{M}(k[\mathbf{x}])$. We claim that

$$\langle \mathrm{LT}(I)\rangle \subset \left\langle \mathrm{LT}(\sqrt{I})\right\rangle \subset \sqrt{\langle \mathrm{LT}(I)\rangle}.$$

The first inclusion is immediate from $I \subset \sqrt{I}$. To establish the second, let x^α be a monomial in $\left\langle \mathrm{LT}(\sqrt{I})\right\rangle$. This means that there is a polynomial $f \in \sqrt{I}$ such that $\mathrm{LT}(f) = x^\alpha$. We know $f^r \in I$ for some $r \geq 0$, and it follows that $x^{r\alpha} = \mathrm{LT}(f^r) \in \langle \mathrm{LT}(I)\rangle$. Thus, $x^\alpha \in \sqrt{\langle \mathrm{LT}(I)\rangle}$. From the definition of Hilbert function we obtain the inequalities

$$\deg\left({}^aHP_{\sqrt{\langle \mathrm{LT}(I)\rangle}}\right) \leq \deg\left({}^aHP_{\langle \mathrm{LT}(\sqrt{I})\rangle}\right) \leq \deg\left({}^aHP_{\langle \mathrm{LT}(I)\rangle}\right).$$

By the result for monomial ideals, the extreme terms are equal and we conclude that ${}^aHP_{\langle \mathrm{LT}(I)\rangle}$ and ${}^aHP_{\langle \mathrm{LT}(\sqrt{I})\rangle}$ have the same degree. The definition of d and Theorem 3.6 imply that the same is true for ${}^aHP_I$ and ${}^aHP_{\sqrt{I}}$, and the proposition is proved. □

Thus, if k is algebraically closed, the degree of the affine Hilbert polynomial is the same for a large collection of ideals defining the same algebraic set; and moreover, the difficulties encountered above vanish because of the Nullstellensatz: $\mathbf{I}(\mathbf{V}(I)) = \sqrt{I}$. More precisely, we have the following theorem that tells us how to compute the dimension in terms of any defining ideal.

3.12. Theorem (The Dimension Theorem) Let $V = \mathbf{V}(I)$ be an affine algebraic set, where $I \subset k[\mathbf{x}]$ is an ideal. If k is algebraically closed, then

$$\dim V = \deg({}^aHP_I).$$

Furthermore, if $>$ is a graded monomial order on $\mathrm{M}(k[\mathbf{x}])$, then

$$\begin{aligned} \dim V &= \deg\left({}^aHP_{\langle \mathrm{LT}(I)\rangle}\right) \\ &= \begin{cases} \text{maximum dimension} \\ \text{of a coordinate subspace in } \mathbf{V}(\langle \mathrm{LT}(I)\rangle). \end{cases} \end{aligned}$$

Finally, the last two equalities hold over any field k when $I = \mathbf{I}(V)$.

In other words, over an algebraically closed field, the dimension of an affine algebraic set $V = \mathbf{V}(I)$ may be computed as follows:

- Compute a Groebner basis for I using a graded monomial order such as grlex or grevlex.
- Compute the maximal dimension d of a coordinate subspace contained in $\mathbf{V}(\langle \mathrm{LT}(I)\rangle) = \mathbf{V}(\langle \mathrm{LT}(G)\rangle)$. (Note that Proposition 1.4 of §1 gives an algorithm for doing this.) Then $\dim V = d$ follows from Proposition 3.9 and Theorem 3.6.

Exercises for §3

1. If an affine algebraic set V consists of a single point P, prove that $\dim V = 0$.
2. Let $I = \langle x^2 + y^2 \rangle \subset \mathbb{R}^2$.

   a. Show carefully that $\deg({}^aHP_I) = 1$.

   b. Show that $\dim \mathbf{V}(I) = 0$.
3. Compute the Hilbert polynomial for each of the following ideals.

   a. $I = \langle x^3y, xy^2 \rangle \subset \mathbb{R}[x, y]$.

   b. $I = \langle x^3yz^5, xy^3z^2 \rangle \subset \mathbb{R}[x, y, z]$.

   c. $I = \langle x^3 - yz^2, y^4 - x^2yz \rangle \subset \mathbb{R}[x, y, z]$.

§4. The Dimension of a Projective Algebraic Set

The discussion of the dimension of a projective algebraic set $V \subset \mathbf{P}^n = \mathbf{P}^n_k$ will parallel what we did in the affine case and, in fact, many of the arguments are the same. We start by studying the projective algebraic set associated with a monomial ideal. This makes sense because every monomial ideal is homogeneous (or graded) with respect to the natural gradation on $k[x_1, ..., x_n, x_{n+1}]$ given by the total degree of polynomials. Thus, a monomial ideal $I \subset k[x_1, ..., x_n, x_{n+1}]$ determines a projective algebraic set $\mathbf{V}_p(I) \subset \mathbf{P}^n$, where we use the subscript p to remind us that we are in projective space.

Let $I = \langle m_1, ..., m_s \rangle \subset k[x_1, ..., x_n]$be a monomial ideal generated by monomials $m_1, ..., m_s$. In $\mathbf{A}^n = \mathbf{A}^n_k$, let $\mathbf{V}(x_{i_1}, ..., x_{i_r})$ be a coordinate subspace of dimension $n - r$ contained in $\mathbf{V}(I) = \bigcap \mathbf{V}(m_i)$. Then:

- $\mathbf{V}_p(x_{i_1}, ..., x_{i_r}) \subset \bigcap_{i=1}^{s} \mathbf{V}_p(m_i) = \mathbf{V}_p(I) \subset \mathbf{P}^{n-1}_k$,
- $\mathbf{V}_p(x_{i_1}, ..., x_{i_r}) = \bigcap_{j=1}^{r} \mathbf{V}_p(x_{i_j})$ may be identified with a copy of $\mathbf{P}^{n-r-1}$ sitting inside $\mathbf{P}^{n-1}$.

If we call $\mathbf{V}_p(x_{i_1}, ..., x_{i_r})$ a *projective linear subspace* of dimension $n - r - 1$, then this leads to the following

4.1. Lemma Let $I \subset k[x_1, ..., x_n, x_{n+1}]$ be a monomial ideal and let $\mathbf{V}_p(I)$ be the projective algebraic set in $\mathbf{P}^n$ defined by I. Then $\mathbf{V}_p(I)$ is a finite union of projective linear subspaces which have dimension one less than the dimension of their affine counterparts.

□

Thus, as in the affine case, we define the *dimension* of a finite union of projective linear subspaces to be the maximum of the dimensions of the subspaces. Then the result of §3 shows that the dimension of the projective algebraic set $\mathbf{V}_p(I)$ of a monomial ideal I is one less than the degree of the polynomial in m counting the number of

monomials not in I of total degree $\leq m$.
Now let $V = \mathbf{V}_p(I) \subset \mathbf{P}^n$ be a projective algebraic set defined by a graded ideal $I \subset k[x_1, ..., x_{n+1}]$. As in the affine case, We want to define and compute $\dim V$ by passing to the monomial ideal $\langle \mathrm{LT}(I)\rangle$ and the coordinate ring $k^g[V]$ of V (see CH.III §3) in a purely algebraic way. We do this as follows.
Fix a monomial order on $k[x_1, ..., x_n, x_{n+1}]$ and consider the monomial ideal $\langle LT(I)\rangle$, then $\dim \mathbf{V}_p(\langle LT(I)\rangle) = d-1$ where $d = \deg\left({}^aHP_{\langle LT(I)\rangle}\right)$. Let us first see what algebraic structure is naturally connected to this number $d-1$.

4.2. Observation (i) The polynomial $h(x) = {}^aHP_{\langle LT(I)\rangle}(x) - {}^aHP_{\langle LT(I)\rangle}(x-1)$ has degree $d-1$ and has positive leading coefficient.
(ii) From §3 we know that for $m \gg 0$

$$\begin{aligned} {}^aHF_{\langle LT(I)\rangle}(m) &= \dim_k \frac{k[x_1, ..., x_n, x_{n+1}]_{\leq m} + I}{I} \\ &= {}^aHP_{\langle LT(I)\rangle}(m). \end{aligned}$$

(iii) ${}^aHF_{\langle LT(I)\rangle}(m) - {}^aHF_{\langle LT(I)\rangle}(m-1) =$

$$= \dim_k \frac{k[x_1, ..., x_n, x_{n+1}]_{\leq m} + I}{k[x_1, ..., x_n, x_{n+1}]_{\leq m-1} + I}.$$

Hence $d-1 = \dim \mathbf{V}_p(\langle LT(I)\rangle)$ is the degree of a polynomial in $\mathbb{Q}[x]$ which measures the "size" of the k-space

$$\frac{k[x_1, ..., x_n, x_{n+1}]_{\leq m} + I}{k[x_1, ..., x_n, x_{n+1}]_{\leq m-1} + I}, \quad \text{for } m \gg 0.$$

The next fact will clarify (iii). First recall that $k[x_1, ..., x_n, x_{n+1}]/I$ has a natural gradation induced by the gradation of $k[x_1, ..., x_n, x_{n+1}]$ which is given by the k-subspaces $k[x_1, ..., x_n, x_{n+1}]_m = \{$ homogeneous of degree $m\}$ (see CH.I. §1), namely,

$$k[x_1, ..., x_n, x_{n+1}] = \bigoplus_{m\geq 0} k[x_1, ..., x_n, x_{n+1}]_m,$$

$$\frac{k[x_1, ..., x_n, x_{n+1}]}{I} = \bigoplus_{m\geq 0} \frac{k[x_1, ..., x_n, x_{n+1}]_m + I}{I}$$

Putting $I_m = I \cap k[x_1, ..., x_n, x_{n+1}]_m$, then $(k[x_1, ..., x_n, x_{n+1}]_m + I)/I \cong k[x_1, ..., x_n, x_{n+1}]_m/I_m$, and we have the following easily verified fact.

4.3. Lemma There is a natural k-space isomorphism:

$$(\Delta) \qquad \frac{k[x_1, ..., x_n, x_{n+1}]_m}{I_m} \xrightarrow{\cong} \frac{k[x_1, ..., x_n, x_{n+1}]_{\leq m} + I}{k[x_1, ..., x_n, x_{n+1}]_{\leq m-1} + I}$$

□

Let $V = \mathbf{V}_p(I) \subset \mathbf{P}^n$ be a projective algebraic set defined by a graded ideal $I \subset k[x_1, ..., x_n, x_{n+1}]$. Using the notation $I_m = I \cap k[x_1, ..., x_n, x_{n+1}]_m$ as before, the *Hilbert function* of I is defined by

$$\begin{aligned} HF_I(m) &= \dim_k \frac{k[x_1, ..., x_n, x_{n+1}]_m}{I_m} \\ &= \dim_k \frac{k[x_1, ..., x_n, x_{n+1}]_m + I}{I} \end{aligned}$$

Strictly speaking, we should call this the projective Hilbert function, but the above terminology is customary in algebraic geometry.
If we combine Lemma 4.3 and Observation 4.2, we see that for any graded ideal $I \subset k[x_1, ..., x_n, x_{n+1}]$, the Hilbert function may be written as a polynomial for m sufficiently large, which is called the *Hilbert polynomial* of I and is denoted $HP_I(x)$ or simply HP_I. It follows that

$$(\nabla) \qquad HP_I(x) = {}^aHP_I(x) - {}^aHP_I(x-1)$$

We then define the dimension of a projective algebraic set in terms of the Hilbert polynomial as follows.

4.4. Definition The *dimension* of a projective algebraic set $V \subset \mathbf{P}^n$, denoted $\dim V$, is the degree of the Hilbert polynomial of the corresponding homogeneous ideal $I = \mathbf{I}_p(V) \subset k[x_1, ..., x_n, x_{n+1}]$, i.e., $\dim V = \deg(HP_{\mathbf{I}_p(V)})$.

Example (i) Since for each m $\dim_k k[x_1, ..., x_n, x_{n+1}]_m = \binom{n+m}{n}$,

we have $\dim \mathbf{P}^n = n$.

Over an *algebraically closed* field k, we can compute the dimension as follows.

4.5. Theorem (The Dimension Theorem) Let $V = \mathbf{V}(I) \subset \mathbf{P}^n$ be a projective algebraic set, where $I \subset k[x_1, ..., x_n, x_{n+1}]$ is a graded ideal. If V is nonempty and k is algebraically closed, then

$$\dim V = \deg(HP_I).$$

Furthermore, for any monomial order on $k[x_1, ..., x_n, x_{n+1}]$, we have

$$\begin{aligned} \dim V &= \deg\left(HP_{\langle LT(I)\rangle}\right) \\ &= \left\{\begin{array}{l} \text{maximum dimension} \\ \text{of a projective coordinate subspace in } \mathbf{V}(\langle LT(I)\rangle). \end{array}\right. \end{aligned}$$

Finally, the last two equalities hold over any field k when $I = \mathbf{I}(V)$.

Proof The first step is to show that I and $\sqrt{I}$ have Hilbert polynomials of the same degree. The proof is similar to the proof of Proposition 3.10 and is left as an exercise.
By the projective Nullstellensatz, we know that $\mathbf{I}(V) = \mathbf{I}(\mathbf{V}(I)) = \sqrt{I}$, and, from here, the proof is identical to the affine case (see Theorem 3.12). □

To compare the dimension of an affine algebraic set and that of its projective closure, we need more preparation.
First, a general version of CH.III. Theorem 6.10.

4.6. Proposition With notation as in CH.III §6, the following properties hold:
(i) For an ideal $I \subset k[x_1, ..., x_n]$, let I^* be the homogenization of I in $k[x_1, ..., x_n, x_{n+1}]$ with respect to x_{n+1}. Then we have the ring epimorphism

$$\begin{array}{rccc} \alpha: & k[x_1, ..., x_n, x_{n+1}]/I^* & \longrightarrow & k[x_1, ..., x_n]/I \\ & [F] & \mapsto & [F_*] \end{array}$$

and $\mathrm{Ker}\alpha = \langle 1 - [x_{n+1}]\rangle$ where $[x_{n+1}] = x_{n+1} + I^*$. Moreover, $[x_{n+1}]$ is a regular element in $k[x_1, ..., x_n, x_{n+1}]/I^*$, and hence $\langle 1 - [x_{n+1}]\rangle$ does not contain any nonzero homogeneous element.

(ii) For each $m \geq 0$, α induces a k-space isomorphism:

$$\frac{k[x_1,...,x_n,x_{n+1}]_m + I^*}{I^*} \xrightarrow{\cong} \frac{k[x_1,...,x_n]_{\leq m} + I}{I}$$

Proof (i) Using CH.III Proposition 5.5(i), the proof is similar to the one given in CH.III Theorem 6.10 combined with CH.III Exercise 8.
(ii) It is clear that α induces the desired k-space homomorphism. If $F \in k[x_1,...,x_n,x_{n+1}]_m$ with $F_* \in I$, then $F = x_{n+1}^r(F_*)^* \in I^*$ for some $r \geq 0$ by CH.III Proposition 5.2.(iii). This shows that the induced map is injective. (Indeed, this can also be obtained by using (i) above because $\langle 1 - [x_{n+1}]\rangle$ does not contain any nonzero homogeneous element.) The surjectivity of the induced map follows easily from the homogenization trick. □

4.7. Theorem (i) If $V \subset \mathbf{P}^n$ is a projective algebraic set and $\mathbf{C}(V) \subset \mathbf{A}_k^{n+1}$ is its affine cone (see CH.III, §2), then

$$\dim\mathbf{C}(V) = \dim V + 1.$$

(ii) Let $I \subset k[x_1,...,x_n]$ be an ideal and let $I^* \subset k[x_1,...,x_n,x_{n+1}]$ be its homogenization with respect to x_{n+1}. Then for $m \geq 0$, we have

$${}^aHF_I(m) = HF_{I^*}(m).$$

There is a similar relation between Hilbert polynomials. Consequently, if $V \subset \mathbf{A}^n$ is an affine algebraic set and $\overline{V} \subset \mathbf{P}^n$ is its projective closure (see CH.III §6), then

$$\dim V = \dim\overline{V}.$$

Proof We will use the subscripts a and p to indicate the affine and projective cases respectively.
(i) Note that the affine cone $\mathbf{C}(V)$ is simply the *affine* algebraic set in $\mathbf{A}^n$ defined by $\mathbf{I}_p(V)$. Further, by CH.III Lemma 2.4 we have $\mathbf{I}_a(\mathbf{C}(V)) = \mathbf{I}_p(V)$. Thus, the dimensions of V and $\mathbf{C}(V)$ are the degrees of $HP_{\mathbf{I}_p(V)}$ and ${}^aHP_{\mathbf{I}_p(V)}$, respectively. Then the equality $\dim\mathbf{C}(V) = \dim V + 1$ follows from the formula (∇) before Definition 4.4.
(ii)By Proposition 4.6(ii) it is clear that ${}^aHF_I(m) = HF_{I^*}(m)$. For the second part of (ii), suppose $V \subset \mathbf{A}^n$. Let $I = \mathbf{I}_a(V) \subset$

$k[x_1, ..., x_n]$ and let $I^* \subset k[x_1, ..., x_n, x_{n+1}]$ be the homogenization of I with respect to x_{n+1}. Then $\overline{V}$ is defined to be $\mathbf{V}_p(I^*) \subset \mathbf{P}^n$. Furthermore, we have $I^* = \mathbf{I}_p(\overline{V})$ by CH.III Proposition 6.7(iv). Then

$$\dim V = \deg({}^aHP_I) = \deg(HP_{I^*}) = \dim \overline{V}$$

follows immediately from the first part of (ii), and the theorem is proved. □

Let $V \subset \mathbf{P}^n$ be a projective algebraic set, and let I be *any* graded ideal of $k[x_1, ..., x_n, x_{n+1}]$ such that $V = \mathbf{V}(I)$. Then $V = \cup_{i=1}^{n+1}(V \cap U_i)$, i.e., V is covered by the open sets $V \cap U_i$, $i = 1, ..., n+1$, which are homeomorphic to affine algebraic sets $\phi_i(\mathbf{V}(I_*))$, where I_* is the dehomogenization of I with respect to x_i (CH.III Theorem 6.12).

4.8. Theorem With notation as above, and assuming that k is algebraically closed, we have:
(i) $\dim \mathbf{V}(I_*) = \dim \mathbf{V}((I_*)^*)$.
(ii) If I is $(\phi_*)^*$-closed, then $\dim \mathbf{V}(I_*) = \dim V$.
(iii) If V is a nonempty projective variety such that $V \cap U_i \neq \emptyset$, then $\dim V = \dim \mathbf{V}(I_*)$.

Proof This follows from Theorem 4.7, CH.III Lemma 6.1(ii), Proposition 6.5 and Proposition 6.9(iii). □

Finally, we point out that

- Some computer algebra systems can compute Hilbert polynomials. REDUCE has a command to find the affine Hilbert polynomial of an ideal, whereas *Macaulay* and *CoCoA* will compute the projective Hilbert polynomial of a homogeneous ideal.

Exercises for §4

1. Give a detailed proof of Theorem 4.5.
2. Let $I \subset k[x_1, ..., x_{n+1}]$ be a graded ideal. Show that $\langle x_1, ..., x_{n+1}\rangle^r \subset I$ for some $r > 0$ if and only if $HP_I = 0$.
3. From above 2. conclude that if $V \subset \mathbf{P}^n$ is an algebraic set and $I = \mathbf{I}(V)$, then $V = \emptyset$ if and only if $HP_I = 0$. Thus, the empty algebraic set in $\mathbf{P}^n$ does not have dimension.

§5. Elementary Properties of Dimension

Let $\mathbf{A}^n = \mathbf{A}_k^n$ and $\mathbf{P}^n = \mathbf{P}_k^n$. In this section we derive several basic properties of the dimension of an algebraic set. We first observe the following.

5.1. Proposition Let V_1 and V_2 be projective or affine algebraic sets over an arbitrary field k. If $V_1 \subset V_2$, then $\dim V_1 \leq \dim V_2$.

Proof Let V_1 and V_2 be affine algebraic sets in $\mathbf{A}^n$. Note that if $V_1 \subset V_2$, then $\mathbf{I}(V_1) \subset \mathbf{I}(V_1)$ and we have the natural k-space homomorphism:

$$\frac{k[x_1, ..., x_n]_{\leq m} + \mathbf{I}(V_2)}{\mathbf{I}(V_2)} \longrightarrow \frac{k[x_1, ..., x_n]_{\leq m} + \mathbf{I}(V_1)}{\mathbf{I}(V_1)}$$

for every $m \geq 0$. This immediately yields the inequality $\dim V_1 \leq \dim V_2$. There is a similar argument for the projective case. □

5.2. Proposition Let k be an algebraically closed field and let $F \in k[x_1, ..., x_n, x_{n+1}]$ be a nonconstant homogeneous polynomial. Then the dimension of the projective hypersurface $\mathbf{V}(F) \in \mathbf{P}^n$ defined by F is

$$\dim \mathbf{V}(F) = n - 1.$$

Proof Fix a monomial order $>$ on $k[x_1, ..., x_n, x_{n+1}]$. Since k is algebraically closed, Theorem 4.5 says the dimension of $\mathbf{V}(F)$ is the maximum dimension of a projective coordinate subspace contained in $\mathbf{V}(\langle \mathrm{LT}(I)\rangle)$, where $I = \langle F\rangle$. One can check that $\langle \mathrm{LT}(I)\rangle = \langle \mathrm{LT}(F)\rangle$, and since $LT(F)$ is a nonconstant monomial, the projective algebraic set $\mathbf{V}(\mathrm{LT}(F))$ is a union of subspaces of $\mathbf{P}^n$ of dimension $n-1$. It follows that $\dim \mathbf{V}(I) = n - 1$. □

Thus, when k is algebraically closed, a hypersurface $\mathbf{V}(F)$ in $\mathbf{P}^n$ always has dimension $n-1$. We leave it as an exercise for the reader to prove the analogous statement for affine hypersurfaces.

It is important to note again that these results are *not valid* if k is not algebraically closed. For $I = \langle F = x^2 + y^2\rangle$ in $\mathbb{R}[x, y]$, we have seen before that $\mathbf{V}(F) = \{(0,0)\} \subset \mathbb{R}^2$ has dimension 0, yet Proposition 5.2 would predict it should be 1.

5.3. Proposition Let V be a nonempty affine or projective algebraic set. Then V consists of finitely many points if and only if $\dim V = 0$.

Proof We will give the proof only in the affine case. Let $>$ be a graded order on $k[x_1, ..., x_n]$. If V is finite, then let a_j, for $j = 1, ..., m_i$, be the distinct elements of k appearing as i-th coordinates of points of V. Then

$$f = \prod_{j=1}^{m_i} (x_i - a_j) \in \mathbf{I}(V)$$

and we conclude that $LT(f) = x_i^{m_i} \in \langle LT(\mathbf{I}(V))\rangle$ where $i = 1, ..., n$. This implies that $\mathbf{V}(\langle LT(\mathbf{I}(V))\rangle) = \{0\}$ and then $\dim V = 0$ by Exercise 1 of §3.
Now, suppose that $\dim V = 0$. Then the affine Hilbert polynomial of $\mathbf{I}(V)$ is a constant c, so that

$$\dim_k (k[x_1, ..., x_n]_{\leq m}/\mathbf{I}(V)_{\leq m}) = c$$

for m sufficiently large (hence $\dim_k(k[x_1, ..., x_n]/\mathbf{I}(V)) = c$). If we also have $m \geq c$, then the classes

$$[1], [x_i], [x_i^2], ..., [x_i^m] \in k[x_1, ..., x_n]_{\leq m}/\mathbf{I}(V)_{\leq m}$$

must be linearly dependent. But a nontrivial linear relation

$$[0] = \sum_{j=0}^{m} a_j \left[x_i^j\right] = \left[\sum_{j=0}^{m} a_j x_i^j\right]$$

yields that $\sum_{j=0}^{m} a_j x_i^j$ is a nonzero polynomial in $\mathbf{I}(V)_{\leq m}$. This polynomial vanishes on V, which implies that there are only finitely many distinct i-th coordinates among the points of V. Since this is true for all $1 \leq i \leq n$, it follows that V must be finite. □

Remark If, in addition, k is algebraically closed, then we see that the five conditions of Theorem 2.4 in §2 are equivalent to $\dim V = 0$. In particular, given any defining ideal I of V, we get a simple criterion for detecting when an algebraic set has dimension 0.

Now that we understand algebraic sets of dimension 0, it is the time to record some interesting properties of positive dimensional algebraic sets, in particular, we will show that a projective algebraic set of dimension > 0 meets every hypersurface (as we have promised in CH.III §1).

5.4. Theorem Let k be an algebraically closed field and let I be a graded ideal of $k[x_1, ..., x_n, x_{n+1}]$. If F is any nonconstant homogeneous polynomial in $k[x_1, ..., x_n, x_{n+1}]$, then

$$\dim\mathbf{V}(I) \geq \dim\mathbf{V}(I + \langle F\rangle) \geq \dim\mathbf{V}(I) - 1.$$

Proof Since $I \subset I+\langle F\rangle$, by Proposition 5.1 (or its proof) we see that $\deg(HP_I) \geq \deg\left(HP_{I+\langle F\rangle}\right)$, and we conclude $\dim\mathbf{V}(I) \geq \dim\mathbf{V}(I+\langle F\rangle)$.
To obtain the other inequality, suppose that F has total degree $r > 0$. Fix a total degree $s \geq r$ and consider the natural linear map

$$\pi : \quad \frac{k[x_1, ..., x_n, x_{n+1}]_s}{I_s} \longrightarrow \frac{k[x_1, ..., x_n, x_{n+1}]_s}{(I + \langle F\rangle)_s}$$

which is obviously surjective. Then $\mathrm{Ker}\pi = (I + \langle F\rangle)_s/I_s = (I_s + \langle F\rangle_s)/I_s$. Since F is a homogeneous polynomial of total degree r, the linear map

$$\begin{array}{cccc} \alpha_F : & \dfrac{k[x_1, ..., x_n, x_{n+1}]_{s-r}}{I_{s-r}} & \longrightarrow & \dfrac{k[x_1, ..., x_n, x_{n+1}]_s}{I_s} \\ & [H] & \mapsto & [FH] \end{array}$$

satisfies $\mathrm{Ker}\pi = \mathrm{Im}\alpha_F$. Hence

$$\begin{aligned} \dim_k \frac{k[x_1, ..., x_n, x_{n+1}]_s}{I_s} &= \dim_k \frac{k[x_1, ..., x_n, x_{n+1}]_s}{(I + \langle F\rangle)_s} + \dim_k(\mathrm{Ker}\pi) \\ &\leq \dim_k \frac{k[x_1, ..., x_n, x_{n+1}]_s}{(I + \langle F\rangle)_s} + \\ &\quad +\dim_k \frac{k[x_1, ..., x_n, x_{n+1}]_{s-r}}{I_{s-r}}. \end{aligned}$$

In terms of Hilbert functions, this becomes:

$$HF_{I+\langle F\rangle}(s) \geq HF_I(s) - HF_I(s - r),$$

whenever $s \geq r$. Thus, if s is sufficiently large we obtain the inequality

$$(*) \qquad HP_{I+\langle F\rangle}(s) \geq HP_I(s) - HP_I(s-r)$$

for Hilbert polynomials.
Suppose that $\deg(HP_I) = d$. It is easy to see that the polynomial on the right hand side in $(*)$ has degree $d-1$. Thus $(*)$ shows that $HF_{I+\langle F\rangle}(s)$ is $\geq$ a polynomial of degree $d-1$ for s sufficiently large, which implies $\deg\left(HP_{I+\langle F\rangle}\right) \geq d-1$ (prove this!). Since k is algebraically closed, we may conclude that $\dim \mathbf{V}(I + \langle F\rangle) \geq \dim \mathbf{V}(I) - 1$, as desired. □

5.5. Corollary Let k be an algebraically closed field and $I \subset k[x_1, ..., x_n, x_{n+1}]$ a graded ideal. Let F be a nonconstant homogeneous polynomial in $k[x_1, ..., x_n, x_{n+1}]$ such that its class in the quotient ring $k[x_1, ..., x_n, x_{n+1}]/I$ is not a zero divisor. Then

$$\dim \mathbf{V}\left(I + \langle F\rangle\right) = \dim \mathbf{V}(I) - 1.$$

□

By induction, Theorem 5.4 may be extended to the case of several polynomials.

5.6. Proposition Let k be an algebraically closed field and let I be a graded ideal in $k[x_1, ..., x_n, x_{n+1}]$. Let $F_1, ..., F_r$ be nonconstant homogeneous polynomials in $k[x_1, ..., x_n, x_{n+1}]$. then

$$\dim \mathbf{V}\left(I + \langle F_1, ..., f_r\rangle\right) \geq \dim \mathbf{V}(I) - r.$$

Remark Here we point out that Theorem 5.4 can fail for affine algebraic sets, even when k is algebraically closed. For example, consider the ideal $I = \langle xz, yz\rangle \subset \mathbb{C}[x, y, z]$. One easily sees that in $\mathbb{C}^3$, we have $\mathbf{V}(I) = \mathbf{V}(z) \cup \mathbf{V}(x, y)$, so that $\mathbf{V}(I)$ is the union of the (x, y)-plane and the z-axis. In paricular, $\mathbf{V}(I)$ has dimension 2. Now, let $f = z - 1 \in \mathbb{C}[x, y, z]$. Then $\mathbf{V}(f)$ is the plane $z = 1$ and it follows that $\mathbf{V}(I + \langle f\rangle) = \mathbf{V}(I) \cap \mathbf{V}(f)$ consists of the single point $(0, 0, 1)$ (check this carefully!), i.e., this intersection has dimension

0. Yet Theorem 5.4 would predict that $\mathbf{V}(I + \langle f \rangle)$ has dimension at least 1.
What goes "wrong" here is that the planes $z = 0$ and $z = 1$ are parallel and, hence, do not meet in affine space. We are missing a component of dimension 1 at infinity. This indicates why dimension theory works better for homogeneous ideals and projective algebraic sets. It is possible to have a version of Theorem 5.4 that is valid for affine algebraic sets, but we will not pursue that here.

5.7. Proposition Let k be algebraically closed.
(i) Let $V \subset \mathbf{P}^n$ be a projective algebraic set of dimension > 0. Then $V \cap \mathbf{V}(F) \neq \emptyset$ for every nonconstant homogeneous polynomial $F \in k[x_1, ..., x_n, x_{n+1}]$. Thus, a positive dimensional projective algebraic set meets every hypersurface in $\mathbf{P}^n$.
(ii) Let $W \subset \mathbf{A}^n$ be an affine algebraic set of dimension > 0. If $\overline{W}$ is the projective closure of W in $\mathbf{P}^n$ with respect to the imbedding ϕ: $\mathbf{A}^n \to U_{n+1} \subset \mathbf{P}^n$, then $W \neq \overline{W}$. Thus, a positive dimensional affine algebraic set always has points at infinity.

Proof (i) Let $V = \mathbf{V}(I)$. Since $\dim V > 0$, Theorem 5.4 shows that $\dim V \cap \mathbf{V}(f) \geq \dim V - 1 \geq 0$. Let us check *carefully* that this guarantees $V \cap \mathbf{V}(f) \neq \emptyset$. If $V \cap \mathbf{V}(F) = \emptyset$, then the projective Nullstellensatz implies that $\langle x_1, ..., x_n, x_{n+1}\rangle^r \subset I + \langle f \rangle$ for some $r \geq 0$. By exercise, it follows that $HP_{I+\langle F \rangle}$ is the zero polynomial. But by the foregoing results, this polynomial cannot be zero when $\dim V > 0$ (we leave the details as an exercise).
(ii) The points at infinity of W are $\overline{W} \cap H_{n+1}^{\infty}$, where $H_{n+1}^{\infty} = \mathbf{V}(x_{n+1})$ is the hyperplane at infinity. By Theorem 4.7, we have $\dim \overline{W} = \dim W > 0$, and then (i) implies that $\overline{W} \cap \mathbf{V}(x_{n+1}) \neq \emptyset$. □

The dimension of the union of two algebraic sets is well-behaved, as stated in the following proposition.

5.8. Proposition If V and W are algebraic sets either both in $\mathbf{A}^n$ or both in $\mathbf{P}^n$, then

$$\dim(V \cup W) = \max(\dim V, \dim W).$$

Proof The proofs for the affine and projective cases are almost identical, so we will give only the affine proof. If k is a finite field, V, W,

and $V \cup W$ are finite and, hence, have dimension 0 by Proposition 5.3. So we may assume that k is infinite.
Let $I = \mathbf{I}(V)$ and $J = \mathbf{I}(W)$, so that $\dim V = \deg({}^{a}HP_I)$ and $\dim W = \deg({}^{a}HP_J)$. By CH.I §4 Proposition 4.3, $\mathbf{I}(V \cup W) = \mathbf{I}(V) \cap \mathbf{I}(W) = I \cap J$. It is more convenient to deal with the product ideal IJ and we note that

$$IJ \subset I \cap J \subset \sqrt{IJ}.$$

Consequently we conclude that

$$\deg\left({}^{a}HP_{\sqrt{IJ}}\right) \leq \deg\left({}^{a}HP_{I\cap J}\right) \leq \deg\left({}^{a}HP_{IJ}\right).$$

But the extreme terms are equal (Proposition 3.11) hence $\dim(V \cup W) = \deg({}^{a}HP_{IJ})$.
Now fix a graded order $>$ on $k[x_1, ..., x_n]$. The results of §3 imply that $\dim V$, $\dim W$, and $\dim(V \cup W)$ are given by the maximal dimension of a coordinate subspace contained in $\mathbf{V}(\langle LT(I)\rangle)$, $\mathbf{V}(\langle LT(J)\rangle)$ and $\mathbf{V}(\langle LT(IJ)\rangle)$ respectively. It is an easy exercise to check that

$$\langle LT(IJ)\rangle = \langle LT(I)\rangle \cdot \langle LT(J)\rangle.$$

This implies

$$\mathbf{V}(\langle LT(IJ)\rangle) = \mathbf{V}(\langle LT(I)\rangle) \cup \mathbf{V}(\langle LT(J)\rangle).$$

Since k is infinite, every coordinate subspace is irreducible. Hence, a coordinate subspace is contained in $\mathbf{V}(\langle LT(IJ)\rangle)$ if and only if it lies in either $\mathbf{V}(\langle LT(I)\rangle)$ or $\mathbf{V}(\langle LT(J)\rangle)$. From this, it follows immediately that $\dim(V \cup W)$ is the maximum of $\dim V$ and $\dim W$.
□

This proposition has the following useful corollary.

5.9. Corollary The dimension of an algebraic set is the largest of the dimensions of its irreducible components.

□

This corollary allows us to reduce computation of dimensions to the case of an irreducible algebraic set.

5.10. Proposition Let k be an algebraically closed field and let $V \subset \mathbf{P}^n$ be a variety.
(i) If $F \in k[x_1, ..., x_n, x_{n+1}]$ is a homogeneous polynomial which does not vanish on V, then $\dim(V \cap \mathbf{V}(F)) = \dim V - 1$.
(ii) If $W \subset V$ is an algebraic set such that $W \neq V$, then $\dim W < \dim V$.

Proof (i) From Theorem 2.2 of CH.III §2, we retain that $\mathbf{I}(V)$ is a prime ideal and $k[V] \cong k[x_1, ..., x_n, x_{n+1}]/\mathbf{I}(V)$ is an integral domain. Since $F \notin \mathbf{I}(V)$, the class of F is nonzero in $k[x_1, ..., x_n, x_{n+1}]/\mathbf{I}(V)$ and, hence, is not a zero divisor. The desired conclusion then follows from Corollary 5.5.
(ii) If W is a proper algebraic subset of V, then we may select $F \in \mathbf{I}(W) - \mathbf{I}(V)$. Thus, $W \subset V \cap \mathbf{V}(F)$, and it follows from (i) and Proposition 5.1 that

$$\dim W \leq \dim(V \cap \mathbf{V}(F)) = \dim V - 1 < \dim V.$$

This completes the proof of the proposition. □

Part (i) of Proposition 5.10 asserts that when V is irreducible and F does not vanish on V, then some component of $V \cap \mathbf{V}(F)$ has dimension $\dim V - 1$. With some more work, it can be shown that *every* component of $V \cap \mathbf{V}(F)$ has dimension $\dim V - 1$. See, for example, Theorem 3.8 in Chapter IV of [Ken] or Theorem 5 of Chapter I, §6 of [Sha].

Exercises for §5

1. Prove the analogous statement of Proposition 5.2 for affine hypersurfaces.
2. Prove the analogous statement of Proposition 5.3 for projective algebraic set.
3. Give a detailed proof of Proposition 5.7(i).
4. Prove that for two polynomial ideals I, J, $\langle \mathrm{LT}(IJ) \rangle = \langle \mathrm{LT}(I) \rangle \langle \mathrm{LT}(J) \rangle$.

§6. Dimension and Algebraic Independence

In §4, we defined the dimension of an affine algebraic set as the degree of the affine Hilbert polynomial. But Hilbert polynomials do not tell the full story. In this section we start a pure algebraic approache to dimension theory.

Let $k[\mathbf{x}] = k[x_1, ..., x_n]$. If $V \subset \mathbf{A}^n = \mathbf{A}_k^n$ is an affine algebraic set, recall that the *coordinate ring* $k[V]$ consists of all polynomial functions on V. This is related to the ideal $\mathbf{I}(V)$ by the natural ring isomorphism $k[V] \cong k[\mathbf{x}]/\mathbf{I}(V)$ discussed in CH.II §4. We can resulte the notion of dimension to $k[V]$. Using the notation as before we first note that for any $m \geq 0$, there is a well-defined linear map

$$\frac{k[\mathbf{x}]_{\leq m}}{\mathbf{I}(V)_{\leq m}} \cong \frac{k[\mathbf{x}]_{\leq m} + \mathbf{I}(V)}{\mathbf{I}(V)} \longrightarrow \frac{k[\mathbf{x}]}{\mathbf{I}(V)} \cong k[V]$$

which is injective. Thus, we may view $k[\mathbf{x}]_{\leq m}/\mathbf{I}(V)_{\leq m}$ as a finite dimensional "piece" of $k[V]$ that approximates $k[V]$ more and more closely as m gets larger. Since the degree of ${}^aHP_{\mathbf{I}(V)}$ measures how fast these finite dimensional approximations are growing, we see that $\dim V$ tells us something about the "size" of $k[V]$ as a k-vector space. On the other hand, we know that $k[V]$ is, as a ring, finitely generated over k, i.e., it is ring-finite over k in the sense of Appendix I.
This suggests that we should be able to reformulate the definition of the dimension of V directly in terms of the ring structure of $k[V]$ over k. To do this, we will use the notion of algebraically independent elements.

6.1. Definition we say that elements $\phi_1, ..., \phi_r \in k[V]$ are ***algebraically independent over*** k if there is no nonzero polynomial p of r variables with coefficients in k such that $p(\phi_1, ..., \phi_r) = 0$ in $k[V]$.

Note that if $\phi_1, ..., \phi_r \in k[V]$ are algebraically independent over k, then the ϕ_i's are distinct and nonzero. Obviously any subset of $\{\phi_1, ..., \phi_r\}$ is also algebraically independent over k.

Example (i) The simplest example of algebraically independent elements occurs when $V = \mathbf{A}^n$. If k is an infinite field, we have $\mathbf{I}(V) = \{0\}$ and, hence, $k[V] = k[\mathbf{x}]$. Here, the elements $x_1, ..., x_n$

are algebraically independent over k since $p(x_1, ..., x_n) = 0$ means that p is the zero polynomial.

(ii) For another example, let V be the twisted cubic in $\mathbb{R}^3$, so that $\mathbf{I}(V) = \langle y - x^2, z - x^3 \rangle$. Let us show that the class of x in $\mathbb{R}[V]$, denoted $[x]$, is algebraically independent over $\mathbb{R}$. Suppose p is a polynomial with coefficients in $\mathbb{R}$ such that $p([x]) = [0]$ in $\mathbb{R}[V]$. This means $[p(x)] = [0]$, so that $p(x) \in \mathbf{I}(V)$. But it is easy to show that $\mathbb{R}[x] \cap \langle y - x^2, z - x^3 \rangle = \{0\}$, which proves that p is the zero polynomial. On the other hand, we leave it to the reader to verify that $[x], [y] \in \mathbb{R}[V]$ are not algebraically independent over $\mathbb{R}$ since $[y] - [x]^2 = [0]$ in $\mathbb{R}[V]$.

From §3 it is clear that the dimension $d = \dim V$ of an affine algebraic set V is related to the *independence* of subsets in $\{x_1, ..., x_n\}$ (modulo $\mathbf{I}(V)$) in the sense that

$$\mathrm{M}(U) \cap \langle \mathrm{LT}(\mathbf{I}(V)) \rangle = \{0\},$$

where for each $U = \{x_{i_1}, ..., x_{i_r}\} \subset \{x_1, ..., x_n\}$, $\mathrm{M}(U)$ denotes the set of the monomials in the polynomial subalgebra $k[U] = k[x_{i_1}, ..., x_{i_r}] \subset k[\mathbf{x}]$, and we have

$$d = \max\left\{ |U| \;\middle|\; U \subset \{x_1, ..., x_n\} \text{ is independent (modulo } \mathbf{I}(V))\right\}.$$

Bearing this in mind, we may relate the dimension of V to the number of algebraically independent elements in the coordinate ring $k[V]$. For $f \in k[\mathbf{x}]$ we write $[f]$ for the class of f in $k[V]$.

6.2. Theorem Let $V \subset \mathbf{A}^n$ be an affine algebraic set. Then the dimension of V equals the maximal number of elements of $k[V]$ which are algebraically independent over k.

Proof We will first show that if $d = \dim V$, then we can find d elements of $k[V]$ which are algebraically independent over k. To do this, let $I = \mathbf{I}(V)$ and consider the ideal of leading terms $\langle LT(I) \rangle$ for some graded order on $k[\mathbf{x}]$. If $U = \{x_{i_1}, ..., x_{i_d}\}$ is independent modulo I, we can now prove that $[x_{i_1}], ..., [x_{i_d}] \in k[V]$ are algebraically independent over k. Let p be a polynomial with coefficients in k such that $p([x_{i_1}], ..., [x_{i_d}]) = [0]$. Then $[p(x_{i_1}, ..., x_{i_d})] = [0]$ in $k[V]$, which shows that $p(x_{i_1}, ..., x_{i_d}) \in I \cap k[U] = \{0\}$ since U is independent modulo I, it follows that $p(x_{i_1}, ..., x_{i_d}) = 0$, and since $x_{i_1}, ..., x_{i_d}$ are variables, we see that p is the zero polynomial. Since $d = \dim V$, we have found the desired number of algebraically independent elements.

Next, we show that if $[f_1], ..., [f_r] \in k[V]$ are algebraically independent, then $r \leq \dim V$. Let N be the largest of the total degrees of $f_1, ..., f_r$ and let $y_1, ..., y_r$ be new variables. If $p \in k[y_1, ..., y_r]$ is a polynomial of total degree $\leq m$, then the polynomial $p(f_1, ..., f_r) \in k[\mathbf{x}]$ has total degree $\leq Nm$. Consider the map

$$(*) \qquad \begin{array}{rccc} \alpha : & k[y_1, ..., y_r]_{\leq m} & \longrightarrow & \dfrac{k[\mathbf{x}]_{\leq Nm}}{I_{\leq Nm}} \\ & p(y_1, ..., y_r) & \mapsto & [p(f_1, ..., f_r)] \end{array}$$

It is clear that α is a well-defined linear map. We claim that α is injective. Indeed, suppose that $p \in k[y_1, .., y_r]_{\leq Nm}$ and $[p(f_1, ..., f_r)] = [0]$ in $k[\mathbf{x}]_{\leq Nm}/I_{\leq Nm}$. Using the map mentioned in the begining of this section, it follows that

$$[p(f_1, ..., f_r)] = p([f_1], ..., [f_r]) = [0] \quad \text{in } k[\mathbf{x}]/I \cong k[V].$$

Since $[f_1], ..., [f_r]$ are algebraically independent and p has coefficients in k, it follows that p must be the zero polynomial. Hence, α is injective.

Comparing dimensions in $(*)$, we see that

$$(**) \qquad {}^aHF_I(Nm) = \dim_k \frac{k[\mathbf{x}]_{\leq Nm}}{I_{\leq Nm}} \geq \dim_k k[y_1, ..., y_r]_{\leq m}.$$

Since $y_1, ..., y_r$ are variables, we know that $\dim_k k[y_1, ..., y_r]_{\leq m} = \binom{r+m}{m}$, which is a polynomial of degree r in m. In terms of the affine Hilbert polynomial, this implies that ${}^aHP_I(Nm)$ and, hence, ${}^aHP_I(m)$ must have degree at least r. Thus, $r \leq \dim V$, which completes the proof of the theorem. □

Note that in the proof of Theorem 6.2, the $d = \dim V$ algebraically independent elements we found in $k[V]$ came from the coordinates.

Now, we prove that isomorphic algebraic sets have the same dimension, as we proposed in the begining of this chapter.

6.3. Corollary Let V and V' be affine algebraic sets which are isomorphic. Then $\dim V = \dim V'$.

Proof From Theorem 4.7 of CH.II it follows that there is a ring isomorphism $\alpha; k[V] \to k[V']$. Then elements $\phi_1, ..., \phi_r \in k[V]$ are algebraically independent over k if and only if $\alpha(\phi_1), ..., \alpha(\phi_r) \in k[V']$ are. We leave the easy proof of this assertion as an exercise. Now, the corollary follows immediately from Theorem 6.2. □

For the final part of the section, we will assume that V is a *variety*, i.e., V is an irreducible algebraic set. Then $\mathbf{I}(V)$ is a prime ideal and hence $k[V]$ is an integral domain. As before we write $k(V)$ for the field of fractions of $k[V]$, which is the *field of rational functions* on V. For elements of $k(V)$, the definition of algebraic independence over k is the same as that given for elements of $k[V]$ in Definition 6.1. We can relate the dimension of V to $k(V)$ as follows.

6.4. Theorem Let $V \subset \mathbf{A}^n$ be an affine variety. Then the dimension of V equals the maximal number of elements of $k(V)$ which are algebraically independent over k.

Proof Let $d = \dim V$. Since $k[V] \subset k(V)$, any d elements of $k[V]$ which are algebraically independent over k will have the same property when regarded as elements of $k(V)$. So it remains to show that if $\phi_1, .., \phi_r \in k(V)$ are algebraically independent, then $r \leq \dim V$. Each ϕ_i is a quotient of elements in $k[V]$, and if we pick a common denominator f, then we can write $\phi_i = \dfrac{[f_i]}{[f]}$ for $1 \leq i \leq r$. Note also that $[f] \neq [0]$ in $k[V]$. We modify the proof of Theorem 6.2 by taking the denominator f into account.

Let N be the largest of the total degree of $f, f_1, ..., f_r$, and let $y_1, ..., y_r$ be new variables. If $p \in k[y_1, ..., y_r]$ is a polynomial of total degree $\leq m$, then

$$f^m \cdot p\left(\frac{f_1}{f}, ..., \frac{f_r}{f}\right)$$

is a polynomial in $k[x_1, ..., x_n]$ of total degree $\leq Nm$. Consider the map

$$\begin{array}{rccc} \beta: & k[y_1,...,y_r]_{\leq m} & \longrightarrow & \dfrac{k[\mathbf{x}]_{\leq Nm}}{I_{\leq Nm}} \\ & p(y_1,...,y_r) & \mapsto & \left[f^m \cdot p\left(\dfrac{f_1}{f},...,\dfrac{f_r}{f}\right)\right] \end{array}$$

Clearly, β is a well-defined linear map. We aim to show that β is injective. Suppose $p \in k[y_1,...,y_r]_{\leq m}$ and $\left[f^m \cdot p\left(\frac{f_1}{f},...,\frac{f_r}{f}\right)\right] = [0]$ in $k[\mathbf{x}]_{\leq Nm}/I_{\leq Nm}$. Using the map mentioned in the begining of the section, it follows that

$$\left[f^m \cdot p\left(\frac{f_1}{f},...,\frac{f_r}{f}\right)\right] = [0] \text{ in } k[\mathbf{x}]/I \cong k[V].$$

However, in $k(V)$, we may write this as

$$[f]^m \cdot p\left(\frac{[f_1]}{[f]},...,\frac{[f_r]}{[f]}\right) = [f]^m \cdot p(\phi_1,...,\phi_r) = [0] \text{ in } k(V).$$

Since $k(V)$ is a field and $[f] \neq [0]$, it follows that $p(\phi_1,...,\phi_r) = [0]$. Then p must be the zero polynomial since $\phi_1,...,\phi_r$ are algebraically independent and p has coefficients in k. Thus, β is injective.
Once we know that β is injective, we get the inequality (**) as in the proof of Theorem 6.2, and from hereon, the proof of Theorem 6.2 shows that dim$V \geq r$. This proves the theorem. □

Let us mention a more general version of Theorem 6.2 resp. Theorem 6.4:

- Let I be an ideal of $k[\mathbf{x}]$. Then deg(${}^aHP_I$) equals the maximal number of elements of $k[\mathbf{x}]/I$ which are algebraically independent over k.
- Let p be a prime ideal of $k[\mathbf{x}]$, and Q the quotient field of the domain $k[\mathbf{x}]/p$. Then deg(${}^aHP_p$) equals the maximal number of elements of Q which are algebraically independent over k.

Recall from field theory that there is a concept of *transcendence degree* which is closely related to what we have been studying. In general, when we have a field K containing k, we have the following definition.

6.5. Definition Let L be a field containing k. Then we say that L has *transcendence degree* d over k and write $d = \text{tr.deg}_k L$, provided that d is the largest number of elements of L which are algebraically independent over k.

If we combine this definition with Theorem 6.4, then for any affine variety V, we have

$$\dim V = \text{tr.deg}_k k(V).$$

Remark (i) Many books on algebraic geometry use this as the definition of the dimension of an affine variety. Of course such a definition is based on the the famous Noether's Normalization Theorem (e.g. see [Mats]). Now it is clear that if we combine Appendix I §3 Theorem 3.1 with the foregoing results, then we can get Noether's Normalization Theorem immediately.

(ii) For an example of transcendence degree, suppose that k is infinite, so that $k(V) = k(x_1, ..., x_n)$ when $V = \mathbf{A}^n$. Since $\mathbf{A}^n$ has dimension n, we conclude that the field $k(x_1, ..., x_n)$ has transcendence dgree n over k, namely, no $n+1$ elements of $k(x_1, ..., x_n)$ can be algebraically independent over k (this is not a trivial result in field theory). For more detail on transcendence degree we refer to Chapters VII and X of [Lan].

(iii) Comparing to Corollary 6.3, we will see in CH.VI §3 that the big advantage of using the above definition is that the dimension of an affine (projective) variety is an invariant for birational equivalence.

Exercises for §6

1. Let $V = \mathbf{V}(I)$ and $W = \mathbf{V}(J)$ be isomorphic affine algebraic sets. Prove ${}^aHP_I = {}^aHP_J$.
2. Give a detailed proof of Corollary 6.3.
3. The notion of transcendence degree is analogous to the idea of the dimension of a vector space. By methods entierly analogous to those for bases of vector spaces, one can prove:
   a. Let $a_1, ..., a_n \in L$, L a finitely generated field extension of k. Then $a_1, ..., a_n$ is a minimal set such that L is algebraic over $k(a_1, ..., a_n)$ if and only if $a_1, ..., a_n$ is a maximal set of algebraically independent elements of L. $\{a_1, ..., a_n\}$ is called a *transcendence basis* of L over k.
   b. Any algebraically independent set may be completed to a

transcendence basis. Any set $\{a_1, ..., a_n\}$ such that L is algebraic over $k(a_1, ..., a_n)$ contains a transcendence basis.

c. $\text{tr.deg}_k L$ = number of elements in any transcendence basis of L over k.

§7. Dimension and Elimination of Variables

If we pay a little more attention to the first part of the proof of Theorem 6.2, it is clear that the only condition needed there is $\mathbf{I}(V) \cap k[U] = \{0\}$. Indeed, this has motivated another description of $\dim\mathbf{V}(I)$ by using a weaker independence condition which, at the same time, yields a quite natural connection between $\dim\mathbf{V}(I)$ and the elimination of variables in I. The exploration of this fact forms the topic of the present section.

let $k[\mathbf{x}] = k[x_1, ..., x_n]$ be the polynomial ring over an *arbitrary* field k of characteristic 0, and let I be an *arbitrary* ideal of $k[\mathbf{x}]$. Instead of playing with $\dim\mathbf{V}(I)$, in this section we deal with the Hilbert polynomial ${}^aHP_I$ which, of course, turns to $\dim\mathbf{V}(I)$ naturally whenever k is algebraically closed.

With notation as in §6, let $U = \{x_{i_1}, ..., x_{i_r}\} \subset \{x_1, ..., x_n\}$ and $k[U] = k[x_{i_1}, ..., x_{i_r}] \subset k[x_1, ..., x_n]$. We say that U is *weakly independent* (modulo I) if $k[U] \cap I = \{0\}$.

From this definition we immediately have

7.1. Lemma If U is independent (modulo I), i.e., $\mathrm{M}(U) \cap \langle \mathrm{LT}(I)\rangle = \{0\}$, then U is weakly independent (modulo I).

□

Puttiing

$$d' = \max\left\{|U| \;\middle|\; U \text{ weakly independent (modulo } I)\right\},$$

we have

7.2. Theorem If I is a proper ideal of $k[\mathbf{x}]$, then $\deg({}^aHP_I) = d'$.

Proof If $d' = 0$, then since I is a proper ideal we have ${}^aHF_I(m) \geq \binom{m+0}{0} = 1.$

Now suppose $d' > 0$, and without loss of generality we let $U = \{x_1, ..., x_{d'}\} \subset \{x_1, ..., x_n\}$ be weakly independent (modulo I). Then the weak independence condition $k[U] \cap I = \{0\}$ and CH.I Lemma 1.6 together yield

$$\begin{aligned} \binom{m+d'}{d'} = \dim_k k[U]_{\leq m} &= \dim_k \frac{k[U]_{\leq m} + I}{I} \\ &\leq \dim_k \frac{k[\mathbf{x}]_{\leq m} + I}{I} \\ &= {}^aHF_I(m). \end{aligned}$$

Thus, for $m \gg 0$ we obtain

$${}^aHP_I(m) \geq \binom{m+d'}{d'} = f(m),$$

where $f(x)$ denotes the polynomial $\binom{x+d'}{d'}$. It follows that

$$\deg({}^aHP_I) \geq \deg f(x) = d'.$$

But by Proposition 3.8 and Lemma 7.1 we also have $\deg({}^aHP_I) \leq d'$. Hence $\deg({}^aHP_I) = d'$, as desired. □

In the algorithmic study of systems of polynomial equations, the Elimination Theory consists of the Elimination (of variables) Step and the Extension (of solutions) Step, both can be obtained in great generality in terms of Groebner bases. We refer the reader to [CLO'] for more details about this theory. What we do here is to point out the natural connection between the dimension theory and the elimination of variables suggested in Theorem 7.2.

7.3. Definition Given $I = \langle f_1, ..., f_s \rangle \subset k[\mathbf{x}]$, the k-th *elimination ideal* I_k is the ideal of $k[x_{k+1}, ..., x_n]$ defined by

$$I_k = I \cap k[x_{k+1}, ..., x_n].$$

Thus, I_k consists of all polynomials in I which eliminate the variables $x_1, ..., x_k$ (the fact that I_k is an ideal of $k[x_{k+1}, ..., x_n]$ is obvious). Note that $I = I_0$ is the 0-th elimination ideal. Also observe that different orderings of the variables lead to different elimination ideals. Using this language, we see that eliminating $x_1, ..., x_k$ means finding *nonzero* polynomials in the k-th elimination ideal I_k. The next theorem expresses that if $\deg(\,{}^aHP_I) = d$ (or $\dim \mathbf{V}(I) = d$ in case k is algebraically closed), then the elimination of any $n - d - j$ $(1 \le j \le n - d)$ variables is always guaranteed; and a solution of the Elimination Step is always realizable by giving a systematic procedure for finding elements of I_k.

7.4. Theorem Let $I \subset k[\mathbf{x}]$ be an ideal.
(i) If $\deg(\,{}^aHP_I) = d$, then for any $U = \{x_{i_1}, ..., x_{i_{d+j}}\} \subset \{x_1, ..., x_n\}$ with $1 \le j \le n - d$, we have $I \cap k[U] \ne \{0\}$. In particular, we have

$$(*) \qquad I \cap k[x_1, ..., x_d, x_h] \ne \{0\}, \ d + 1 \le h \le n.$$

(ii) If G is a Groebner basis of I with respect to the lexicographic ordering $>_{lex}$ such that $x_1 >_{lex} >_{lex} x_2 >_{lex} \cdots >_{lex} x_n$, then the set

$$G_k = G \cap k[x_{k+1}, ..., x_n]$$

is a Groebner basis of the k-th elimination ideal I_k with respect to $>_{lex}$.

Proof (i) This follows from Theorem 7.2.
(ii) Suppose that $G = \{g_1, ..., g_s\}$. If $f \in I_k = I \cap k[x_{k+1}, ..., x_n]$, then

$$f = h_1g_1 + \cdots + h_sg_s, \text{ with } \operatorname{md}(f) \ge \operatorname{md}(h_ig_i) \text{ whenever } h_ig_i \ne 0.$$

Since we are using $>_{lex}$, if some $\mathrm{LM}(g_i)$ contains one of $x_1, ..., x_k$ as a divisor, then $h_ig_i = 0$. Thus, after reordering $\{g_1, ..., g_s\}$ if necessary, we obtain

$$f = h_1g_1 + \cdots + h_rg_r, \text{ with } g_j \in k[x_{k+1}, ..., x_n], \ j = 1, ..., r, \ r \le s.$$

This shows that $\{g_1, ..., g_r\}$ is a Groebner basis of I_k with respect to $>_{lex}$. □

Indeed, the elimination property (∗) in Theorem 7.4(i) characterizes $\dim \mathbf{V}(I)$ when I is a *prime* ideal (or equivalently when $\mathbf{V}(I)$ is an affine variety). To see this, let $k(\mathbf{x}) = k(x_1, ..., x_n)$ be the field of rational functions in n variables $x_1, ..., x_n$. For any subset $U = \{x_{i_1}, ..., x_{i_d}\} \subset \{x_1, ..., x_n\}$ with $1 \le d \le n$, we write $k[\mathbf{x}_d]$ for the polynomial subring $k[x_{i_1}, ..., x_{i_d}] \subset k[x_1, ..., x_n]$, write $k(\mathbf{x}_d)$ for the field $k(x_{i_1}, ..., x_{i_d})$ of rational functions in d variables $x_{i_1}, ..., x_{i_d}$, and write $k(\mathbf{x}_d)[x_{h_1}, ..., x_{h_{n-d}}]$ for the subring of $k(\mathbf{x})$ generated by $k(\mathbf{x}_d)$ and $\{x_{h_1}, ..., x_{h_{n-d}}\} = \{x_1, ..., x_n\} - U$. One may check that this ring is indeed the localization of $k[\mathbf{x}]$ at the multiplicative subset $S_d = k[\mathbf{x}_d] - \{0\}$ in the sense of Appendix II §1. So we also write $S_d^{-1}k[\mathbf{x}]$ for this ring.

With notation as above, if I is an ideal of $k[\mathbf{x}]$, we have the *extension ideal* I^e which is the ideal generated by I in $S_d^{-1}k[\mathbf{x}]$, i.e.,

$$I^e = \left\{ \frac{a}{s} \;\middle|\; a \in I, s \in S_d \right\}.$$

Conversely, for any ideal $J \subset S_d^{-1}k[\mathbf{x}]$, we have the *contraction ideal* J^c in $k[\mathbf{x}]$ which is defined as

$$J^c = J \cap k[\mathbf{x}].$$

7.5. Lemma With notation as above, let I be an ideal of $k[\mathbf{x}]$ and J an ideal of $S_d^{-1}k[\mathbf{x}]$. Let $U = \{x_{i_1}, ..., x_{i_d}\} \subset \{x_1, ..., x_n\}$, $1 \le d \le n$.
(i) $J^{ce} = J$. $I \subseteq I^{ec}$ and the equality holds if I is a prime ideal.
(ii) U is weakly independent (modulo I) if and only if $I \cap S_d = \emptyset$ if and only if $I^e \ne S_d^{-1}k[\mathbf{x}]$.

Proof Exercise. □

7.6. Proposition Let I be an ideal of $k[\mathbf{x}]$. With notation as before, suppose that $\deg({}^aHP_I) = d$. Then there exists a subset $U = \{x_{i_1}, ..., x_{i_d}\} \subset \{x_1, ..., x_n\}$ such that

(a) $k[\mathbf{x}_d] \cap I = \{0\}$, i.e., U is weakly independent (modulo I), and

(b) for any $x_j \in \{x_1, ..., x_n\} - U$,

$$I \cap k[\mathbf{x}_d][x_j] \ne \{0\}.$$

Consequently, $I^e \ne S_d^{-1}k[\mathbf{x}]$ and $S_d^{-1}k[\mathbf{x}]/I^e$ is a finite dimensional vector space over $k(\mathbf{x}_d)$, i.e., $\deg({}^aHP_{I^e}) = 0$.

Proof The existence of U with properties (a) and (b) follows from Theorem 7.2.
Since U is weakly independent (modulo I), $I^e \neq S_d^{-1}k[\mathbf{x}]$ by Lemma 7.5. To see that $S_d^{-1}k[\mathbf{x}]/I^e$ is a finite dimensional vector space over $k(\mathbf{x}_d)$, for each $h_j \in \{h_1, ..., h_{n-d}\} = \{1, ..., n\} - \{i_1, ..., i_d\}$, let $P_j(x_{h_j})$ be a nonzero polynomial in $I \cap k[\mathbf{x}_d][x_{h_j}]$, i.e., a polynomial in x_{h_j} with coefficients in $k[\mathbf{x}_d]$, and assume $\deg P_j(x_{h_j}) = k_j$. Then it is not hard to see that $S_d^{-1}k[\mathbf{x}]/I^e$ is spanned as a vector space over $k(\mathbf{x}_d)$ by

$$\left\{x_{h_1}^{a_1} \cdots x_{h_{n-d}}^{a_{n-d}}\right\}_{0 \leq a_j < k_j,\ 1 \leq j \leq n-d}.$$

□

7.7. Proposition With notation as before, let $U = \{x_{i_1}, ..., x_{i_d}\} \subset \{x_1, ..., x_n\}$ and let J be a proper ideal of $S_d^{-1}k[\mathbf{x}]$.
(i) $J^c = J \cap k[\mathbf{x}]$ is a proper ideal of $k[\mathbf{x}]$ and $\deg({}^aHP_{J^c}) \geq d$.
(ii) If furthermore $S_d^{-1}k[\mathbf{x}]/J$ is a finite dimensional vector space over $k(\mathbf{x}_d)$, then $\deg({}^aHP_{J^c}) = d$.

Proof Since J is a proper ideal of $S_d^{-1}k[\mathbf{x}]$, J^c is a proper ideal of $k[\mathbf{x}]$ by Lemma 7.5. Again by lemma 7.5 we claim that U is weakly independent (modulo I). It follows from Theorem 7.2 that $\deg({}^aHP_{J^c}) \geq d$. This proves (i).
To prove (ii), we only have to show that $\deg({}^aHP_{J^c}) \leq d$. Without loss of generality we may assume that $U = \{x_1, ..., x_d\}$. Since J is a proper ideal, $S_d^{-1}k[\mathbf{x}]/J$ is a $k(\mathbf{x}_d)$-vector space of dimension ≥ 1. We may assume that a $k(\mathbf{x}_d)$-basis $\{s_1, ..., s_t\}$ of $S_d^{-1}k[\mathbf{x}]/J$ consists of monomials in $x_{d+1}, ..., x_n$ and that $s_1 = 1$. Working with this basis, there exist $P \in k[\mathbf{x}_d]$ and $q_{vj}^u \in k[\mathbf{x}_d]$, $d+1 \leq u \leq n$, $1 \leq v, j \leq t$, such that

$$(\bullet) \qquad P x_u s_v = \sum_{j=1}^{t} q_{vj}^u s_j.$$

Now consider the k-subspace of $S_d^{-1}k[\mathbf{x}]/J$:

$$M = \sum_{j=1}^{t} k[\mathbf{x}_d][P^{-1}]s_j,$$

where $k[\mathbf{x}_d][P^{-1}]$ is the subring of $S_d^{-1}k[\mathbf{x}]$ generated by $k[\mathbf{x}_d]$ and $\{P^{-n} \mid n \geq 0\}$ (which is indeed the localization of $k[\mathbf{x}_d]$ at the multiplicative subset $\{1, P, P^2, ...\}$ in the sense of Appendix II §1). Put

$$T = \max_{d+1 \leq u \leq n,\ 1 \leq v,j \leq t} \left\{\deg P + 1,\ \deg q_{vj}^{u}\right\}.$$

It is clear that each element $D = \sum_{j=1}^{t} F_j s_j$ of M belongs to a k-subspace of the form:

$$F_m M = \left\{P^{-m}\sum_{j=1}^{t} h_j s_j \;\middle|\; h_j \in k[\mathbf{x}_d],\ \deg h_j \leq mT\right\} \subset M,\ m \geq 0.$$

Clear, $F_m M \subset F_{m+1} M$ and

$$\dim_k F_m M = t \cdot \binom{mT + d}{d}$$

which is a polynomial in m and has degree d (see CH.I §1).
Since $k[\mathbf{x}]J^c = k[\mathbf{x}]/(J \cap k[\mathbf{x}]) \cong (k[\mathbf{x}] + J)/J \subset S_d^{-1}k[\mathbf{x}]/J$ and $(k[\mathbf{x}]_{\leq m} + J^c)/J^c \cong (k[\mathbf{x}]_{\leq m} + J)/J$, it follows from §3 that we will finish the proof of (ii) by showing that

$$(\bullet\bullet) \qquad \frac{k[\mathbf{x}]_{\leq m} + J}{J} \subset F_m M, \quad m \geq 0.$$

Indeed, let $D = x_1^{\alpha_1} \cdots x_d^{\alpha_d} x_{d+1}^{\beta_1} \cdots x_n^{\beta_{n-d}}$ be a monomial in $(k[\mathbf{x}] + J)/J$ such that $|\alpha| + |\beta| \leq m$, where $|\alpha| = \alpha_1 + \cdots + \alpha_d$, $|\beta| = \beta_1 + \cdots + \beta_{n-d}$. If we start with x_{d+1} in the foregoing $(\bullet)$:

$$Px_{d+1} = Px_{d+1}s_1 = \sum_{j=1}^{t} q_{1j}^{d+1} s_j,$$

then we obtain

$$P^{|\beta|}D = \sum_{j=1}^{t} x_1^{\alpha_1} \cdots x_d^{\alpha_d} q_j^{\beta} s_j$$

with $q_j^{\beta} \in k[\mathbf{x}_d]$ and $\deg q_j^{\beta} \leq |\beta| T$. Thus $\deg\left(x_1^{\alpha_1} \cdots x_d^{\alpha_d} q_j^{\beta}\right) \leq |\alpha| + |\beta| T \leq mT$, and it follows that

$$\begin{aligned} D &= \sum_{j=1}^{t} \frac{x_1^{\alpha_1}\cdots x_d^{\alpha_d} q_j^{\beta}}{P^{|\beta|}} s_j \\ &= \sum_{j=1}^{t} \frac{P^{|\alpha|} x_1^{\alpha_1}\cdots x_d^{\alpha_d} q_j^{\beta}}{P^{|\alpha|+|\beta|}} s_j \end{aligned}$$

with

$$\begin{aligned} \deg\left(P^{|\alpha|} x_1^{\alpha_1}\cdots x_d^{\alpha_d} q_j^{\beta}\right) &\le |\alpha|\deg P + |\alpha| + |\beta| T \\ &= |\alpha|(\deg P + 1) + |\beta| T \\ &\le (|\alpha| + |\beta|)T \\ &\le mT, \end{aligned}$$

i.e., $D \in F_{|\alpha|+|\beta|}M \subset F_m M$, as desired. □

Combining Proposition 7.6 and Proposition 7.7 the following results are obtained.

7.8. Theorem Let $U = \{x_{i_1}, ..., x_{i_d}\} \subset \{x_1, ..., x_n\}$ and let J be a proper ideal of $S_d^{-1}k[\mathbf{x}]$. Then the following are equivalent:
(i) $J \cap k[\mathbf{x}_d][x_j] \neq \{0\}$ for every $x_j \in \{x_1, ..., x_n\} - U$;
(ii) $S_d^{-1}k[\mathbf{x}]/J$ is a finite dimensional vector space over $k(\mathbf{x}_d)$, i.e., $\deg({}^aHP_J) = 0$;
(iii) $\deg({}^aHP_{J^c}) = d$.

□

7.9. Corollary Let I and U be as in Proposition 7.6. Then $\deg({}^aHP_I) = d = \deg({}^aHP_{I^{ec}})$.

□

7.10. Theorem Let I be an ideal of $k[\mathbf{x}]$ and $U = \{x_{i_1}, ..., x_{i_d}\} \subset \{x_1, ..., x_n\}$ be such that $I \cap k[\mathbf{x}_d] = \{0\}$. Suppose that

(**) $k[\mathbf{x}]/I$ is S_d-torsionfree, i.e., $s \in S_d = k[\mathbf{x}_d] - \{0\}$, $f \in k[\mathbf{x}]$, $s \cdot f \in I$ implies $f \in I$.

Then $\deg({}^aHP_I) = d$ if and only if $S_d^{-1}k[\mathbf{x}]/I^e$ is a finite dimensional vector space over $k(\mathbf{x}_d)$, i.e., $\deg({}^aHP_{I^e}) = 0$.

Every prime ideal p of $k[\mathbf{x}]$ such that $p \cap k[\mathbf{x}_d] = \{0\}$ does satisfy the condition (**).

□

Remark The notion of weak independence modulo a polynomial ideal $I \subset k[x_1, ..., x_n]$ was introduced by W. Gröbner in his book *Algebraic geometrie*, I, II (Bibliographisches Institut, Mannheim, 1968, 1970). Its usefulness in algebraic geometry was realized after the combination with the algorithmic techniques of Groebner bases in order to compute the dimension $\dim \mathbf{V}(I)$ of the affine algebraic set $\mathbf{V}(I)$ determined by the ideal I, or the degree of the Hilbert polynomial of the $k[x_1, ..., x_n]$-module $k[x_1, ..., x_n]/I$. The notion of (strong) independence modulo a polynomial ideal I was introduced by H. Kredel and V. Weispfenning in the article *Computing dimension and independent sets for polynomial ideals* (In: Computational Aspects of Commutative Algebra, a special issue of the Journal of Symbolic Computation, L. Robbiano ed., Academic Press, 1989, 97–113) as a *key* link between the weak independence modulo I and a Groebner basis of I.

Exercises for §7

1. If $U = \{x_{i_1}, ..., x_{i_d}\} \subset \{x_1, ..., x_n\}$ is weakly independent (modulo I), where I is an ideal of $k[\mathbf{x}]$, is U necessarily independent (modulo I)?
2. Consider the system of polynomial equations
$$x^2 + y + z = 1,$$
$$x + y^2 + z = 1,$$
$$x + y + z^2 = 1.$$
let I be the ideal generated by $f_1 = x^2 + y + z - 1$, $f_2 = x + y^2 + z - 1$, $f_3 = x + y + z^2 - 1$ in $\mathbb{R}[x, y, z]$. Find the elimination ideals I_1 and I_2.
3. Prove Lemma 7.5.
4. Prove that there is a one-to-one correspondence between the prime ideals Q of $S_d^{-1}k[\mathbf{x}]$ and the prime ideals p of $k[\mathbf{x}]$ such that $U = \{x_{i_1}, ..., x_{i_d}\} \subset \{x_1, ..., x_n\}$ is weakly independent (modulo p).
5. In the proof of Proposition 7.7 check that $M = \cup_{m \geq 0} F_m M$ and $F_n M \subset F_{m+1} M$, $m \geq 0$.

§8. The Krull Dimension of an Affine k-Algebra

This section is devoted to a purely algebraic characterization of dimension of an affine algebraic set.

Let k be a field and $k \subset A$ a commutative ring extension in the sense of Appendix I.

8.1. Definition If A is ring-finite over k in the sense of Appendix I, i.e., there exist a finitely many $a_1, ..., a_n \in A$ such that $A = k[a_1, ..., a_n]$, then we say that A is an *affine k-algebra*. If furthermore A is a domain, we say that A is an *affine domain over k*.

From the definition it is clear that every affine k-algebra is a homomorphic image of the affine k-algebra $k[\mathbf{x}] = k[x_1, ..., x_n]$ of polynomials in n variables. Conversely, for any ideal I of $k[\mathbf{x}]$, $k[\mathbf{x}]/I$ is an affine k-algebra. In this section, we discuss the relation between the Krull dimension (see the definition below) of $k[\mathbf{x}]/I$ and $\dim \mathbf{V}(I)$ for an ideal I of $k[\mathbf{x}]$.

We start with an arbitrary commutative ring A. A finite sequence of $n + 1$ prime ideals of A

$$p_0 \supset p_1 \supset \cdots \supset p_n$$

is called a *prime chain* of length n.

8.2. Definition (i) If p is any prime ideal of A, the *height* of p, denoted $\mathrm{ht}(p)$, is defined to be

$$\mathrm{ht}(p) = \sup\{\text{lengths of the prime chains with } p = p_0\}.$$

(ii) The *Krull dimension* of A, denoted K.dim$A$, is defined to be

$$\mathrm{K.dim}A = \sup\{\mathrm{ht}(p) \mid p \text{ runs over all prime ideals of } A\}.$$

Example (i) A field has Krull dimension 0. If A is a PID(principal ideal domain), then K.dim$A = 1$.

The Krull dimension is a very important algebraic invariant in commutative algebra. We refer the reader to any commutative algebra text book (e.g., [Mats]) for the generalities concerning Krull dimension.

From here on, for an ideal $I \subset k[\mathbf{x}]$, we discuss the relation between $\dim\mathbf{V}(I)$ and K.dim$k[\mathbf{V}(I)]$. However, as in §7 our discussion is nore general, namely, we will work on an arbitrary field k and will deal with the Hilbert polynomial ${}^aHP_I$. Theorem 3.12 allows to replace $\deg({}^aHP_I)$ by $\dim\mathbf{V}(I)$ everywhere the first appears, at least in the case where k is algebraically closed.

The first result characterizes prime ideals p in $k[\mathbf{x}]$ with $\deg({}^aHP_p) = 0$.

8.3. Theorem Let p be a prime ideal of $k[\mathbf{x}]$. Then p is a maximal ideal if and only if $\deg({}^aHP_p) = 0$.

Proof First let p be maximal. Then we have the field extension $k \subset L = k[\mathbf{x}]/p$. Clearly, L is ring-finite over k. It follows from Theorem 3.1 of Appendix I that L is finite dimensional over k. Hence $\deg({}^aHP_p) = 0$ by §3.
Conversely, suppose $\deg({}^aHP_p) = 0$. Then $k[\mathbf{x}]/p$ is finite dimensional over k. To prove p is maximal it is sufficient to prove that $k[\mathbf{x}]/p$ is a field. To this end, take any nonzero element $[f] \in k[\mathbf{x}]/p$ with $f \in k[\mathbf{x}]$, and consider the k-linear map

$$\begin{array}{rccc} \varphi_{[f]}: & k[\mathbf{x}]/p & \longrightarrow & k[\mathbf{x}]/p \\ & [g] & \mapsto & [f]\cdot[g] = [gf] \end{array}$$

Since p is a prime ideal and $[f] \neq 0$, we see that $\varphi_{[f]}$ is injective. But $k[\mathbf{x}]/p$ is finite dimensional over k, it follows that $\varphi_{[f]}$ is also surjective. This means that $[f]\cdot[g] = [1]$ for some $[g] \in k[\mathbf{x}]/p$, i.e., $[f]$ is invertible, as desired. □

To handle a prime ideal p with $\deg({}^aHP_p) > 0$, let $U = \{x_{i_1}, ..., x_{i_d}\} \subset \{x_1, ..., x_n\}$ be such that U is weakly independent (modulo p). With notation as in §7, it follows from Lemma 7.5(i) and §7 exercise 4 that the extension ideal $p^e \subset S_d^{-1}k[\mathbf{x}]$ has some nice properties. Here we record one of them:

8.4. Lemma p^e is a prime ideal of $S_d^{-1}k[\mathbf{x}]$ and $p^{ec} = p$.

□

8.5. Lemma (Comparing to Proposition 5.10) Let p, q be prime ideals of $k[\mathbf{x}]$. Suppose $p \subset q$ and $p \neq q$. Then $\deg({}^aHP_q) < \deg({}^aHP_p)$.

Proof By the assumption it is clear that $\deg({}^aHP_q) \le \deg({}^aHP_p)$. Suppose that $\deg({}^aHP_p) = \deg({}^aHP_q) = d$, and that $U = \{x_1, ..., x_d\}$ is weakly independent (modulo q). Then U is also weakly independent (modulo p). If we put $S_d = k[\mathbf{x}_d] - \{0\}$ as before, then it follows from Lemma 8.4 and Theorem 7.10 that p^e and q^e are prime ideals of $S_d^{-1}k[\mathbf{x}]$ with $\deg({}^aHP_{p^e}) = \deg({}^aHP_{q^e}) = 0$. Thus, p^e, q^e are maximal ideals of $S_d^{-1}k[\mathbf{x}]$ by Theorem 8.3. But $p^e \subset q^e$, and then $p^e = q^e$. It follows from Lemma 8.4 that $p = q$, a contradiction. This shows that we must have $\deg({}^aHP_q) < \deg({}^aHP_p)$. □

8.6. Corollary Let

$$p_0 \supset p_1 \supset \cdots \supset p_m$$

be a prime chain in $k[\mathbf{x}]$. Then $m \le \deg({}^aHP_{p_m}) - \deg({}^aHP_{p_0})$.

□

8.7. Proposition Let p be a prime ideal of $k[\mathbf{x}]$. Then
(i) $\mathrm{K.dim}(k[\mathbf{x}]/p) \le \deg({}^aHP_p)$.
(ii) $\mathrm{ht}(p) \le n - \deg({}^aHP_p)$.

Proof (i) In Corollary 8.6, put $p_m = p$. Then by the definition of $\mathrm{K.dim}(k[\mathbf{x}]/p)$ we obtain the desired inequality.
(ii) In Corollary 8.6, put $p_0 = p$. Then by the definition of $\mathrm{ht}(p)$ we obtain the desired inequality. □

8.8. Proposition Let p be a prime ideal of $k[\mathbf{x}]$. If $\deg({}^aHP_p) = d$, then there exists a prime chain

$$p_0 \supset p_1 \supset \cdots \supset p_d$$

such that $p_d = p$.

Proof Assume $\{x_1, ..., x_d\}$ is weakly independent (modulo p). We use induction on d. If $d = 0$, then p is maximal by Theorem 8.3, and $p = p_0$ is the desired prime chain.
Suppose $d > 0$. Put $S_{d-1} = k[\mathbf{x}_{d-1}] - \{0\}$ as before, and consider $p^e \subset S_{d-1}^{-1}k[\mathbf{x}]$. Then since $p \cap k[\mathbf{x}_d] = \{0\}$, it is easy to see that

$$p^e \cap k(\mathbf{x}_{d-1})[x_d] = \{0\}.$$

This means $\deg({}^aHP_{p^e}) > 0$, i.e., p^e is not maximal in $S_{d-1}^{-1}k[\mathbf{x}]$. Let Q be a maximal ideal in $S_{d-1}^{-1}k[\mathbf{x}]$ such that $p^e \subset Q$. Then Q^c is a prime ideal of $k[\mathbf{x}]$ and $p = p^{ec} \subset Q^c$ by Lemma 8.4. By the maximality of Q we have $\deg({}^aHP_Q) = \deg({}^aHP_{Q^{ce}}) = 0$. It follows from Theorem 7.10 that $\deg({}^aHP_{Q^c}) = d - 1$, in particular, $p \neq Q^c$. By the induction hypothesis we have a prime chain $p_0 \supset p_1 \supset \cdots \supset p_{d-1} = Q^c$, and consequently,

$$p_0 \supset p_1 \supset \cdots \supset p_{d-1} = Q^c \supset p$$

is the desired prime chain. □

Combining Proposition 8.7 and Proposition 8.8, we have the following

8.9. Theorem For any prime ideal $p \subset k[\mathbf{x}]$, we have

$$\deg({}^aHP_p) = \text{K.dim}(k[\mathbf{x}]/p).$$

□

The next result deals with the height of a prime ideal. It is not needed for the main result of this section (we mention it here only for completeness), so we refer the reader to [BW] pp.325–327 (or [Mats] pp.89–92) for a proof.

8.10. Theorem Let p be a prime ideal of $k[\mathbf{x}]$ with $\deg({}^aHP_p) = d$, where $0 \leq d \leq n$. Then there exists a prime chain

$$p_0 \supset p_1 \supset \cdots \supset p_{n-d}$$

such that $p_0 = p$, in particular, we have $\mathrm{ht}(p) = n - \deg({}^{a}HP_p)$.

□

Now, let $A = k[\mathbf{x}]/I$ be an affine k-algebra, where I is an ideal of $k[\mathbf{x}]$, and let $\sqrt{I}$ be the radical of I. Then the following fact is clear:

$$\begin{aligned} \mathrm{K.dim}A &= \sup\left\{\mathrm{ht}(p) \;\middle|\; p \text{ runs over all primes in } k[\mathbf{x}]/I\right\} \\ &= \sup\left\{\mathrm{ht}(p) \;\middle|\; p \text{ runs over all primes in } k[\mathbf{x}]\big/\sqrt{I}\right\} \end{aligned}$$

Putting $V = \mathbf{V}(I)$, then V has a (unique) decomposition into irreducible components: $V = \cup_{i=1}^{s} V_i$, where $V_i \neq V_j$ whenever $i \neq j$ (see CH.II). Furthermore, if we put $p_i = \mathbf{I}(V_i)$, then we have the following

8.11. Proposition With notation as above,
(i) if p is a prime ideal containing I, then $p \supset p_i$ for some i.
(ii) $\{p_1, ..., p_s\}$ is the set of primes which are minimal in the set of all primes containing I (or equivalently, $\sqrt{I}$).

Proof (i) is easy. (ii) follows from Exercise 4 of CH.III §6. □

8.12. Corollary With notation as above,

$$\mathrm{K.dim}A = \max\left\{\mathrm{K.dim}(k[\mathbf{x}]/p_i) \;\middle|\; i = 1, ..., s\right\}.$$

□

If we combine Corollary 5.9, Theorem 8.9 and the above corollary, then we have the following

8.13. Theorem With notation as before, if k is algebraically closed, then

$$\dim \mathbf{V}(I) = \mathrm{K.dim}(k[\mathbf{x}]/I)$$

□

§9. The Topological Dimension of an Affine Algebraic Set

Let $V \subset \mathbf{A}^n = \mathbf{A}_k^n$ be an affine algebraic set. Recall from CH.II that we have defined the Zariski topology on V by taking algebraic subsets of V as closed sets. In this section, we give a short discussion on the relation between the topological dimension of V and $\dim V$.

9.1. Definition If X is a topological space, then the *topological dimension* of X, denoted $\text{t.dim}X$, is defined to be the supremum of all integers n such that there exists a chain

$$Z_0 \subset Z_1 \subset \cdots \subset Z_n$$

of distinct irreducible closed subsets of X.

9.2. Proposition Let $V \subset \mathbf{A}^n$ be an affine algebraic set and $k[V]$ the coordinate ring of V. Then $\text{t.dim}V = \text{K.dim}k[V]$, where the latter is the Krull dimension of $k[V]$ defined in the last section.

Proof From CH.II §4 we know that the closed irreducible subsets of V correspond to prime ideals of $k[\mathbf{x}]$ containing $\mathbf{I}(V)$. These in turn correspond to prime ideals of $k[V] = k[\mathbf{x}]/\mathbf{I}(V)$. Hence, $\text{t.dim}V = \text{K.dim}k[V]$. □

Summing up, so far in this chapter, we have proved the following.

9.3. Theorem Let V be an affine variety and $k[V]$ its coordinate ring. If k is algebraically closed, then

$$\dim V = \deg\left({}^aHP_{\mathbf{I}(V)}\right) = \text{tr.deg}_k k(V) = \text{K.dim}k[V] = \text{t.dim}V.$$

□

Thus, for an affine variety V, we have $\deg\left({}^aHP_{\mathbf{I}(V)}\right)$ as a combinitorical (algebra) invariant, $\text{K.dim}k[V]$ as a ring theoretical (ideals) invariant, $\text{tr.deg}_k k(V)$ as a field theoretical invariant and $\text{t.dim}V$ as a topological invariant, moreover the interpretation of dimension in terms of the tangent space is part of the local theory in CH.VI but it definitely has a differential (derivatives, etc...) aspect. This evidence of "mathematical unity" makes dimension theory esthetically satisfying, establishing how essential notions are at the core of seemingly different kinds of mathematics.

CHAPTER VI
An Introduction to Local Theory

Standing on the world you see a part of it, your neighborhood. Move about and you discover other parts of it; these are glued in your memory so as to obtain your own map of the world. Did you ever detect for yourself (or even find a proof) that the world is round? How? From childhood on we have learned to observe complex objects by looking at specific points, edges, faces, shaded areas, bumps,... and then compose a global picture. Most of the geometrical objects we are studying now are too complex to be visualized, even with the use of sophisticated computers the images or at best approximations and we cannot see more than three dimensional shapes. Again abstract mathematics follows closely the way our thinking and observing have evolved.

Indeed, we observed in CH.III §3 that a nonconstant polynomial does not define a function on a projective variety. For a nonempty projective varierty Y, any rational function $\frac{\phi}{\psi}$ in $k(Y)$ is a "local functioin" in the sense that it defines a function on an open subset of Y, i.e., on a neighborhood of some point(s). More precisely, such a nonempty projective variety $Y \subset \mathbf{P}^n_k$ may be covered by a finite number of nonempty open subsets $Y \cap U_i$ which may be identified with affine varieties. If $I = \mathbf{I}(Y)$ then the latter varieties are given as $\mathbf{V}(I_*)$ by means of the continuous mappings φ_i: $U_i \to \mathbf{A}^n = \mathbf{A}^n_k$, where I_* is the dehomogenization of I with respect to x_i, and moreover, $K(Y) \cong K(\mathbf{V}(I_*))$. Conversely, any nonempty affine variety $V \subset \mathbf{A}^n$ with $I = \mathbf{I}(V)$ is homeomorphic to the *open subset* $Y \cap U_{n+1}$ of the projective variety $Y = \mathbf{V}(I^*) \subset \mathbf{P}^n$, where I^* is the homogenization of I with respect to x_{n+1}, and moreover, $K(V) \cong K(Y)$. This shows that in both cases, although the differrent open subsets $Y \cap U_i$, $1 \le i \le n+1$, are generally not projective algebraic sets, each of them (if not empty) has a "function field" which is isomorphic to $k(Y)$.

In view of CH.II Theorem 2.1 and CH.III Proposition 2.2, the above remark and the discussion in CH.II §6 and CH.III §3 suggest that the following approach may be useful.

- Consider algebraic sets in the unified setting of quasi-varieties.
- Study the geometry of an algebraic set by using the rational-like functions on quasi-varieties.
- Construct mappings between quasi-varieties by using rational-like functions on a quasi-variety.

This leads to the "local theory" in algebraic geometry. Roughly speaking, the main idea in such a local theory is to view each quasi-variety Y as beeing covered by a number of "basic" open affine varieties having the same function fields and which are "glued" together by rational mappings on quasi-varieties. As a result one may classify all varieties up to "birational equivalence", and under this birational equivalence, one may study a point by taking any of its open neighborhood in an arbitrary ambient space without losing any geometric property.

Once the "local-global" principle has been understood, a pure algebraic realization of the local theory is the study of local rings, and in the study of local rings an algorithmic theory has been developed. We refer to any book of Commutative Algebra for generalities on local rings, and we refer to the book *Using Algebraic Geometry* (Springer-Verlag, 1998) written by D. Cox, J. Little and D. O'shea for computation in local rings in terms of Groebner basis. In this chapter we will provide only one example using Groebner bases for checking the birationality of a rational mapping (§5).

The discussion in this chapter is based on the following key observation:

- Any nonempty open subset of a (affine or projective) variety Y is dense and irreducible in Y with respect to the Zariski topology on Y (see CH.II §1, CH.III §2).

As a consequence of this fact, rational functions in $k(x_1, ..., x_n)$, or more generally, rational functions in $k(Y)$, where Y is a (affine or projective) variety, are analytic-like functions on open subsets of Y. (See CH.II Lemma 6.7.) It is this property that gives rise to the typical philosophy of algebraic geometry: ***studying points and classifying varieties by means of regular functions and rational mappings on Zariski open sets.***

In this chapter, k will be an *algebraically closed* field.

§1. Regular Functions on Quasi-Varieties

In this section, we first introduce the notion of the regular function on a Zariski open subset by generalizing the properties of a rational function discussed in CH.II and CH.III.

We start with a quasi-affine variety $Y \subset \mathbf{A}^n = \mathbf{A}^n_k$.

1.1. Definition A function f: $Y \to k$ is *regular at a point* $P \in Y$ if there is an open neighborhood U with $P \in U \subseteq Y$, and polynomials $g, h \in k[x_1, ..., x_n]$, such that $h(Q) \neq 0$ for all $Q \in U$, and $f = \frac{g}{h}$ on U. We say that f is *regular on* Y if it is regular at every point of Y.

Example (i) Let Y be a quasi-affine variety in $\mathbf{A}^n_k$. Then every polynomial $f \in k[x_1, ..., x_n]$ is regular on Y. More generally, if V is an affine variety containing Y, then every polynomial function $[f] \in k[V]$ is regular on Y.

(ii) Let V be an affine variety, $k(V)$ the field of rational functions of V. From CH.II. §6 it follows that any $\frac{\phi}{\psi} \in k(V)$ is regular on the open subset

$$U_{\frac{\phi}{\psi}} = \left\{ P \in V \;\middle|\; \frac{\phi}{\psi} \text{ is defined at } P \right\}.$$

1.2. Lemma **A regular function is continuous (where k is identified with A_k^1 endowed with its Zariski topology).**

Proof Let f be a regular function on Y. It is enough to show that f^{-1} of a closed set is closed. A closed set of $\mathbf{A}_k^1$ is a finite set of points, so it is sufficient to show that $f^{-1}(a) = \{P \in Y \mid f(P) = a\}$ is closed for any $a \in k$. This can be done locally as follows.
Let U be an open set on which f may be represented as $\frac{g}{h}$, with $g, h \in k[x_1, ..., x_n]$, and h nowhere 0 on U. Then

$$f^{-1}(a) \cap U = \left\{ P \in U \;\middle|\; \frac{g(P)}{h(P)} = a \right\}.$$

But $\frac{g(P)}{h(P)} = a$ if and only if $(g - ah)(P) = 0$. So $f^{-1}(a) \cap U = \mathbf{V}(g - ah) \cap U$ which is closed in U.
By the definition of regular functions on a quasi-affine variety it is easy to see that $Y = \cup_{\alpha\in\Lambda} U_\alpha$ where the U_α's are open subsets and on each U_α, f is represented by a rational function. Thus $f^{-1}(a) = \cup_{\alpha\in\Lambda}(U_\alpha \cap f^{-1}(a))$ with each $U_\alpha \cap f^{-1}(a)$ closed in U_α. By the definition of open sets

$$U_\alpha = Y - V_\alpha \text{ where } V_\alpha \text{ is closed in } Y,$$

and by the definition of closed sets

$$U_\alpha \cap f^{-1}(a) = U_\alpha \cap Z_\alpha \text{ where } Z_\alpha \text{ is closed in } Y.$$

We claim that $f^{-1}(a) = \cap_{\alpha\in\Lambda}(V_\alpha \cap Z_\alpha)$, and that in turn leads to the result that $f^{-1}(a)$ is closed. To see this, let $P \in f^{-1}(a)$. If $P \in U_\alpha$, then $P \in U_\alpha \cap f^{-1}(a) \subset Z_\alpha$; and if $P \notin U_\alpha$, then $P \in Y - U_\alpha = V_\alpha$, so that $P \in Z_\alpha \cup V_\alpha$ for all $\alpha \in \Lambda$. Conversely, let $P \in Z_\alpha \cup V_\alpha$ for all $\alpha \in \Lambda$. From the fact that $Y = \cup_{|\alpha\in\Lambda} U_\alpha$ it follows that $P \in U_\alpha$ for some U_α. Then $P \notin V_\alpha$, and hence $P \in Z_\alpha$, $P \in Z_\alpha \cap U_\alpha$ $= f^{-1}(a)\mathcal{U}_\alpha \subset \{^{-\infty}(\dashv)$. Therefore, the desired equality holds. □

Next, we consider a quasi-projective variety $Y \subset \mathbf{P}^n = \mathbf{P}^n_k$.

1.3. Definition A function $f\colon Y \to k$ is *regular at a point* $P \in Y$ if there is an open neighborhood U with $P \in U \subseteq Y$, and homogeneous polynomials $g, h \in k[x_1, ..., x_n, x_{n+1}]$, of the same degree, such that $h(P) \neq 0$ for all $P \in U$, and $f = \frac{g}{h}$ on U. (Note that in this case, even though g and h are not functions on $\mathbf{P}^n$, their quotient is a well-defined function whenever $h \neq 0$, since they are homogeneous of the same degree.) We say that f is *regular on* Y if it is regular at every point .

As in the affine case, a regular function on a quasi-projective variety is necessarily continuous.

Example Let V be a projective variety, $k(V)$ the field of rational functions on V (CH.III §3). As in the affine case, one may check that any $\frac{\phi}{\psi} \in k(V)$ is regular on the open subset

$$U_{\frac{\phi}{\psi}} = \left\{ P \in V \;\middle|\; \frac{\phi}{\psi} \text{ is defined at } P \right\}.$$

(See CH.III §3 Exercise 3 for the open set $U_{\frac{\phi}{\psi}}$.)

1.4. Corollary (Comparing with CH.II Lemma 6.7) An important consequence of the continuity of regular functions is the fact that if f and f' are regular functions on a quasi-affine (or quasi-projective) variety X, and if $f = f'$ on some nonempty open subset $U \subseteq X$, then $f = f'$ everywhere.

Proof Indeed, $(f - f')^{-1}(0) \supset U$ implies $(f - f')(0)$ is closed and dense in X (CH.II Proposition 1.3), hence equal to X. □

1.5. Definition Let k be a fixed algebraically closed field. A ***variety*** X over k (or simply a *variety*) is an affine, quasi-affine, projective, or quasi-projective variety as defined before.

If Y is a closed subset of $X \subset V$, where V is an affine or projective variety, we say that Y is *irreducible* if Y is not the union of two proper closed subsets. Y is then also a variety, for if $\overline{Y}$ is the closure of Y in V, it is easy to varify that $\overline{Y}$ is irreducible in V and that

$Y = \overline{Y} \cap X$, so Y is open in $\overline{Y}$. Such a Y is called a *closed subvariety* of X.

Next let us define several rings of functions.

(a) Ring of global functions

Put

$$\mathcal{O}(Y) = \{\text{all regular functions on } Y\}.$$

Then $\mathcal{O}(Y)$ is a commutative ring containing k, with the usual addition and multiplication of functions, the additive group $(\mathcal{O}(Y), +)$ has the 0 function as its identity and the multiplicative semigroup $(\mathcal{O}(Y), \cdot)$ has the constant function 1 as its identity. $\mathcal{O}(Y)$ is called the ring of *global functions* on Y.
Obviously, if U is any nonempty open subset of Y, then by restricting functions to U, $\mathcal{O}(Y) \subset \mathcal{O}(U)$.

(b) Local ring at a point

Let $P \in Y$. Put

$$\mathcal{G}_P = \left\{ \begin{array}{ll} \text{pair } (U, f), & \text{where} \\ & U \text{ is an open subset of } Y \text{ containing } P, \\ & f \text{ is a regular function on } U. \end{array} \right\}$$

Because of Corollary 1.4, we may define an equivalence relation $\sim$ on $\mathcal{G}_P$:

$$(U, f) \sim (V, g) \Longleftrightarrow f = g \text{ on } U \cap V.$$

We denote the quotient set $\mathcal{G}_P / \sim$ by $\mathcal{O}_{P,Y}$ (or simply $\mathcal{O}_P$ if no confusion is possible), and denote the equivalence class of (U, f) by $[U, f]$. An element in $\mathcal{O}_{P,Y}$ is called a *germ* of regular functions on Y near P, and $\mathcal{O}_{P,Y}$ is called the *local ring* of Y at P. (The interested reader may compare the notion of a *germ* here with the notion appearing in differential geometry or complex analytic geometry.)

If the addition and multiplication on $\mathcal{O}_{P,Y}$ are defined by putting

$$\begin{aligned} [U, f] + [V, g] &= [U \cap V, f + g] \\ [U, f] \cdot [V, g] &= [U \cap V, f \cdot g], \end{aligned}$$

then we have

1.6. Lemma With operations defined as before, $\mathcal{O}_{P,Y}$ is a local ring (in the sense of CH.II §6) containing k, and its unique maximal ideal is

$$\mathcal{M}_{P,Y} = \left\{ [U,f] \in \mathcal{O}_{P,Y} \;\middle|\; f(P) = 0 \right\}.$$

The additive group $(\mathcal{O}_{P,Y}, +)$ has $[Y, 0]$ as its identity and the multiplicative semigroup $(\mathcal{O}_{P,Y}, \cdot)$ has $[Y, 1]$ as its identity. Moreover, $\mathcal{O}_{P,Y}/\mathcal{M}_{P,Y} \cong k$.

Proof The fact that $\mathcal{O}_{P,Y}$ is a ring containing k for the given addition and multiplication may be checked directly. Since we have a ring epimorphism

$$\begin{array}{ccc} \mathcal{O}_{P,Y} & \longrightarrow & k \\ [U,f]) & \mapsto & f(P) \end{array}$$

with kernel $\mathcal{M}_{P,Y}$, it follows that $\mathcal{O}_{P,Y}/\mathcal{M}_{P,Y} \cong k$. If $f(P) \neq 0$, then $\frac{1}{f}$ defines a regular function on some neighborhood of P. Hence $\mathcal{O}_{P,Y}$ is a local ring by CH.II §6 Exercise 3. □

(c) Field of rational functions

Put

$$\mathcal{G}_Y = \left\{ \begin{array}{ll} \text{pair } (U,f), & \text{where} \\ & U \neq \emptyset \text{ is an open subset of } Y \text{ and} \\ & f \text{ is a regular function on } U. \end{array} \right\},$$

and define an equivalence relation $\sim$ on $\mathcal{G}_Y$ via Corollary 1.4:

$$(U,f) \sim (V,g) \Longleftrightarrow f = g \text{ on } U \cap V.$$

We denote the quotient set $\mathcal{G}_Y/\sim$ by $\mathcal{R}(Y)$, and denote the equivalence class of (U,f) by $[U,f]$. The elements of $\mathcal{R}(Y)$ are called *rational functions* on Y.

If the addition and multiplication on $\mathcal{R}(Y)$ are defined by putting

$$\begin{aligned} [U,f] + [V,g] &= [U \cap V, f + g] \\ [U,f] \cdot [V,g] &= [U \cap V, f \cdot g], \end{aligned}$$

then we have

1.7. Lemma (i) With the operations defined above, $\mathcal{R}(Y)$ is a field containing k, and the additive group $(\mathcal{R}(Y),+)$ has $[Y,0]$ as its identity and the multiplicative semigroup $(\mathcal{R}(Y),\cdot)$ has $[Y,1]$ as its identity.
(ii) If U is a nonempty open subset of Y, then $\mathcal{R}(Y)=\mathcal{R}(U)$.

Proof (i) May be checked directly; note that for $[U,f]\in\mathcal{R}(Y)$ with $f\neq 0$, $U\neq\emptyset$ and f is regular on U. Considering the open subset $V=U-f^{-1}(0)\neq\emptyset$ (note that $f\neq 0$), then $[V,\frac{1}{f}]$ is the inverse of $[U,f]$.
(ii) Straightforward. □

So far we have defined, for any variety Y, the ring of *global functions* $\mathcal{O}(Y)$, the local ring $\mathcal{O}_{P,Y}$ at a point $P\in Y$, and the function field $\mathcal{R}(Y)$. By restricting functions we obtain the maps

$$\mathcal{O}(Y)\to\mathcal{O}_{P,Y}\to\mathcal{R}(Y),$$

which in fact are injective by Corollary 1.4. So in our later discussion we may always regard $\mathcal{O}(Y)$ and $\mathcal{O}_{P,Y}$ as subrings of $\mathcal{R}(Y)$ which may also be replaced, if it is necessary, by $\mathcal{R}(U)$ where U is any nonempty open subset of Y.

Exercises for §1

1. Show that the following two properties hold for a variety X.
   a. Every family of closed subsets of a variety has a minimal member.
   b. If $X=\cup_{\alpha\in\Lambda}U_\alpha$, U_α open in X, then $X=\cup_{i=1}^{n}U_{\alpha_i}$ for a finite number of $U_{\alpha_1},\dots,U_{\alpha_n}\in\{U_\alpha\}_{\alpha\in\Lambda}$.
2. Let X be a variety. Show that for any $z=[U,f]\in\mathcal{R}(X)$,

$$U_z=\left\{P\in U\;\middle|\;z(P)\neq 0\right\}$$

   is an open subset in U.
3. Show that the operations we defined on $\mathcal{R}(Y)$, $\mathcal{O}(Y)$, and $\mathcal{O}_{P,Y}$ are well defined.
4. we have seen that $\mathcal{R}(Y)=\mathcal{R}(U)$ for a nonempty open U in Y. Is it also true that $\mathcal{O}_{P,U}=\mathcal{O}_{P,Y}$?

5. Show that $\mathcal{O}_{P,Y}$ is exactly the localization of the coordinate ring of Y at the maximal ideal of P if Y is an affine variety. (See Appendix II §1.)

§2. Morphisms

We define mappings between varieties (in the sense of Definition 1.5). Before doing this, let us recall from CH.II §§3–4 that if we have a polynomial mapping α: $V \to W$ for affine varieties, then α is continuous with respect to the Zariski topology and α induces a k-linear ring homomorphism α_*:

$$\begin{array}{ccc} k[W] & \xrightarrow{\alpha_*} & k[V] \\ \phi & \mapsto & \phi \circ \alpha \end{array}$$

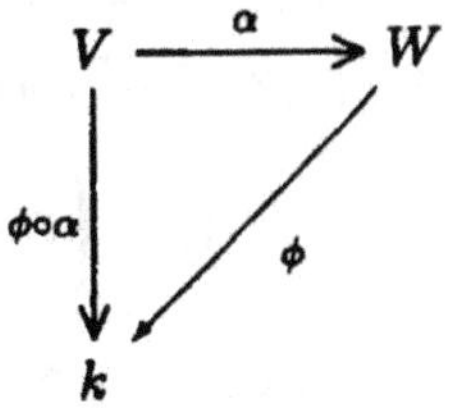

Moreover, α is a polynomial isomorphism if and only if α_* is a ring isomorphism; and we may algorithmically check whether α is a polynomial isomorphism or not.

2.1. Definition If X, Y are two varieties, a *morphism* φ: $X \to Y$ is a continuous map such that for every open set $V \subset Y$, and for every regular function f: $V \to k$, the function $f \circ \varphi$: $\varphi^{-1}(V) \to k$ is regular.

Clearly the composition of two morphisms is a morphism, so we have the notion of isomorphism: an *isomorphism* φ: $X \to Y$ of two varieties is a morphism which admits an inverse morphism ψ: $Y \to X$ with $\psi \circ \varphi = 1_X$ and $\varphi \circ \psi = 1_Y$. If X and Y are isomorphic varieties, then we write $X \cong Y$.

Example (i) Let V and W be affine varieties. Any polynomial mapping ϕ: $V \to W$ is a morphism. If ϕ is a polynomial isomorphism, then it is an isomorphism as defined above.

(ii) Let X be an affine variety, $f \in \mathcal{O}(X)$. Let ϕ: $X \to \mathbf{A}^1$ be the function defined by f: $\phi(P) = f(P)$ for $P \in X$. Then ϕ is a morphism of varieties.

(iii) More generally, let $X \subset \mathbf{A}^n$ be any variety. Let $f_1, ..., f_s$ be regular functions on X. Then

$$\begin{array}{rcl} \phi: X & \longrightarrow & \mathbf{A}^s \\ P & \mapsto & (f_1(P), ..., f_s(P)) \end{array}$$

defines a morphism from X to $\mathbf{A}^s$. For example, let $V = \mathbf{V}(f_1, ..., f_s) \subset \mathbf{A}^n$ be an affine variety. If V has a rational parametrization given by the rational functions $r_1 = \frac{g_1}{h_1}, ..., r_n = \frac{g_n}{h_n} \in k(t_1, ..., t_m)$, then $r_1, ..., r_n$ are regular functions on the open subset $U = \cap_{i=1}^n V_i \subset \mathbf{A}^n$, where

$$V_i = \left\{ P \in \mathbf{A}^m \;\middle|\; \frac{g_i}{h_i}(P) \neq 0 \right\},$$

and we have a morphism from U to V defined as above.

(iv) From CH.I §3 we retain that the parametrization of the unit circle $x^2 + y^2 = 1$ does not cover the whole circle—it has an exceptional point, namely $(-1, 0)$. However, there is a good remedy for the parametrization of a conic if we pass to the projective space, as we already pointed out in CH.III §1. Indeed, every nonhomogeneous conic

$$W = \mathbf{V}\left(ax^2 + bxy + cy^2 + dx + ey + f\right) \subset \mathbf{A}^2_{\mathbb{R}}$$

with $a, b, c, d, e, f \in \mathbb{R}$ corresponds to a quadratic curve

$$V = \mathbf{V}\left(AX^2 + 2BXY + CY^2 + 2DXZ + 2EYZ + FZ^2\right) \subset \mathbf{P}^2_{\mathbb{R}}.$$

It is well known (also see Example (iii) of §6 later) that if the above quadratic curve is nondegenerate then V is equivalent to the curve $C = \mathbf{V}(XZ - Y^2)$ under the projective change of coordinates $(X, Y, Z) \mapsto (X - Z, Y, X + Z)$. Thus, from CH.III §6 we obtain

$$W = C \cap \left(\mathbf{P}^2_{\mathbb{R}} - \mathbf{V}(Z)\right), \quad \text{by taking } Z = 1.$$

Moreover, C has a parametrization given as follows.

$$\phi_2 : \quad \mathbf{P}^1_{\mathbb{R}} \longrightarrow C \subset \mathbf{P}^2_{\mathbb{R}}$$

$$(x_1, x_2) \mapsto (x_1^2, x_1x_2, x_2^2)$$

If we write the homogeneous coordinates in $\mathbf{P}^2_{\mathbb{R}}$ as $(Z_{(2,0)}, Z_{(1,1)}, Z_{(0,2)})$, then the equation $XZ = Y^2$ has the form

$$Z_{(1,1)}Z_{(1,1)} = Z_{(2,0)}Z_{(0,2)}.$$

Now, for each $P \in C$, put

$$\begin{cases} Z_{(2,0)} = x_1 \\ Z_{(1,1)} = x_2 \end{cases} \quad \text{if } Z_{(0,2)} = 0,\ Z_{(2,0)} \neq 0,$$

$$\begin{cases} Z_{(0,2)} = x_1 \\ Z_{(1,1)} = x_2 \end{cases} \quad \text{if } Z_{(2,0)} = 0,\ Z_{(0,2)} \neq 0,$$

then it is easy to see that this defines a morphism and it is the inverse mapping of ϕ_2. This shows that ϕ_2 is an isomorphism. It is clear that this parametrization does not lose any information concerning W.

The rational map ϕ_2 is called the *2-uple embedding* of $\mathbf{P}^1$ in $\mathbf{P}^2$. Later in §4 we will generalize this example to the higher dimensional case.

(v) Let E be a d-dimensional subspace of a projective space $\mathbf{P}^n$, determined by $n-d$ linearly independent linear equations $L_1 = L_2 = \cdots = L_{n-d} = 0$, where the L_i are linear forms (i.e., homogeneous polynomials of degree 1). The mapping

$$\pi : \quad \mathbf{P}^n - E \longrightarrow \mathbf{P}^{n-d-1}$$

$$P \mapsto (L_1(P), ..., L_{n-d}(P))$$

is a morphism defined on the variety $\mathbf{P}^n - E$. This morphism is called a *projection* with centre at E. Consequently, π determines a morphism, again denoted π, $X \to \mathbf{P}^{n-d-1}$ where X is any variety in $\mathbf{P}^n$ with $X \cap E = \emptyset$. The geometric meaning of a projection is the following. As a model of $\mathbf{P}^{n-d-1}$ we take any $(n-d-1)$-dimensional subspace $H \subset \mathbf{P}^n$ disjoint from E. A unique $(d+1)$-dimensional subspace E_P passes through any point $P \in \mathbf{P}^n - E$ and

E. This subspace intersects H in a unique point, namely $\pi(P)$. If X intersects E but is not contained in it, then the projection is defined on the nonempty open subset $X - E$.

Comparison with CH.III Theorem 6.12 leads to the following

2.2. Proposition Let $U_i \subset \mathbf{P}_k^n$ be the open set defined by the equation $x_i \neq 0$. Then the mapping φ_i: $U_i \to \mathbf{A}^n$ of (Ch.III §5) is an isomorphism of varieties. Consequently, if $V \subset \mathbf{P}^n$ is a projective variety defined by a hogeneous ideal $I \subset k[x_1, ..., x_{n+1}]$, then φ_i induces an isomorphism $V \cap U_i \to \mathbf{V}(I_*) \subset \mathbf{A}^n$ whenever $V \cap U_i \neq \emptyset$, where I_* is the dehomogenization of I with respect to x_i. If $V \subset \mathbf{A}^n$ is an affine variety defined by an ideal $I \subset k[x_1, ..., x_n]$, then φ induces an isomorphism $\mathbf{V}(I^*) \cap U_i \to V$, where I^* is the homogenization of I with respect to x_i.

Proof We have already shown that it is a homeomorphism, so we need only check that the regular functions are the same on any open set. On U_i the regular functions are locally quotients of homogeneous polynomials in $x_1, ..., x_n, x_{n+1}$ of the same degree. On $\mathbf{A}^n$ the regular functions are locally quotients of polynomials in $y_1, ..., y_n$. Clearly these notions are identified by using homogenization and dehomogenization of polynomials (Ch.III). □

Remark Note that an isomorphism is necessarily bijective and bicontinuous, but a bijective bicontinuous morphism need not be an isomorphism (see Exercise 5).

With notation as in §1, a given morphism φ: $X \to Y$ corresponds to ring homomorphisms:

$$G_\varphi : \mathcal{O}(Y) \longrightarrow \mathcal{O}(X)$$

$$f \mapsto f \circ \varphi$$

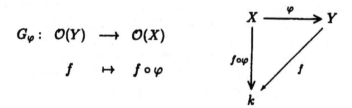

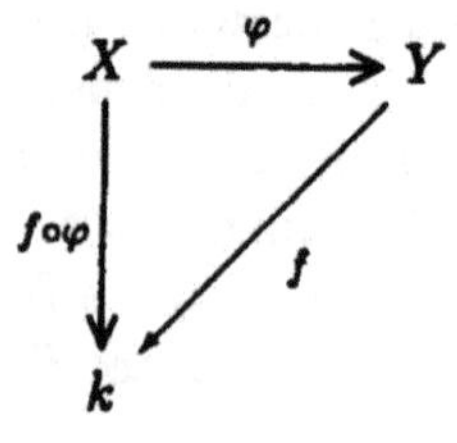

$$
\begin{array}{rccc}
L_\varphi : & \mathcal{O}_{\varphi(P),Y} & \longrightarrow & \mathcal{O}_{P,X} \\
& [U,f] & \mapsto & [\varphi^{-1}(U), f\circ\varphi]
\end{array}
\qquad
\begin{array}{ccc}
\varphi^{-1}(U) & \xrightarrow{\varphi} & U \\
{\scriptstyle f\circ\varphi}\downarrow & \swarrow{\scriptstyle f} & \\
k & &
\end{array}
$$

where $\varphi(P) \in U$.

Naturally, one might also expect that φ induce the ring homomorphism

$$
\begin{array}{rccc}
R_\varphi : & \mathcal{R}(Y) & \longrightarrow & \mathcal{R}(X) \\
& [U,f] & \mapsto & [\varphi^{-1}(U), f\circ\varphi]
\end{array}
\qquad
\begin{array}{ccc}
\varphi^{-1}(U) & \xrightarrow{\varphi} & U \\
{\scriptstyle f\circ\varphi}\downarrow & \swarrow{\scriptstyle f} & \\
k & &
\end{array}
$$

where $U \neq \emptyset$ is open in Y.

However, it is easy to see that unless $\varphi(X)$ is *dense* in Y, R_φ may not exist (we shall come back to this point in the next section). Nevertheless, we do have the following proposition.

2.3. Proposition Let φ: $X \to Y$ be an isomorphism of varieties. Then φ induces k-linear ring isomorphisms: G_φ: $\mathcal{O}(Y) \xrightarrow{\cong} \mathcal{O}(X)$, L_φ: $\mathcal{O}_{P,X} \xrightarrow{\cong} \mathcal{O}_{\varphi(P),Y}$, R_φ: $\mathcal{R}(X) \xrightarrow{\cong} \mathcal{R}(Y)$ defined as above. In particular, if $\mathcal{M}_{P,X}$ is the maximal ideal of $\mathcal{O}_{P,X}$ then $L_\varphi(\mathcal{M}_{P,X}) = \mathcal{M}_{\varphi(P),Y}$, the maximal ideal of of $\mathcal{O}_{\varphi(P),Y}$. Thus we may say that $\mathcal{O}(X)$, $\mathcal{O}_{P,X}$, and $\mathcal{R}(X)$ are *invariants* of the variety X (and the point P) up to isomorphism.

□

Now, we want to relate $\mathcal{O}(Y)$, $\mathcal{O}_{P,V}$, and $\mathcal{R}(Y)$ to the affine coordinate ring $k[Y]$, the local ring at a point and the function field $k(Y)$ for an affine variety Y (resp. to the homogeneous coordinate ring $k^g[Y]$, the local ring at a point and the function field $k(Y)$ for a projective variety Y), which were introduced earlier in CH.II (resp.

in CH.III). To this end, let us record the "local-global" property of a rational function in $\mathcal{R}(Y)$ which will be frequently used:

- If $[U, f] \in \mathcal{R}(Y)$ and W is any nonempty open subset of U, then $[W, f] = [U, f]$ in $\mathcal{R}(Y)$. Moreover, if f is represented by some rational function $\frac{g}{h} \in k(x_1, ..., x_n)$ on a nonempty open subset $W \subset U$, then $\left[W, \frac{g}{h}\right] = [U, f]$ in $\mathcal{R}(Y)$.

2.4. Theorem Let $Y \subset \mathbf{A}^n$ be an affine variety with affine coordinate ring $k[Y]$ and function field $k(Y)$. Then:
(i) $\mathcal{R}(Y) \cong k(Y)$
(ii) There is an one-to-one correspondence between Y and $\{\mathcal{O}_{P,Y}\}_{P\in Y}$.
(iii) $\mathcal{O}(Y) \cong k[Y]$.

Proof (i) For each $\frac{\phi}{\psi} \in k(Y)$, let $U_{\frac{\phi}{\psi}}$ be the opensubset of V defined in CH.II §6. Thus we have a well-defiend ring homomorphism

$$\begin{array}{rccc} \alpha: & k(Y) & \longrightarrow & \mathcal{R}(Y) \\ & \frac{\phi}{\psi} & \mapsto & \left[U_{\frac{\phi}{\psi}}, \frac{\phi}{\psi}\right] \end{array}$$

Let $[U, f] \in \mathcal{R}(Y)$. Then f is regular on the open subset $U \subset Y$. Suppose that f is represented on an open subset $W \subset U$ by a rational function $\frac{g}{h} \in k(x_1, ..., x_n)$, then $\left[W, \frac{g}{h}\right] = [U, f]$ and $h \notin \mathbf{I}(Y)$. Thus, $\alpha\left(\frac{[g]}{[h]}\right) = \left[W, \frac{g}{h}\right] = [U, f]$. This shows that α is surjective.
Finally, suppose $\alpha\left(\frac{\phi}{\psi}\right) = [Y, 0]$, then on some nonempty open subset $U \subset Y$ the function $\frac{\phi}{\psi}$ is the 0 function, and hence $\frac{\phi}{\psi} = 0$ on Y by Corollary 1.4. It follows from Ch.II Lemma 6.7(i) that $\frac{\phi}{\psi} = 0$ in $k(Y)$. This shows that α is also injective. Therefore, α is an isomorphism.
(ii) It is clear that under the isomorphism in (i) the local ring at a point $P \in Y$ defined in $k(Y)$ (CH.II §6) is isomorphic to the local ring of Y at P defined in §1, and the maximal ideal is mapped to the maximal ideal. Now (ii) follows from CH.II Theorem 6.14.
(iii) It is clear that under the isomorphism in (i) we have

$$k[Y] \subset \mathcal{O}(Y) \subset \bigcap_{P\in Y} \mathcal{O}_{P,Y}.$$

On the other hand, by CH.II Proposition 6.10 we have (in $k(Y)$):

$$\bigcap_{P\in Y} \mathcal{O}_{P,Y} = k[Y].$$

It follows from the proof of (i) that $k[Y] \cong \mathcal{O}(Y)$.
This finishes the proof of the theorem. □

Now let $Y \subset \mathbf{P}^n$ be a projective variety, $I = \mathbf{I}(Y)$, and let $k^g[Y]$ be the (homogeneous) coordinate ring of Y. Suppose $Y \cap U_i \neq \emptyset$, where $U_i \subset \mathbf{P}^n$ is the open set $x_i \neq 0$, $1 \leq i \leq n+1$. Then by CH.III Proposition 6.9 and Theorem 6.11 we know that $\overline{\mathbf{V}(I_*)} = Y$, and the dehomogenization map from $k[x_1, ..., x_{n+1}]$ to $k[x_1, ..., x_n]$ induces two isomorphisms of rings α: $k[\mathbf{V}(I_*)] \cong k[Y]/\langle 1 - [x_i]\rangle$, and F_α: $k(Y) \cong k(\mathbf{V}(I_*))$. If we look at the isomorphism F_α and put

$$\begin{aligned} k^g[Y][x_i^{-1}]_{(0)} &= \left\{ \frac{[F]}{[x_i^n]} \;\middle|\; F \text{ is a homogeneous of degree } n \geq 0 \right\} \\ &\subset k(Y), \end{aligned}$$

then it is easy to verify that $k^g[Y][x_i^{-1}]_{(0)}$ is a subring of $k(Y)$ containing k and we obtain the following

2.5. Lemma The dehomogenization map

$$k[x_1, ..., x_{n+1}] \xrightarrow{\alpha_*} k[x_1, ..., x_n]$$

of CH.III §5 induces a ring isomorphism

$$\begin{aligned} k^g[Y][x_i^{-1}]_{(0)} &\longrightarrow k[\mathbf{V}(I_*)] \\ \frac{[F]}{[x_i^n]} &\mapsto [F_*] \end{aligned}$$

□

2.6. Theorem With notation as before, let $Y \subset \mathbf{P}^n$ be a projective variety with coordinate ring $k^g[Y]$ and the function field $k(Y)$. Then the following hold.

(i) $\mathcal{R}(Y) \cong k(Y)$, and if $Y \cap U_i \neq \emptyset$ then $\mathcal{R}(Y) \cong k(\mathbf{V}(I_*))$, where I_* is the dehomogenization of $I = \mathbf{I}(Y)$ with respect to x_i.
(ii) If $P \in Y \cap U_i$, then $\mathcal{O}_{P,Y} \cong \mathcal{O}_{\varphi_i(P),\mathbf{V}(I_*)}$, where $\varphi_i \colon U_i \to \mathbf{A}^n$ is as before, and I_* is the same as in (i).
(iii) $\mathcal{O}(Y) = k$.

Proof (i) By Definition 1.3 the proof of the first isomorphism is similar to the proof of Theorem 2.4(i), and the second isomorphism follows from CH.III Theorem 6.11.
(ii) Under the first isomorphism of (i), the local ring at $P \in Y$ defined in CH.III §3 is isomorphic to the local ring at P defined in §1. Then the desired isomorphism of (ii) follows from CH.III Theorem 6.11.
(iii) Let $f \in \mathcal{O}(Y)$ be a global regular function. Since $Y = \bigcup_{i=1}^{n+1}(Y \cap U_i)$, if $Y \cap U_i \neq \emptyset$, then f is regular on $Y \cap U_i \cong \mathbf{V}(I_*)$ by Proposition 2.2 and Proposition 2.3. It follows from Theorem 2.4(iii) and Lemma2.5 that $f \in k[\mathbf{V}(I_*)] \cong k^g[Y][x_i^{-1}]_{(0)}$. So we conclude that f can be written as $\frac{G_i}{x_i^{n_i}}$ where $G_i \in k^g[Y]$ is homogeneous of degree n_i. Thinking of $\mathcal{O}(Y)$, $\mathcal{R}(Y)$ and $k^g[Y]$ as subrings of the quotient field L of $k^g[Y]$, this means that $x_i^{n_i} f \in k^g[Y]_{n_i}$, the n_i-th homogeneous component of the graded ring $k^g[Y]$, for each i. Now choose $N \geq \sum n_i$. Then $k^g[Y]_N$ is spanned as a k-space by monomials of degree N in $x_1, ..., x_{n+1}$, and in any such monomial , at least one x_i occurs to a power $\geq n_i$. Thus we have $k^g[Y]_N \cdot f \subset k^g[Y]_N$. Iterating, we have $k^g[Y]_N \cdot f^q \subset k^g[Y]_N$ for all $q > 0$. In particular, $x_1^N f^q \in k^g[Y]$ for all $q > 0$. This shows that the subring $k^g[Y][f]$ of L (generated by $k^g[Y]$ and f) is contained in $x_1^{-N} k^g[Y]$, which is a finitely generated $k^g[Y]$-module. Since $k^g[Y]$ is a Noetherian ring (see Appendix I §1), $k^g[Y][f]$ is a finitely generated $k^g[Y]$-module, and therefore f is integral over $k^g[Y]$ (see Apendix I). This means that there are elements $a_1, ..., a_m \in k^g[Y]$ such that

$$f^m + a_1 f^{m-1}, ..., +a_m = 0.$$

Since f has degree 0, we may replace the a_i by their homogeneous components of degree 0, and still have a valid equation. But $k^g[Y]_0 = k$, hence $a_i \in k$, and so f is algebraic over k. Since k is algebraically closed, it follows that $f \in k$, as desired. □

Remark It follows from Proposition 2.3 and Theorem 2.4 that $k[Y]$

is an invariant up to isomorphism of affine varieties. However, for a projective variety Y, $k^g[Y]$ is not an invariant: it depends on the imbedding of Y in projective space. For example, let $X = \mathbf{P}^1$, and let Y be the 2-uple imbedding of $\mathbf{P}^1$ in $\mathbf{P}^2$ given in the previous example. Then $X \cong Y$. But $k^g[X] \not\cong k^g[Y]$.

2.7. Theorem Let X be any variety and let Y be an affine variety. Then there is a natural bijective mapping of sets

$$\alpha: \quad \mathrm{Hom}(X,Y) \xrightarrow{\cong} \mathrm{Hom}(k[Y], \mathcal{O}(X))$$

where the Hom on the left refers morphisms of varieties, and the Hom on the right refers to k-linear ring homomorphisms.

Proof As before, any morphism φ: $X \to Y$ induces a k-linear ring homomorphism G_φ: $\mathcal{O}(Y) \to \mathcal{O}(X)$ via the commutative diagram

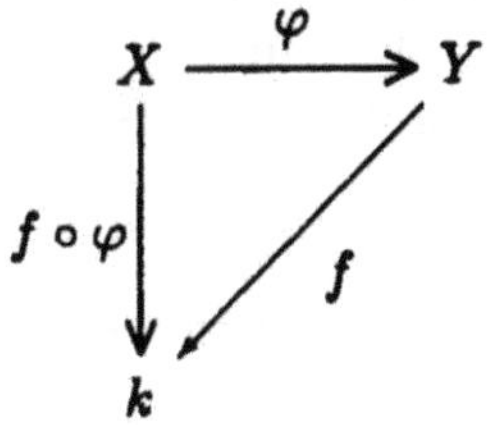

But we have seen in Theorem 2.4 that $\mathcal{O}(Y) \cong k[Y]$, so we obtain a homomorphism $k[Y] \to \mathcal{O}(X)$. This defines α.
Conversely, suppose a k-linear ring homomorphism β: $k[Y] \to \mathcal{O}(X)$ is given. Assume that Y is a closed subset of $\mathbf{A}^n$, so that $k[Y] = k[x_1, ..., x_n]/\mathbf{I}(Y)$. Let $\overline{x_i}$ be the image of x_i in $k[Y]$, and consider the elements $\xi_i = \beta(\overline{x_i}) \in \mathcal{O}(X)$. These are global functions on X, that may be used to define a mapping ψ: $X \to \mathbf{A}^n$ by $\psi(P) = (\xi_1(P), ..., \xi_n(P))$ for $P \in X$.
we show next that the image of ψ is contained in Y. Since $Y = \mathbf{V}(\mathbf{I}(Y))$, it is sufficient to show that for any $P \in X$ and any $f \in \mathbf{I}(Y)$, $f(\psi(P)) = 0$. We have:

$$f(\psi(P)) = f(\xi_1(P), ..., \xi_n(P)).$$

Now f is a polynomial, and β is a k-linear ring homomorphism, so we have

$$f(\xi_1(P), ..., \xi_n(P)) = \beta(f(\overline{x_1}, ..., \overline{x_n}))(P) = 0$$

since $f \in \mathbf{I}(Y)$. So ψ defines a map from X to Y, which iduces the given homomorphism β.
To complete the proof, we must show that ψ is a morphism. This is a consequence of the following lemma. □

2.8. Lemma Let X be any variety, and let $Y \subset \mathbf{A}^n$ be an affine variety. A map of sets ψ: $X \to Y$ is a morphism if and only if $x_i \circ \psi$ is a regular function on X for each i, where $x_1, ..., x_n$ are the coordinate functions on $\mathbf{A}^n$.

Proof If ψ is a morphism, the $x_i \circ \psi$ must be regular functions by definition of a morphism. Conversely, suppose the $x_i \circ \psi$ are regular. Then for any polynomial $f = f(x_1, ..., x_n)$, $f \circ \psi$ is also regular on X. Since $Y = \mathbf{V}(\mathbf{I}(Y))$, and since regular functions are continuous, we see that ψ^{-1} takes closed sets to alosed sets, so ψ is continuous. Finally, since regular functions on open subsets of Y are locally quotients of polynomials, $g \circ \psi$ is regular for any regular function g on any open subset of Y. Hence ψ is a morphism. □

Exercises for §2

1. Show that any regular function on a quasi-projective variety is continuous.
2. Show that any conic in $\mathbf{A}_k^2$ is isomorphic either to $\mathbf{A}_k^1$ or $\mathbf{A}_k^1 - \{0\}$ (cf. CH.II §4 Exercises 2–4).
3. Show that $\mathbf{A}_k^1$ is not isomorphic to any proper open subset of itself.
4. If an affine variety is isomorphic to a projective variety, then it consists of only one point.
5. Show that $(x_1, ..., x_n) \mapsto (x_1, ..., x_n, 0)$ defines an isomorphism of $\mathbf{P}^{n-1}$ with $H_\infty^{n+1} \subset \mathbf{P}^n$ (see CH.III §1). If a variety V in $\mathbf{P}^n$ is contained in H_∞^{n+1}, V is isomorphic to a variety in $\mathbf{P}^{n-1}$. Any projective variety is isomorphic to a closed subvariety $V \subset \mathbf{P}^m$ (for some m) such that V is not contained in any hyperplane in $\mathbf{P}^m$.
6. A morphism whose underlying map on the topological space is a homeomorphism need not be an isomorphism.
   a. (Compare with CH.II §4 Exercise 2) For example, let φ: $\mathbf{A}_k^1 \to \mathbf{A}_k^2$ be defined by $t \mapsto (t^2, t^3)$. Show that φ defines a bijective bicontinuous morphism of $\mathbf{A}_k^1$ onto the curve

$y^2 = x^3$, but that φ is not an isomorphism.

b. For another example, let the characteristic of the base field k be $p > 0$, and define a map φ: $\mathbf{A}^1_k \to \mathbf{A}^1_k$ by $t \mapsto t^p$. Show that φ is bijective and bicontinuous but not an isomorphism. This is called the *Frobenius morphism.*

7. Prove Proposition 2.3. (First prove that G_φ, L_φ, and R_φ are well-defined.)

§3. Rational Maps

In the last two sections, we have associated to each variety X its function field $\mathcal{R}(X)$ which contains "global" and "local" information of X in the sense that $\mathcal{O}(X) \subset \mathcal{R}(X)$ and $\mathcal{O}_{P,X} \subset \mathcal{R}(X)$ for every $P \in X$. If X and Y are isomorphic varieties, then $\mathcal{R}(X) \cong \mathcal{R}(Y)$ (Proposition 2.3). Now we come back to the question posed in the end of CH.II §6:

- If X and Y are varieties such that $\mathcal{R}(Y) \xrightarrow{\cong} \mathcal{R}(X)$, is there an isomorphism $X \to Y$?

Note that the field of rational functions $\mathcal{R}(X)$ is a "local invariant", namely, if U is any nonempty open subset of X then $\mathcal{R}(X) = \mathcal{R}(U)$ (Lemma 1.7), and if Y is any variety containing a nonempty open subset V which is isomorphic to U as varieties, then $\mathcal{R}(X) = \mathcal{R}(U) \cong \mathcal{R}(V) = \mathcal{R}(Y)$ (Proposition 2.3). So it is more natural to ask

- If X and Y are varieties such that $\mathcal{R}(X) \cong \mathcal{R}(Y)$, can we find nonempty open subsets U and V in X and Y, respectively, such that $U \cong V$? (If this can be done, we do obtain most if not all information about X from Y and vice versa because open subsets are dense.) More generally, if we replace $\operatorname{Hom}(k[Y], \mathcal{O}(X))$ in Theorem 2.7 by $\operatorname{Hom}(\mathcal{R}(Y), \mathcal{R}(X))$, the set of k-linear field homomorphisms, is there an analogue of Theorem 2.7?

The answer to the above question leads to the notion of birational equivalence of varieties. That is the topic of this section.

Let X and Y be varieties and U a nonempty open subset of X. As we have observed in §2, if φ: $U \to Y$ is a morphism such that $\varphi(U)$ is dense in Y, then φ induces a k-linear homomorphism of fields R_φ: $\mathcal{R}(Y) \to \mathcal{R}(U) = \mathcal{R}(X)$ with $R_\varphi(f) = \varphi \circ f$ for $f \in \mathcal{R}(Y)$, i.e., we have the following correspondence:

$$\left\{\begin{array}{l}\text{pair } (U,\varphi) \text{ with } U \text{ open}\\ \text{and } \varphi: U \to Y \text{ a morphism,}\\ \varphi(U) \text{ is dense in } Y\end{array}\right\} \longrightarrow \left\{\begin{array}{l}k\text{-linear}\\ \text{field homomorphisms:}\\ \mathcal{R}(Y) \to \mathcal{R}(U) = \mathcal{R}(X)\end{array}\right\}$$

To see that this correspondence is well-defined, we have to understand the following

- in what sense are two pairs (U,φ) and (V,ψ) identifiable?

This leads to the following generalization of Corollary 1.4. (and definitions).

3.1. Lemma Let X and Y be varieties, let φ and ψ be two morphisms from X to Y. Suppose that there is a nonempty open subset $U \subset X$ such that $\varphi|_U = \psi|_U$, then $\varphi = \psi$.

Proof We may assume that $Y \subset \mathbf{P}^n_k$ for some n. Then by composing with the inclusion morphism $Y \to \mathbf{P}^n$, we reduce the problem to the case $Y = \mathbf{P}^n$. We consider the product $\mathbf{P}^n \times \mathbf{P}^n$, which has a structure of projective variety given by its Segre embedding (CH.III §7). The morphism φ and ψ determine a map $\varphi \times \psi$: $X \to \mathbf{P}^n \times \mathbf{P}^n$, which in fact is a morphism. Let $\Delta = \{(P,P) \mid P \in \mathbf{P}^n\}$ be the *diagonal* subset of $\mathbf{P}^n \times \mathbf{P}^n$. It is defined by the equations $\{x_iy_j = x_jy_i \mid i,j = 1,...,n+1\}$ and so is a closed subset of $\mathbf{P}^n \times \mathbf{P}^n$ (see Appendix II). By the hypothesis $(\varphi \times \psi)(U) \subseteq \Delta$. But U is dense in X, and Δ is closed, so $(\varphi \times \psi)(X) \subseteq \Delta$. This states that $\varphi = \psi$. □

3.2. Definition Let X, Y be varieties. A *rational map* φ: $X \cdots \to Y$ is an equivalence class of pairs (U,φ_U) where U is a nonempty open subset of X, φ_U is a morphism of U to Y, and where (U,φ_U) and (V,φ_V) are equivalent if φ_U and φ_V agree on $U \cap V$. The rational map φ is *dominant* if for some (and hence every) pair (U,φ_U), the image of φ_U is dense in Y.

Note that the lemma implies that the relation on pairs (U,φ_U) as described above is an equivalence relation. Note also that a rational map $\varphi\colon X \cdots \to Y$ is in general *not* a map of the set X to Y. Clearly the composition of two dominant rational maps is , a dominant rational map. In this way we arrive at:

3.3. Definition A *birational map* is a rational map $\varphi\colon X \cdots \to Y$ which admits an inverse, namely a rational map $\psi\colon Y \cdots \to X$ such that $\psi \circ \varphi = 1_X$ and $\varphi \circ \psi = 1_Y$ as rational maps. If such a birational map exists for X and Y, we say that X and Y are *birationally equivalent.*

The first result concerning rational maps is

3.4. Proposition Let $\varphi\colon X \cdots \to Y$ be a rational map.
(i) If φ is dominant, then φ induces the k-linear field homomorphism $R_\varphi\colon \mathcal{R}(Y) \to \mathcal{R}(X)$ as defined before.
(ii) If φ is birational, then there exists nonempty open subsets $U_0 \subset X$ and $V_0 \subset Y$ such that $\varphi\colon U_0 \xrightarrow{\cong} V_0$. Consequently, $\mathcal{R}(Y) \cong \mathcal{R}(X)$.

Proof (i) This follows from the foregoing remark.
(ii) Let φ be represented by (U, φ_U) and its inverse, say ψ, be represented by (V, ψ_V). Then we have

$$\begin{array}{ccc} \psi_V^{-1}(U) & \xrightarrow{\ \psi_V\ } & U \\ \cap\big\downarrow & & \big\downarrow \varphi_U \\ Y & \xrightarrow[\ 1_Y\ ]{} & Y \end{array} \qquad \varphi_U \circ \psi_V = 1_Y\Big|_{\psi_V^{-1}(U)}$$

Put

$$U_0 = \varphi_U^{-1}\left(\psi_V^{-1}(U)\right), \quad V_0 = \psi_V^{-1}\left(\varphi_U^{-1}(V)\right),$$

$$\varphi_{U_0} = \varphi_U\Big|_{U_0} : \ U_0 \longrightarrow \psi_V^{-1}(U).$$

If $P \in \psi^{-1}(U)$, then $\varphi_U(\psi_V(P)) = P$. It follows that $\psi_V^{-1}(U) \subset \psi_V^{-1}(\varphi_U^{-1}(V)) = V_0$. This shows that $\varphi_{U_0}\colon U_0 \to V_0$ is a morphism.

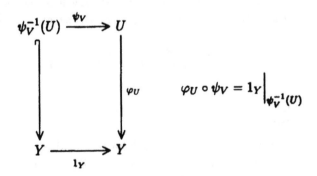

Similarly, ψ_{V_0}: $V_0 \to U_0$ is a morphism. Since φ_{U_0} is clearly the inverse morphism of ψ_{V_0}, we obtain $U_0 \cong V_0$, as desired. □

Secondly, let us consider the opposite problem, i.e., given a k-linear ring homomorphism θ: $\mathcal{R}(Y) \to \mathcal{R}(X)$, we want to find a dominant rational map φ: $X \cdots \to Y$ inducing the given ring homomorphism. We need some lemmas to the effect that on any variety, the open affine subsets form a base for the Zariski topology. At this stage, working with varieties and morphisms of varieties, we say that a variety is *affine* if it is isomorphic to an affine variety. For instance, if U is an open subset of a variety X which is isomorphic to an affine variety, then we say that U is an *open affine subvariety* of X.

3.5. Lemma Let $V \subset \mathbf{A}^n$ be an affine variety, and let $f \in k[\mathbf{x}] = k[x_1, ..., x_n]$, $f \neq 0$. Let

$$V_f = \left\{P \in V \,\middle|\, f(P) \neq 0\right\} = V - \mathbf{V}(f),$$

an open subset of V. Then
(i) V_f is an affine variety with coordinate ring $k[V_f]$ which is isomorphic to a subring of $k(V)$ (which is indeed the localization of $k[V]$ at the multiplicative subset $\{1, f, f^2, ...\}$ in the sense of APP.II §1):

$$k[V][f^{-1}] = \left\{\frac{g}{f^m} \,\middle|\, g \in k[V], m \geq 0\right\}.$$

(ii) $V = \bigcup_{f \in k[\mathbf{x}]} V_f$.

Proof Consider $I = \mathbf{I}(V)$, and let $\widetilde{I}$ be the ideal in $k[x_1, ..., x_{n+1}]$ generated by I and by $x_{n+1}f - 1$. Put $V' = \mathbf{V}(\widetilde{I}) \subset \mathbf{A}^{n+1}$. We first look at the ring homomorphisms

$$\begin{array}{ccc} k[x_1, ..., x_{n+1}] & \xrightarrow{\pi} & k[x_1, ..., x_{n+1}]/\widetilde{I} \\ \downarrow \varphi & \nearrow \psi & \\ k[V][f^{-1}] & & \end{array}$$

where π is the natural ring homomorphism and

$$\varphi\left(\sum x_1^{\alpha_1}\cdots x_n^{\alpha_n}x_{n+1}^{\alpha_{n+1}}\right)=\frac{\sum x_1^{\alpha_1}\cdots x_n^{\alpha_n}}{f^{\alpha_{n+1}}}$$

$$\psi\left(\frac{g}{f^m}\right)=\frac{g}{f^m},\quad \text{(note that } f \text{ is invertible in } k[x_1,..,x_{n+1}]/\widetilde{I}.)$$

(Check that φ and ψ are well-defined ring homomorphisms.) Since $\mathrm{Ker}\pi=\widetilde{I}\subset\ker\varphi$, it follows from the basic homomorphism theorem that there exists a ring homomorphism $\overline{\psi}$: $k[x_1,...,x_{n+1}]/\widetilde{I}\to k[V][f^{-1}]$, and one may check that $\overline{\psi}$ is indeed the inverse of φ. Hence $k[V][f^{-1}]\cong k[x_1,...,x_{n+1}]/\widetilde{I}$, in particular, $\widetilde{I}$ is a prime ideal, so V' is a variety (note that k is algebraically closed).
On the other hand, consider the projection π: $(x_1,...,x_n,x_{n+1})\mapsto(x_1,...,x_n)$ from $\mathbf{A}^{n+1}$ to $\mathbf{A}^n$. If $P\in V'=\mathbf{V}(I,x_{n+1}f-1)$, then $\pi(P)\in V$ and $f(\pi(P))\neq 0$. This means that $\pi(V')\subset V_f$. Thus, π induces a polynomial map α: $V'\to V_f$ which is easily seen to be bijective. To show that α is an isomorphism, it is sufficient to show that α^{-1} is a morphism. But for $P\in V_f$, if $P=(a_1,...,a_n)$, then $\alpha^{-1}(P)=(a_1,...,a_n,\frac{1}{f(P)})$, so the fact that α^{-1} is a morphism on V_f follows from Lemma 2.8. This proves (i).
To prove (ii), take $P\in V$. If $P=(a_1,...,a_n)$ with $a_i\neq 0$, then $P\in V_{x_i}$; If $P=(0,0,...,0)$ is the origin, then $P\in V_f$ where f is any polynomial in $k[x_1,...,x_n]$ with nonzero constant term. It follows from (i) that V is covered by the open affine varieties of the form V_f. □

3.6. Proposition For any variety Y, the open affine subvarieties of Y form a basis of the Zariski topology on Y.

Proof We must show for any point $P\in Y$ and any open set U containing P, that there exists an open affine set V with $P\in V\subset U$. First, since U is also a variety, we may assume $U=Y$. Secondly, since any variety is covered by quasi-affine varieties (CH.III §6), we may assume that Y is quasi-affine in $\mathbf{A}^n$, namely, Y is an open subset of an affine variety, say V. Then $Y=V-W$ where W is a closed subset of V. But $W=\mathbf{V}(I)$ for some ideal I of $k[x_1,...,x_n]$, it follows that

$$\begin{aligned} Y &= V - W \\ &= V - \bigcap_{f \in I} \mathbf{V}(f) \\ &= \bigcup_{f \in I} (V - \mathbf{V}(f)) \\ &= \bigcup_{f \in I} V_f \end{aligned}$$

where every V_f is affine by Lemma 3.5. □

The former result is fundamental for the local theory in algebraic geometry. It reduces many problems on a variety to problems on an affine variety.

3.7. Lemma With notation as in §2, let φ: $X \to Y$ be a morphism of varieties.
(i) If $\varphi(X)$ is dense in Y, then the induced ring homomorphism G_φ: $\mathcal{O}(Y) \to \mathcal{O}(X)$ is injective.
(ii) If X and Y are affine, then $\varphi(X)$ is dense in Y if and only if the induced ring homomorphism G_φ: $\mathcal{O}(Y) \to \mathcal{O}(X)$ is injective.

proof (i) If $\varphi(X)$ is dense in Y, then for any $f \in \mathcal{O}(Y)$ such that $f \circ \varphi = 0$, we have $f^{-1}(0) \supset \varphi(X)$, and hence $f^{-1}(0) = Y$, i.e., $f = 0$. This shows that G_φ is injective.
(ii) Now suppose that X and Y are affine and G_φ is injective. We want to show that $\varphi(X) \cap V \neq \emptyset$ for any nonempty open subset $V \subset Y$. By Lemma 3.4 we may assume that $V = V_f$ where f is a regular function on Y with $f(P) \neq 0$ for all $P \in V$. If $\varphi(X) \cap V = \emptyset$, then $G_\varphi(f) = f \circ \varphi = 0$, a contradiction. This shows that $\varphi(X)$ must be dense in Y. □

We are now ready to prove the main result of this section.

3.8. Theorem For any two varieties X and Y, the foregoing construction defines a bijection between
(i) the set of dominant rational maps from X to Y, and
(ii) the set of k-linear ring homomorphisms from $\mathcal{R}(Y)$ to $\mathcal{R}(X)$.

Furthermore, this correspondence defines a bijection

$$\left\{\begin{array}{c}\text{varieties}\\ \text{and}\\ \text{dominant rational maps } (U,\varphi_U)\end{array}\right\}$$

$$\updownarrow$$

$$\left\{\begin{array}{c}\text{finitely generated field extensions}\\ \text{of } k \text{ and}\\ k\text{-lnear field homomorphisms}\end{array}\right\}$$

where R_{φ_U} is the induced field isomorphism: $\mathcal{R}(Y) \to \mathcal{R}(X)$ with $R_{\varphi_U}(f) = f \circ \varphi_U$ for $f \in \mathcal{R}(Y)$.

Proof In view of Proposition 3.4 it suffices to construct an inverse to the mapping (i) $\to$ (ii). Let θ: $\mathcal{R}(Y) \to \mathcal{R}(X)$ be a k-linear ring homomorphism. We wish to define a dominant rational map from X to Y. By Proposition 3.6, Y is covered by affine varieties, so we may assume Y is affine. Let $k[Y]$ be its affine coordinate ring, and let $y_1, ..., y_n$ be generators for $k[Y]$ as a finite ring extension of k (Appendix I). Then $\theta(y_1), ..., \theta(y_n)$ are rational functions on X. We may find an open set $U \subset X$ such that the functions $\theta(y_i)$ are all regular on U. Then θ defines an injective k-linear homomorphism of rings $k[Y] \to \mathcal{O}(U)$. By Theorem 2.7 this corresponds to a morphism φ: $U \to Y$, yielding a dominant rational map from X to Y (Lemma 3.7). It is easy to see that this defines a set map (ii) $\to$ (i) which is inverse to the one defined by Proposition 3.4.

To see that we have a bijection as stated, we need only check that for any variety Y, $\mathcal{R}(Y)$ is finitely generated over k, and conversely, if K/k is a finitely generated field extension, then $K = \mathcal{R}(Y)$ for some Y. If Y is a variety, then $\mathcal{R}(Y) = \mathcal{R}(U)$ for any open affine subset (Lemma 1.7), so we may assume Y affine. Then by Theorem 2.4, $\mathcal{R}(Y) = k(Y)$ is a finitely generated field extension of k. On the other hand, let K be a finitely generated field extension of k. Let $y_1, ..., y_n \in K$ be a set of generators, i.e., $K = k(y_1, ..., y_n)$, and let $B = k[y_1, ..., y_n]$ be the subring of K generated by $\{y_1, ..., y_n\}$ over k. Then B is a quotient of the polynomial ring $k[x_1, ..., x_n]$, so $B \cong k[Y]$ for some variety $Y \subset \mathbf{A}^n$. Then $K \cong k(Y)$ and so we are done. □

3.9. Corollary For any two varieties X, Y the following conditions are equivalent:
(i) X and Y are birationally equivalent;
(ii) There are open subsets $U \subset X$ and $V \subset Y$ with U isomorphic to V;
(iii) $\mathcal{R}(Y) \cong \mathcal{R}(X)$ as fields.

Proof (i) $\Rightarrow$ (ii) $\Rightarrow$ (iii) follows from Proposition 3.4. (iii) $\Rightarrow$ (ii) follows from Theorem 3.7. (ii) $\Rightarrow$ (i) is clear. □

3.10. Corollary Let Y be a variety. Then the following statements are equivalent:
(i) $\mathcal{R}(Y)$ is a purely transcendental extension of k, i.e., $\mathcal{R}(Y) \cong k(t_1, ..., t_n)$ for some n, where the latter is the rational function field in n variables;
(ii) There exists open subset $U \subset Y$ such that U is isomorphic to an open set of $\mathbf{A}^n$.

□

From Proposition 3.6 we retain that a variety X has a basis for the Zariski topology consisting of "standard" open affine varieties V_f. But we also know that $\mathcal{R}(X) = \mathcal{R}(V_f) \cong k(V_f)$. It follows from Corollary 3.9 that:

- Every variety X is birational to an affine variety.
- Each point $P \in X$ is contained in some V_f, and all V_f are "glued" together by birational maps. (See CH.III §6 Example (iii) for an illustration of this principle.)

Furthermore, as an illustration of the notion of birational correspondence, we will show that every variety is birational to a hypersurface.

3.11. Theorem Any variety X is birational to a hypersurface Y in $\mathbf{P}^d$ for some $d \geq 0$.

Proof Proposition 3.6 entails that we only need to consider the affine case. Let X be an affine variety in $\mathbf{A}^n$, and $k(X)$ the function field of X. Suppose $\dim X = d$. By Ch.V Theorem 6.4 we know that $d = \text{tr.deg}_k k(X)$, i.e., there are $a_1, ..., a_d \in k(X)$ which are algebraically independent over k, and $k(X)$ is a finite algebraic extension of $k(a_1, ..., a_d)$. Since k is algebraically closed, it is a perfect field. Hence $k(X)$ is a finite separable field extension of $k(a_1, ..., a_d)$. Then by the theorem of the primitive element (see Appendix I Theorem 4.4) we may select an element $y \in K$ such that $K = k(a_1, ..., a_d, y)$. Now y is algebraic over $k(a_1, ..., a_d)$, so it satisfies a polynomial equation with coefficients which are rational functions in $a_1, ..., a_d$. Clearing denominators, we obtain an irreducible polynomial $f(t_1, ..., t_d, t)$ with coefficients in k such that $f(a_1, ..., a_d, y) = 0$. This defines a hypersurface in $\mathbf{A}^{d+1}$ with function field isomorphic to K, which, according to Corollary 3.9, is birational to X. Its projective closure is the required hypersurface $Y = \mathbf{V}(f^*) \subset \mathbf{P}^{d+1}$ (see CH.III).

□

Exercises for §3

1. Let (U, φ_U) and (V, φ_V) represent the same rational map φ: $X \cdots \to Y$. Show that $\varphi_V(V)$ is dense in Y if $\varphi_U(U)$ is dense in Y.
2. Show that any variety X is the union of a finite number of open affine subvarieties. (Hint: Use §1 Exercise 1.)
3. Use Lemma 3.4 and CH.I §4 Exercise 2 to show that if P, Q are points in a variety X, then there exists an affine open set $V \subset X$ which contains P and Q.
4. Is Lemma 3.6(ii) still true if Y is not affine?

§4. Examples of Rational Varieties

Varieties satisfying the equivalent conditions of Corollary 3.10 are called *rational varieties*. CH.II §1 entails that the existence of a rational parametrization, for an affine algebraic set, forces the set to be irreducible. On the he opposite side, the condition (ii) of Corollary 3.10 is equivalent to say that the variety Y *can be parametrized by* n *independent variables*. As we have pointed out in CH.I §3, most elementary applications of algebraic geometry to other mathematical branches and scientific subjects are related to rational varieties. With an eye to applications it is useful to see some more examples of rational varierties.

For convenience, in this section we let $k = \mathbb{C}$, $\mathbf{A}^n = \mathbf{A}^n_k$, $\mathbf{P}^n = \mathbf{P}^n_k$.

First, all examples we considered in CH.I §3 are examples of rational varieties.

Another example of rational variety is the curve $\mathbf{V}(x^2-y^3, y^2-z^3) \subset \mathbf{A}^3$. This may be seen as follows. The curve V may be parametrized by the mapping

$$\begin{array}{cccc} \phi: & \mathbf{A}^1 & \longrightarrow & V \\ & t & \mapsto & (t^9, t^6, t^4) \end{array}$$

Note that ϕ is not a polynomial isomorphism (CH.II §4 Exercise 6). But ϕ induces a birational map $\mathbf{A}^1_{\mathbb{R}} - \{0\} \to V - \{(0,0,0)\}$ where ϕ^{-1} is defined as $(x, y, z) \mapsto \dfrac{x}{z^2}$.

From §2 we derive that every nondegenerate quadratic curve C in $\mathbf{P}^2$ is a rational variety. We now generalize the rational map ϕ: $\mathbf{P}^1 \to C \subset \mathbf{P}^2$ to the higher dimensional case.

For given $n, d > 0$, let $t_1, t_2, ..., t_{N+1}$ be all monomials of degree d in the $n+1$ variables $x_1, ..., x_{n+1}$, where $N = \left(\begin{array}{c} n+d \\ n \end{array}\right) - 1$ (see CH.I §1), i.e., $t_i = x_1^{\alpha_1} \cdots x_{n+1}^{\alpha_{n+1}}$, $(\alpha_1, ..., \alpha_{n+1}) \in \mathbb{Z}^{n+1}_{\geq 0}$, $\alpha_1 + \cdots + \alpha_{n+1} = d$. Consider the mapping (note that all t_i are homogeneous)

$$\begin{array}{cccc} \phi_d: & \mathbf{P}^n & \longrightarrow & \mathbf{P}^N \\ & P & \mapsto & (t_1(P), ..., t_{N+1}(P)) \end{array}$$

This is called a *d-uple embedding* of $\mathbf{P}^n$ in $\mathbf{P}^N$. If we denote the homogeneous coordinates in $\mathbf{P}^N$ by

$$Z_{(\alpha_1,...,\alpha_{n+1})}, \quad \alpha_1 + \cdots + \alpha_{n+1} = d,$$

then each $\phi_d(P)$ with $P \in \mathbf{P}^n$ satisfies the equations

$$(*) \quad \left\{ \begin{array}{l} Z_{(\alpha_1,...,\alpha_{n+1})} Z_{(\beta_1,...,\beta_{n+1})} = Z_{(\gamma_1,...,\gamma_{n+1})} Z_{(\eta_1,...,\eta_{n+1})} \\ \text{where } \alpha_1 + \beta_1 = \gamma_1 + \eta_1, ..., \alpha_{n+1} + \beta_{n+1} = \gamma_{n+1} + \eta_{n+1}, \end{array} \right.$$

We write V_d for the algebraic set defined by (*) in $\mathbf{P}^N$, then $\mathrm{Im}\phi_d \subset V_d$. It is easy to see that ϕ_d is not surjective. However, we do have the following

4.1. Proposition ϕ_d: $\mathbf{P}^n \to V_d$ is an isomorphism of varieties. Hence V_d is a rational variety.

Proof That ϕ_d is a morphism is clear. We construct an inverse of ϕ_d as follows. For any $Q \in V_d$, if some homogeneous coordinate $Z_{(\alpha_1,\dots,\alpha_{n+1})} \neq 0$, we may easily derive from (*) that at least one of the $Z_{(0,\dots,0,d,0,\dots,0)}$ is different from 0. Without loss of generality we may put $Z_{(d,0,\dots,0)} \neq 0$. Now define:

$$(**)\qquad \begin{cases} Z_{(d,0,\dots,0)} & = & x_1, \\ Z_{(d-1,1,0,\dots,0)} & = & x_2, \\ Z_{(d-1,0,1,0,\dots,0)} & = & x_3, \\ \vdots & & \\ Z_{(d-1,0,\dots,0,1)} & = & x_{n+1}. \end{cases}$$

If $Z_{(\beta_1,\dots,\beta_{n+1})}$ is a coordinate of Q such that $\beta_j = 0$ for $j < i \leq n+1$, i.e., β_i is the first nonzero component in $(\beta_1,\dots,\beta_{n+1})$, then $Z_{(\beta_1,\dots,\beta_{n+1})} = Z_{(0,\dots,0,\beta_i,\beta_{i+1},\dots,\beta_{n+1})}$. Let $\beta_i = d - s$ where $0 \leq s < d$. From the equation

$$Z_{(0,\dots,0,d-s,\beta_{i+1},\dots,\beta_{n+1})} Z_{(d,0,\dots,0)} = \\ Z_{(1,\dots,0,d-s-1,\beta_{i+1},\dots,\beta_{n+1})} Z_{(d-1,0,\dots,0,1,0,\dots,0)}$$

it is clear that we may assume that $\beta_1 \neq 0$ and $\beta_1 \neq d, d-1$. Thus, there is some $\beta_i \neq 0$ and from the equation

$$Z_{(\beta_1,\dots,\beta_{n+1})} Z_{(d,0,\dots,0)} = Z_{(\beta_1+1,0,\dots,0,\beta_i-1,\dots,\beta_{n+1})} Z_{(d-1,0,\dots,0,1,0,\dots,0)}$$

we may inductively derive that (**) indeed defines the inverse of ϕ_d. It is not difficult to see that this inverse mapping is a morphism. It follows that ϕ_d is an isomorphism. □

The significance of a d-uple embedding lies in the fact that if $F = \sum a_{(\alpha_1,\dots,\alpha_{n+1})} x_1^{\alpha_1} \cdots x_{n+1}^{\alpha_{n+1}}$ is a homogeneous polynomial of degree d in $k[x_1,\dots,x_{n+1}]$ and if H is the hypersurface $\mathbf{V}(F) \subset \mathbf{P}^n$, then $\phi_d(H)$ is the intersection of $\phi_d(\mathbf{P}^n)$ and the hyperplane with the equation $\sum a_{(\alpha_1,\dots,\alpha_{n+1})} Z_{(\alpha_1,\dots,\alpha_{n+1})} = 0$ in $\mathbf{P}^N$. Therefore, a d-uple embedding makes it possible to reduce the study of certain problems connected with hypersurfaces to the case of hyperplanes.

Another important class of birational maps is given by the Segre embedding constructed in CH.III §7.

4.2. Proposition Let S: $\mathbf{P}^m \times \mathbf{P}^n \to V_{m,n} \subset \mathbf{P}^N$ be the Segre embedding of $\mathbf{P}^m \times \mathbf{P}^n$, where $N = (m+1)(n+1) - 1$. Then the Segre variety is birationally equivalent to $\mathbf{P}^{m+n+1}$, hence a rational variety.

Proof Let U_m denote the open subset $\mathbf{P}^{m+n+1} - \mathbf{V}(x_1, ..., x_{m+1})$ and U_n denote the open subset $\mathbf{P}^{m+n+1} - \mathbf{V}(y_1, ..., y_{n+1})$. Putting $U = U_m \cap U_n$, the Segre embedding S induces an isomorphism $U \to V_{m,n}$. □

The simplest Segre variety is $V_{1,1} = \mathbf{V}(Z_{00}Z_{11} - Z_{01}Z_{10}) \subset \mathbf{P}^3$. Since we have been working over $k = \mathbb{C}$, any nondegenerate (or nonsingular, see next section for the definition) quadric is projectively equivalent to the quadric

$$\sum_{i=0}^{3} x_i^2 = 0.$$

(We refer to CH.III §4 Example (ii) and the Examples given in §6 later for more details about the quadrics.) We now set

$$\begin{cases} Z_{00} = x_0 + ix_1 \\ Z_{11} = x_0 - ix_1 \\ Z_{01} = -(x_2 + ix_3) \\ Z_{10} = x_2 - ix_3 \end{cases}$$

which amounts to a projective change of coordinates in $\mathbf{P}^3$. Then the given quadric becomes $Z_{00}Z_{11} = Z_{01}Z_{10}$. Thus, we have obtained the following

4.3. Proposition Every nondegenerate quadric in $\mathbf{P}^3_{\mathbb{C}}$ is projectively (or birationally) equivalent to the Segre variety $V_{1,1}$, hence it is a rational variety.

□

Note that the Segre variety $V_{1,1}$ is given by the single equation $Z_{00}Z_{11} = Z_{01}Z_{10}$, and the Veronese variety W_1 (see CH.III §7) is determined by this equation and the extra equation $Z_{01} = Z_{10}$. We now set

$$\begin{cases} x = Z_{00} \\ y = Z_{11} \\ u = Z_{01} - Z_{10} \\ v = Z_{10} \end{cases}$$

which amounts to a projective change of coordinates of $\mathbf{P}^3$. The equations of W_1 now take the form

$$\begin{cases} xy = v^2, \\ u = 0. \end{cases}$$

Thus W_1 appears as a nondegenerate conic in the plane $u = 0$. All nondegenerate conics in this plane are projectively equivelant since we are working over $\mathbb{C}$. Also, any projective change of coordinates of this plane can be extended to all of $\mathbf{P}^3$ by fixing u. Thus any non-degenerate conic in any plane of $\mathbf{P}^3$ is projectively (or birationally) equivalent to the Veronese variety W_1.

Exercises for §4

1. Show that $\mathbf{V}(y^3 - x^4 - x^3) \subset \mathbf{A}^2_{\mathbb{R}}$ is a rational variety (see CH.I §3 Exercise 1).
2. Show that the d-uple embedding ϕ_d is well-defined and ϕ_d^{-1} is a morphism.
3. Note that the d-uple embedding ϕ_d induces a ring homomorphism:

$$\begin{array}{rccc} \phi_{d*}: & k[y_1, ..., y_{N+1}] & \longrightarrow & k[x_1, ..., x_{n+1}] \\ & F & \mapsto & F(t_1, ..., t_{N+1}), \end{array}$$

   which sends each y_i to t_i, show that
   a. Kerϕ_{d*}, denoted I, is a graded prime ideal of $k[y_1, ..., y_{N+1}]$;
   b. The image Im$\phi_d = \mathbf{V}(I)$;
4. Show that the Segre embedding S: $\mathbf{P}^m \times \mathbf{P}^m \to V_{m,m}$ induces an isomorphism from $\mathbf{P}^m$ to the Veronese variety W_m (see Appendix II).

§5. An Algorithm Checking the Birationality of Rational Maps

The examples of rational varieties mentioned in the last section prompt the following question:

- Let φ: $X \cdots \to Y$ be a rational map represented by (U, φ_U) where U is an open subset of X and φ_U: $U \to Y$ is given by rational functions:

$$\varphi_U(P) = \left(\frac{f_1}{g_1}(P), \cdots, \frac{f_m}{g_m}(P)\right).$$

  Is there an algorithm to determine whether φ_U is birational or not?

This section is devoted to providing a definite answer to the above question which is inspired by the algorithmic criterion for polynomial isomorphisms (see CH.IV §5). We refer the reader to (Li Zhong-Tang, A Groebner basis criterion for birational equivalence of affine varieties, *Journal of Pure and Applied Algebra*, 123(1998)) for a detailed proof of this criterion.

First, We reduce the question to the following situation. Let A^m and $\mathbf{A}^n$ be affine spaces, $\frac{f_1}{g_1}, \ldots, \frac{f_n}{g_n} \in k(x_1, \ldots, x_n)$, and $g = \prod_{i=1}^n g_i$. Then $\left(\frac{f_1}{g_1}, \ldots, \frac{f_n}{g_n}\right)$ defines a rational map φ: $\mathbf{A}^m \cdots \to \mathbf{A}^n$ which is represented by (U, φ_U) with $U = V_g = \{P \in \mathbf{A}^m \mid g(P) \neq 0\}$. Let $V = \mathbf{V}(h_1, \ldots, h_s)$ be a nonempty variety in $\mathbf{A}^m$. Suppose $V \cap V_g \neq \emptyset$, and let U_g denote this open subset of V. Then φ restricts to a rational map

$$\begin{array}{rcccl} \varphi|_V : & V & \cdots \to & \overline{\varphi(U_g)} \subset \mathbf{A}^n \\ & P & \mapsto & \left(\frac{f_1}{g_1}(P), \cdots \frac{f_n}{g_n}(P)\right) \end{array}$$

where $\overline{\varphi(U_g)}$ denotes the Zariski closure of $\varphi(U_g)$ in $\mathbf{A}^n$.
In order to obtain an algorithmic criterion for the birationality of φ, we introduce the lexicographic order in $k[x, x_1, \ldots, x_m, y_1, \ldots, y_n]$ by fixing

$$x > x_1 > \cdots > x_m > y_1 > \cdots > y_n,$$

and let

$$\begin{aligned} J &= \langle h_1, ..., h_s, g_1 y_1 - f_1, ..., g_n y_n - f_n, 1 - gx \rangle \\ &\subset k[x, x_1, ..., x_m, y_1, ..., y_n]. \end{aligned}$$

The following two preliminary results are obtained.

a. Let $\tilde{J} = J \cap k[y_1, ..., y_n]$. Then $\mathbf{V}(\tilde{J}) = \overline{\varphi(U_g)}$.

b. Let G be a Groebner basis (resp. the reduced Groebner basis) of the ideal J with respect to $>$. Then $G \cap k[y_1, ..., y_n]$ is a Groebner basis (resp. the reduced Groebner basis) of the ideal $\mathbf{I}(\overline{\varphi(U_g)}) \subset k[y_1, ..., y_n]$ for the order induced by $>$ in $k[y_1, ..., y_n]$. In particular, $\overline{\varphi(U_g)} = \mathbf{A}^n$ if and only if $G \cap k[y_1, ..., y_n] = \emptyset$.

5.1. Proposition With notation as above, let G be the reduced Groebner basis of J with respect to $>$. The following statements are equivalent:

(i) φ: $V \cdots \to \overline{\varphi(U_g)}$ is birational.

(ii) φ has a local left inverse defined by a rational parametrization.

(iii) J contains polynomials of the type $b_i(Y)x_i - a_i(Y)$, $i = 1, ..., m$, where $b_i(Y), a_i(Y) \in k[y_1, ..., y_n]$, and $b_i(Y)$ are nonzero on $\overline{\varphi(U_g)}$ for $i = 1, ..., m$.

(iv) G contains the following two sets

$$\begin{aligned} G_{\text{inverse}} &= \{b_{m,0}(Y)x_m - b_{m,1}(Y), b_{m-1,0}(Y)x_{m-1} - b_{m-1,1}(Y, x_m), \\ &\quad ..., b_{1,0}(Y)x_1 - b_{1,1}(Y, x_m, x_{m-1}, ..., x_2)\} \\ G_{\text{image}} &= \emptyset \text{ or} \\ G_{\text{image}} &= \{Q_1(Y), ..., Q_m(Y)\}, \end{aligned}$$

where for $i = 1, ..., m$,

$$\begin{aligned} &b_{i,0}(Y) \in k[y_1, ..., y_n],\ b_{i,0}(Y) \notin J, \\ &b_{i,1}(Y, x_m, x_{m-1}, ..., x_{i+1}) \in k[y_1, ..., y_n, x_m, x_{m-1}, ..., x_{i+1}], \end{aligned}$$

Furthermore, if the above equivalent conditions hold, then G_{image} is the reduced Groebner basis of $\mathbf{I}(\overline{\varphi(U_g)})$ with respect to the induced order in $k[y_1, ..., y_n]$, and the local inverse of φ is given by $x_i = \dfrac{d_i(Y)}{c_i(Y)}$, obtained by solving the system of equations defined by the polynomials in G_{inverse} successively (first solve for x_m, then substitute x_m and solve for x_{m-1}, etc.).

□

§6. Nonsingular Points in Algebraic Sets

Given a point P on a plane curve C, we are usually concerned with the problem whether P is a smooth point or a singular point (e.g. a node, a cusp, isolated, n-fold, etc.). In this section, we define singularity for points in an algebraic set and prove that this definition is independent of the choice of its ambient space. Moreover, we will give a purely algebraic characterization of the singularity of a point P by using its associated local ring, as we have announced earlier.

Because of the connection between algebraic sets and analytic subvarieties (see the remark given in CH.III §6), the notion of nonsingular variety in algebraic geometry corresponds to the notion of manifold in topology. Over the complex numbers, for example, the nonsingular varieties are those which in the "usual" topology are complex manifolds. Accordingly, the most natural (and historically earliest) definition of nonsingularity uses the derivatives of the functions defining the variety.

We start with a hypersurface $V = \mathbf{V}(f)$ in the affine space $\mathbf{A}^n = \mathbf{A}^n_k$, where $f \in k[x_1, ..., x_n]$ is a polynomial. Let $P = (a_1, ..., a_n) \in V$. After an affine change of coordinates (if necessary) we may *always* assume $P = (0, ..., 0)$. Recall that if $n = 2$, then the *tangent line* of V at P is given by the equation

$$\frac{\partial f}{\partial x_1}(P)x_1 + \frac{\partial f}{\partial x_2}(P)x_2 = 0$$

and P is a singular point if and only if $\frac{\partial f}{\partial x_1}(P) = \frac{\partial f}{\partial x_2}(P) = 0$. For example, the origin $(0, 0)$ is a node on the curve $y^2 = x^3 + x^2$; and the origin $(0, 0)$ is a cusp on the curve $y^2 = x^3$.

Generally, for $n \geq 2$ we call

$$T_P V = \mathbf{V}\left(\sum_{i=1}^{n} \frac{\partial f}{\partial x_i}(P)x_i\right) \subset A^n$$

the *tagent space* of V at P.

6.1. Definition A point $P \in V$ is called a *nonsingular point* if there exists some i such that $\frac{\partial f}{\partial x_i}(P) \neq 0$; otherwise, P is called a *singular point*.

Observation Note that $T_P(V)$ is a linear k-space. From the definition it is clear that for a nonsingular point P of V, we have $\dim_k T_P V = n - 1$ if $f \neq 0$, and $\dim_k T_P(V) = n$ otherwise.

If $P \in V$ is nonsingular, then clearly P is nonsingular in some irreducible component of V (see CH.II §2 Exercise 4). So we may further assume that V is *irreducible*.
We write $\mathcal{S}(V)$ for the set of singular points in V. Then the next result shows that "most" of the points in V are nonsigular.

6.2. Theorem With notation as above, $\mathcal{S}(V)$ is a proper closed subset of V. Hence $V - \mathcal{S}(V)$, the set of nonsigular points in V is a dense open subset of V.

Proof Since V is irreducible we may assume that f is irreducible. It is easy to see that

$$\mathcal{S}(V) = \mathbf{V}\left(f, \frac{\partial f}{\partial x_1}, \dots, \frac{\partial f}{\partial x_n}\right)$$

is a closed subset in V. Suppose $V = \mathbf{V}(f) = \mathcal{S}(V)$. Then the restriction of each $\frac{\partial f}{\partial x_i}$ on V is 0. By the Nullstellensatz $\frac{\partial f}{\partial x_i} \in I(V) = (f)$. However, $\deg(\frac{\partial f}{\partial x_i}) < \deg(f)$. It follows that $\frac{\partial f}{\partial x_i} = 0$, $i = 1, \dots, n$. We claim that this is impossible by looking at two cases.
If $\mathrm{char} k = 0$, this is possible only when x_i does not appear in f. Hence f is a constant in k, a contradiction.
If $\mathrm{char} k = p > 0$, this is possible only when x_i appears in f in the form of $x_i^{\alpha p}$. Hence we may write f as $f = \sum a_{\alpha_1 \cdots \alpha_n} x_1^{p\alpha_1} \cdots x_n^{p\alpha_n}$. Since k is algebraically closed the equation $a_{\alpha_1 \cdots \alpha_n} = x^p$ has a solution, say $a_{\alpha_1 \cdots \alpha_n} = s^p_{\alpha_1 \cdots \alpha_n}$ for some $s_{\alpha_1 \cdots \alpha_n} \in k$. Thus, putting

$g = \sum s_{a_1 \cdots a_n} x_1^{a_1} \cdots x_n^{a_n}$ we have $g^p = f$, contradicting to the irreducibility of f. □

Hereafter we let $V = \mathbf{V}(f_1, ..., f_s)$ be an arbitrary *affine variety* in A^n, where $f_i \in k[x_1, ..., x_n]$, and $P = (a_1, ..., a_n) \in V$. Define

$$T_P V = \mathbf{V}\left(\sum_{i=1}^{n} \frac{\partial f_j}{\partial x_i}(P) x_i \;\middle|\; j = 1, ..., s\right)$$

to be the *tangent space* of V at P.

6.3. Proposition Consider the function $V \to I\!N$ defined by putting $P \mapsto \dim_k T_P V$, then for any $r \in I\!N$ the subset

$$S(r) = \{P \in V \mid \dim_k T_P V \geq r\} \subseteq V$$

is closed with respect to the Zariski topology on V.

Proof. $P \in S(r)$ if and only if the Jacobian matrix

$$\begin{pmatrix} \frac{\partial f_1}{\partial x_1}(P) & \frac{\partial f_1}{\partial x_2}(P) & \cdots & \frac{\partial f_1}{\partial x_n}(P) \\ \vdots & \vdots & \vdots & \vdots \\ \frac{\partial f_s}{\partial x_1}(P) & \frac{\partial f_s}{\partial x_2}(P) & \cdots & \frac{\partial f_s}{\partial x_n}(P) \end{pmatrix}$$

has rank $\leq n - r$ if and only if every $(n - r + 1) \times (n - r + 1)$ minor is 0. Since every minor is a polynomial in P, it follows that $S(r)$ is an algebraic set in V. □

6.4. Corollary There exists $r \in I\!N$ and a dense open subset $V_0 \subset V$ such that $\dim_k T_P V \geq r$ for all $P \in V$ and $\dim_k T_P V = r$ whenever $P \in V_0$.

Proof Put $r = \min\{\dim_k T_P V \mid P \in V\}$. Then it is obvious that

$$S(r) = V, \qquad S(r+1) \subset V.$$

Hence $S(r) - S(r+1) = \{P \in V \mid \dim_k T_P V = r\}$ is a nonempty open set of V. □

6.5. Definition Let r be as in Corolary 6.4. If $P \in V$ and $\dim_k T_P V = r$ then P is called a *nonsingular point*; otherwise P is called a *singular point*.

As for a hypersurface we write $\mathcal{S}(V)$ for the set of singular points in V. Then from the above we see that $\mathcal{S}(V)$ is a proper closed subset of V and hence $V - \mathcal{S}(V)$, the set of nonsingular points in V, is a dense open subset of V. This means again that "most" of the points in V are nonsingular.

Now, we claim that

- For an affine variety V and and $P \in V$, the singularity of P is intrinsic in the sense that it does not depend on the embedding of the open subset containing P in an ambient space.
- For an affine variety V, the number r defined in Corollary 6.4 is nothing but $\text{tr.deg}_k k(V) = \dim V$.

Let $P = (a_1, ..., a_n) \in V$. After an appropriate affine change of coordinates we may assume $P = (0, 0, ..., 0)$. Let $\mathbf{I}(P)$ be the ideal of P in $k[V]$ as defined in CH.II Lemma 6.13, i.e., $\mathbf{I}(P) = \{[g] \in k[V] \mid g(P) = 0\}$, and let $\Omega = \langle x_1, ..., x_n \rangle \subset k[\mathbf{x}] = k[x_1, ..., x_n]$. Then it is clear that $\mathbf{I}(P) = \Omega/\mathbf{I}(V)$.
Furthermore, let $\mathcal{O}_{P,V}$ be the local ring of V at P with maximal ideal $\mathcal{M}_{P,V}$ as before. Adopting notation as in CH.II §6, it follows from CH.II Lemma 6.13 and CH.II §6 Exercise 6(b) that

$$(*) \qquad \begin{cases} S_\Omega^{-1}k[\mathbf{x}]/S_\Omega^{-1}\mathbf{I}(V) & \cong \quad S_{\mathbf{I}(P)}^{-1}k[V] = \mathcal{O}_{P,V} \\ \\ S_\Omega^{-1}\Omega/S_\Omega^{-1}\mathbf{I}(V) & \cong \quad S_{\mathbf{I}(P)}^{-1}\mathbf{I}(P) = \mathcal{M}_{P,V} \end{cases}$$

6.6. Theorem With notation as above, in particular, $P = (0, ..., 0)$, the following hold.
(i) There is a k-space isomorphism $(T_P V)^* \cong \mathcal{M}_{P,V}/\mathcal{M}_{P,V}^2$, where $(T_P V)^*$ is the dual space of $T_P V$ consisting of linear functions.
(ii) If $f \in k[\mathbf{x}]$ satisfies $f(P) \neq 0$ and V_f is the open affine subvariety as defined in Lemma 3.5, then $T_P V_f \cong T_P V$ as k-spaces.

Proof (i) Let $(k^n)^*$ be the dual space of k^n. If $f \in k[x_1, ..., x_n]$, and if we write

$$f_P^{(1)} = \sum_{i=1}^{n} \frac{\partial f}{\partial x_i}(P)x_i,$$

then it is clear $f_P^{(1)} \in (k^n)^*$. This defines a k-linear map $\Omega \to (k^n)^*$ which, in turn, defines a k-linear map (check it!)

$$\begin{array}{rccc} d: & S_\Omega^{-1}\Omega & \longrightarrow & (k^n)^* \\ & \dfrac{g}{h} & \mapsto & \dfrac{1}{h(P)} g_P^{(1)} \end{array}$$

Since $\{d(x_i) \mid i = 1, ..., n\}$ is the standard k-basis of $(k^n)^*$, it follows that d is surjective. Moreover, note that since $P = (0, 0, ..., 0)$, we have

$$\begin{aligned} g_P^{(1)} = 0 \quad &\text{if and only if} \quad g \text{ starts with a term of degree 2} \\ &\text{if and only if} \quad g \in \Omega^2. \end{aligned}$$

This shows that $\text{Ker} d = (S_\Omega^{-1}\Omega)^2$, and hence

$$S_\Omega^{-1}\Omega/(S_\Omega^{-1}\Omega)^2 \cong (k^n)^*.$$

Now the inclusion map $T_PV \subset k^n$ induces the restriction linear map $(k^n)^* \xrightarrow{v} (T_PV)^*$ which is surjective. Hence we obtain the following composition linear map

$$D: \quad S_\Omega^{-1}\Omega \xrightarrow{d} (k^n)^* \xrightarrow{r} (T_PV)^*$$

Since both of r and d are surjective, so is D. We conclude $\text{Ker} D = (S_\Omega^{-1}\Omega)^2 + S_\Omega^{-1}\mathbf{I}(V)$. In fact

$$\begin{aligned} \frac{g}{h} \in \text{Ker} D \quad &\text{implies} \quad g_P^{(1)}\Big|_{T_PV} = 0 \\ &\text{implies} \quad g_P^{(1)} = \textstyle\sum_j a_j f_{jP}^{(1)} \text{ for some } f_j \in \mathbf{I}(V),\ a_j \in k \\ &\text{implies} \quad g - \textstyle\sum_j a_j f_j \in \Omega^2 \\ &\text{implies} \quad g \in \Omega^2 + \mathbf{I}(V)(S_\Omega^{-1}\Omega)^2 + S_\Omega^{-1}\mathbf{I}(V). \end{aligned}$$

On the other hand, it is clear that $(S_\Omega^{-1}\Omega)^2+S_\Omega^{-1}\mathbf{I}(V)\subset \mathrm{Ker}D$. This proves the conclusion. Combining with $(*)$ above, we obtain

$$\begin{aligned}\frac{\mathcal{M}_{P,V}}{\mathcal{M}_{P,V}^2} &\cong \left(\frac{S_\Omega^{-1}}{S_\Omega^{-1}\mathbf{I}(V)}\right)\Big/\left(\frac{S_\Omega^{-1}}{S_\Omega^{-1}\mathbf{I}(V)}\right)^2\\ &\cong S_\Omega^{-1}\Omega/\left((S_\Omega^{-1}\Omega)^2+S_\Omega^{-1}\mathbf{I}(V)\right)\\ &\cong (T_PV)^*\end{aligned}$$

as desired.

(ii) By Lemma 3.5 we have $V_f\cong \mathbf{V}(\mathbf{I}(V),x_{n+1}f-1)\subset \mathbf{A}^{n+1}$, $\mathbf{I}(V_f)=\langle \mathbf{I}(V),x_{n+1}f-1\rangle$. Hence

$$T_PV_f=\mathbf{V}\begin{pmatrix}\displaystyle\sum_{i=1}^{n+1}\frac{\partial f_j}{\partial x_i}(P)x_i,\ f_j\in\mathbf{I}(V),\\ \\ \displaystyle\sum_{i=1}^{n+1}\frac{\partial G}{\partial x_i}(P)x_i,\ G=x_{n+1}f-1\end{pmatrix},$$

and the k-space isomorphism $T_PV_f\to T_PV$ follows. □

Remark Note that in the above proof of (i) if we restrict d to Ω, then we also arrive at the k-space isomorphism $\dfrac{\mathbf{I}(P)}{\mathbf{I}(P)^2}\cong (T_PV)^*$.

Recall from Proposition 2.3 that if $\varphi\colon X\to Y$ is an isomorphism of varieties with $\varphi(P)=Q$, then φ induces the isomorphism of local rings $L_\varphi\colon \mathcal{O}_{Q,Y}\to\mathcal{O}_{P,X}$ with $L_\varphi(\mathcal{M}_{Q,Y})=\mathcal{M}_{P,X}$. This property plus the above theorem yields the following corollary.

6.7. Corollary Up to isomorphism, the singularity is intrinsic. More precisely, if $P\in V_0\subset V$ and $Q\in W_0\subset W$, where V_0 and W_0 are open subsets of the affine varieties V and W respectively, and if $\varphi\colon V_0\to W_0$ is an isomorphism with $\varphi(P)=Q$, then there exists k-space isomorphism $T_PV\cong T_QW$, and consequently $\dim_kT_PV=\dim_kT_QW$.

In particular, if V is birational to W, then they have the same number r as defined in Corollary 6.4 which defines the singularity of points in V and W.

□

Of course it is desirable to have nonsingularity at every point for nonsingular varieties; this is evident from the following theorem.

6.8. Theorem For any affine variety V, let r be the number defined in Corollary 6.4. Then

$$r = \text{tr.deg}_k k(V) = \dim V$$

which is equal to $\dim_k T_P V$ for any nonsingular point $P \in V$.

Proof From Corollary 6.4 we retain that if $V = \mathbf{V}(f) \subset \mathbf{A}^n$ is a hypersurface, where f is an irreducible polynomial, then $r = n - 1$. On the other hand, from CH. V §5 Exercise 1 and CH.V §6 we retain that $\dim V = \text{tr.deg}_k k(V) = n - 1$. Hence, for a hypersurface $V = \mathbf{V}(f) \subset \mathbf{A}^n$, we have $r = n - 1 = \text{tr.deg}_k k(V)$.
On the other hand, every affine variety V is birational to a hypersurface (Theorem 3.11). Hence the required equality follows from the above corollary. □

Summing up:

- Since any nonempty variety is birational to an open affine subvariety, we are ready to define the nonsingularity for a point $P \in Y$ where Y is any variety: $P \in Y$ is *nonsingular* if P is a nonsingular point in some affine open subset $Y_0 \subset Y$ (by the foregoing discussion this notion is independent of the choice of Y_0). If every point of Y is nonsingular, then we say that Y is nonsingular; otherwise Y is called singular. Foregoing results also establish that "most" points of a variety are nonsingular.
- We may now also define the *dimension* for any variety Y by: $\dim Y = \text{tr.deg}_k \mathcal{R}(Y)$.

Finally, we give a purely algebraic characterization of nonsingularity. This expresses that the singularity is an intrinsic property.

6.9. Definition Let A be a Noetherian local ring with maximal ideal $\mathfrak{m}$ and residue field $k = A/\mathfrak{m}$. A is called a *regular local ring* if $\dim_k \frac{\mathfrak{m}}{\mathfrak{m}^2} = \text{K.dim}A$, where the latter is the Krull dimension of A defined in CH.V §7.

6.10. Theorem Let V be an affine variety in $\mathbf{A}^n$ and $P \in V$. Then P is nonsingular if and only if the local ring $\mathcal{O}_{P,V}$ is a regular local ring.

Proof As before we may assume $P = (0, 0, ..., 0)$. Let Ω be the maximal ideal of P in $k[x_1, ..., x_n]$, $I = \mathbf{I}(V) = \langle f_1, ..., f_s \rangle$, and $\mathbf{I}(P) = \Omega/I$ the maximal ideal of P in $k[V]$. If we consider the linear map as defined before:

$$\begin{array}{rcl} \theta: \ k[x_1, ..., x_n] & \longrightarrow & k^n \\ f & \mapsto & \left(\dfrac{\partial f}{\partial x_1}(P), ..., \dfrac{\partial f}{\partial x_n}(P)\right) \end{array}$$

then it is easy to see that

$$\theta(I) = k\text{-Span}\,(\theta(f_1), ..., \theta(f_s)).$$

Hence the rank of the Jacobian matrix

$$J = \begin{pmatrix} \dfrac{\partial f_1}{\partial x_1}(P) & \cdots & \dfrac{\partial f_1}{\partial x_n}(P) \\ \vdots & & \vdots \\ \dfrac{\partial f_s}{\partial x_1}(P) & \cdots & \dfrac{\partial f_s}{\partial x_n}(P) \end{pmatrix}$$

is just the k-dimension of $\theta(I)$. Thus, using the isomorphism θ': $\Omega/\Omega^2 \rightarrow k^n$ induced by θ, we have $\dim_k(I + \Omega^2)/\Omega^2 = \dim_k\theta(I) = \text{rank}J$. Finally, the classical isomorphism $(\Omega/\Omega^2)/((I + \Omega^2)/\Omega^2) \cong \Omega/(I + \Omega^2)$ plus the foregoing result yields

$$\begin{aligned} n = \dim_k \frac{\Omega}{\Omega^2} &= \dim_k \frac{I + \Omega^2}{\Omega^2} + \dim_k \frac{\Omega}{I + \Omega^2} \\ &= \text{rank}J + \dim_k \frac{\mathbf{I}(P)}{\mathbf{I}(P)^2} \\ &= \text{rank}J + \dim_k T_P V \end{aligned}$$

where the last equality follows from the remark before Corollary 6.7.

Now recall from CH.II §6 that $S^{-1}_{\mathbf{I}(P)}k[V] = \mathcal{O}_{P,V}$, it follows that the Krull dimension of $\mathcal{O}_{P,V}$ is equal to the height of $\mathbf{I}(P)$ in $k[V]$. On the other hand, from CH.V §7 Theorem 7.13 we know that every maximal ideal in $k[x_1, ..., x_n]$ has height n. Since $\mathbf{I}(V)$ is a prime ideal, it follows that K.dim$\mathcal{O}_{P,V}$ = K.dim$k[V]$ = dim$V$. Put dim$V$ = r. Then $\mathcal{O}_{P,V}$ is a local ring of dimension r. Since $\mathcal{O}_{P,V}/\mathcal{M}_{P,V} = k[V]/\mathbf{I}(P) = k$ where $\mathcal{M}_{P,V}$ is the maximal ideal of $\mathcal{O}_{P,V}$ (CH.II §6), it follows that $\mathcal{O}_{P,V}$ is regular if and only if

$$\dim_k \frac{\mathcal{M}_{P,V}}{\mathcal{M}^2_{P,V}} = \dim_k \frac{\mathbf{I}(P)}{\mathbf{I}(P)^2} = r.$$

But this is also equivalent to rank$J = n - r$, i.e., equivalent to the nonsingularity of P. □

Historically, the above result is at the origin of the local ring theory in commutative algebra.

Remark Although we have defined the singularity for any variety in the "local" sense, and we can always study the singularity of a point in a projective variety by using the affine-projective machinery developed in CH.III, it is also convenient to have an analogue of Definition 6.1 in the projective case (see the examples below), we leave this work as an exercise (Exercise 4 below) to the reader (but we will deal with it in detail for a projective plane curve in CH.VII §1).

Example (i) Let $V = \mathbf{V}(F) \subset \mathbf{P}^n$ be a quadric defined by

$$(1) \qquad F = \sum_{i,j=1}^{n+1} a_{ij}x_ix_j = 0, \; a_{ij} \in k$$

over a field k with char$k \neq 2$. Hence $a_{ij} = a_{ji}$, $i, j = 1, ..., n+1$. Since

$$\frac{\partial F}{\partial x_i} = 2\sum_{j=1}^{n+1} a_{ij}x_j, \quad i = 1, ..., n+1,$$

V has a singular point if and only if the system of equations

$$\sum_{j=1}^{n+1} a_{ij}x_j = 0, \quad i = 1, ..., n+1,$$

has a nontrivial solution (Exercise 4) if and only if $\det(a_{ij}) = 0$. Since from CH.III §4 Example (ii) we know that V is projectively equivalent to the quadric defined by

$$\sum_{i=1}^{n+1} a_i x_i^2 = 0, \quad a_i \in k, \tag{2}$$

it follows that the determinant of coefficients of (2) is $\prod_{i=1}^{n+1} a_i$ and V is nonsingular if and only if $a_i \neq 0$ for $i = 1, ..., n+1$. If V is nonsingular, then again from CH.III §4 Example (ii) we know that if k is algebraically closed then V is projectively equivalent to the quadric defined by

$$\sum_{i=1}^{n+1} x_i^2 = 0. \tag{3}$$

This implies that:

- If k is algebraically closed and $\operatorname{char} k \neq 2$, there is, up to a projective change of coordinates (hence a birational equivalence), exactly one nonsingular quadric in $\mathbf{P}_k^n$ and its equation is (3).

(ii) Now suppose $P = (c_1, ..., c_n, c_{n+1})$ is a singular point of the quadric V given by (1) in the last example and that $Q = (d_1, ..., d_n, d_{n+1})$ is any other point of V. Consider the line L passing through P and Q. Let λ, μ be any two elements of k. Then

$$\begin{aligned}
&\sum_{i,j=1}^{n+1} a_{ij}(\lambda c_i + \mu d_i)(\lambda c_j + \mu d_j) \\
&= \sum_{i=1}^{n+1} (\lambda c_i + \mu d_i) \sum_{j=1}^{n+1} a_{ij}(\lambda c_J + \mu d_j) \\
&= \sum_{i=1}^{n+1} (\lambda c_i + \mu d_i) \sum_{j=1}^{n+1} a_{ij} \cdot \mu d_j \quad \left(\text{since } \sum_{j=1}^{n+1} a_{ij} c_j = 0\right) \\
&= \sum_{i,j=1}^{n+1} \lambda\mu a_{ij} c_i d_j + \mu^2 \sum_{i,j=1}^{n+1} a_{ij} d_i d_j.
\end{aligned}$$

The first term here is equal to $\lambda\mu\sum_{i,j=1}^{n+1} a_{ij}c_jd_i$ since the a_{ij} are symmetric in their indices, and $\sum_{j=1}^{n+1} a_{ij}c_j = 0$ for each i. The second term is zero since Q is on V. This shows that

- The line L joining P and Q lies entirely in V.

(iii) Let C be a curve in $\mathbf{P}_k^2$ defined by a homogeneous polynomial $F \in k[x_1, x_2, x_3]$. Suppose that k is algebraically closed and that $\deg F > 1$. We claim that

- If C contains a line L, then C has a singular point.

To see this, suppose that L is defined by $G = \sum_{i=1}^{3} a_ix_i = 0$. It follows from CH.III §4 Example (iii) that F may be written as $F = G \cdot H$ for some polynomial $H \in k[x_1, x_2, x_3]$. Then

$$\frac{\partial F}{\partial x_i} = G \cdot \frac{\partial H}{\partial x_i} + H \cdot \frac{\partial G}{\partial x_i}$$

for $i = 1, 2, 3$. Let $Q = (c_1, c_2, c_3)$ be a point on the intersection of $\mathbf{V}(G)$ and $\mathbf{V}(H)$ (CH.V Proposition 5.7). Obviously $\frac{\partial F}{\partial x_i}(Q) = 0$ for $i = 1, 2, 3$ and hence Q is a singular point of C.

In conclusion, Examples (i)–(iii) yield the following:

- Let C be a conic in $\mathbf{P}^2$ where k is an algebraically closed field of char$k \neq 2$. Then the following statements are equivalent:
  1. C has a singular point;
  2. $\det(a_{ij}) = 0$, where (a_{ij}) is the matrix of C;
  3. C consists of two (possible coincident) straight lines.

Exercises for §6

1. Let L be a line passing through $P = (a_1, ..., a_n) \in V = \mathbf{V}(f) \subset \mathbf{A}^n$. Suppose that the parametrized form of L is

$$L: \quad x_i = a_i + b_it, \quad i = 1, ..., n.$$

   Prove that $L \subset T_PV$ if and only if $t = 0$ is a multiple root of $g(t) = f(..., a_i + b_it, ...)$.

2. Find all singular points of the following curves in $\mathbf{A}^2$ resp. surfaces in $\mathbf{A}^3$.
   a. $y^2 = x^3 - x$.
   b. $x^2y^2 + +x^2 + y^2 + 2xy(x + y + 1) = 0$.

c. $xy^2 = z^2$.
d. $xy + x^3 + y^3 = 0$.

3. Let $V = \mathbf{V}(f) \subset \mathbf{A}^n$ be a hypersurface which is not a hyperplane. Show that P is a singular point in $\mathbf{V}(f) \cap T_P V$.
4. Let $V = V(F) \subset \mathbf{P}^n$ be a hypersurface defined by a homogeneous polynomial $F \in k[x_1, ..., x_n, x_{n+1}]$. For $P \in V$, we say that P is *singular* if

$$\frac{\partial F}{\partial x_i}(P) = 0, \quad i = 1, ..., n+1;$$

otherwise it is *simple*.
Show that the singularity defined for P above is consistent with Definition 5.1 in the affine case. (Hint: For $f \in k[x_1, ..., x_n]$ consider the homogenization $F = f^*$ in $k[x_1, ..., x_n, x_{n+1}]$ with respect to x_{n+1}, and use Euler's theorem given in CH.I §1 Exercise 6.)

CHAPTER VII
Curves

This chapter is devoted to a more detailed introduction of the study of plane algebraic curves by using the methods developed in previous chapters, in particular, the local method introduced in CH.VI is essential for understanding singularities on curves, intersections of curves and the elementary function theory on curves. In CH.IV §5, CH.V and CH.VI §5 we have seen the power of Groebner basis in algebraic geometry. But there are other computational techniques prominently present in the theory of curves, some of are algebraic nature and some of are analytical type. We point out that there have been several algorithms for parametrizing a curve and for computing the linear space $\mathcal{L}(D)$ associated to a divisor D on a curve (see §5), in certain special cases. We will not introduce all techniques in this text but (as in CH.I §3), we refer the reader to the special issue (2,3)23(1997) of *Journal of Symbolic Computation* for some detailed discussion on this topic.

§1. Nonsingular Curves

We assume that k is an algebraically closed field.

Let $\mathbf{A}^2 = \mathbf{A}_k^2$, and write $k[x,y]$ for its coordinate ring. If $f \in k[x,y]$ is a *nonconstant* polynomial, then (up to a nonzero constant) the algebraic set $\mathbf{V}(f)$ is called an *affine plane curve* (i.e., if $\lambda \in k$ is a nonzero constant, then $\mathbf{V}(f)$ and $\mathbf{V}(\lambda f)$ repersent the same curve). the *degree* of a plane curve is the degree of a defining polynomial for the curve. A plane curve of degree one is a *line*; so we speak of "the line $ax + by + c$", or "the line $ax + by + c = 0$".

Let $C = \mathbf{V}(f)$ be a plane curve of degree n in $\mathbf{A}^2$. We provide an algebraic way to define the tangent line of C at a point P.
Let L be a line passing $P = (a,b) \in C$, which is parametrized as

$$\begin{cases} x = a + ct \\ y = b + dt \end{cases}$$

where t is a variable. Put $g(t) = f(a + ct, b + dt)$. Then $t = 0$ is a root of g. Expanding $g(t)$ in a formal MacLaurin series at 0, we have

$$g(t) = g(0) + \frac{g'(0)}{1!}t + \cdots + \frac{g^{(n)}(0)}{n!}t^n.$$

If $g^{(m)}(0) \neq 0$, and $g^{(l)}(0) = 0$ for $l < m$, then 0 is a root of multiplicity m for $g(t)$, and we say that *L intersects C at $P = (a,b)$ with multiplicity m*. It is easy to show that this definition is independent of particular parametrization of the line L.

1.1. Proposition Put $\nabla f = \left(\frac{\partial f}{\partial x}, \frac{\partial f}{\partial y}\right)$ (i.e., the gradient vector).
(i) If $\nabla f(P) \neq (0,0)$, then there is a *unique* line through P which intersects C with multiplicity ≥ 2.
(ii) If $\nabla f(P) = (0,0)$, then *every* line through P intersects C with multiplicity ≥ 2.

Proof Note that $t = 0$ is a root of multiplicity ≥ 2 if and only if $g'(0) = 0$ if and only if

$$(*) \qquad \frac{\partial f}{\partial x}(P) \cdot c + \frac{\partial f}{\partial y}(P) \cdot d = 0.$$

Hence, if $\nabla f(P) = (0,0)$, then L always intersects C at P with multiplicity ≥ 2. This proves (ii).

Now suppose $\nabla f(P) \neq (0,0)$. Then the solution space of $(*)$ with indeterminates c and d is 1-dimensional. Thus, there is $(c_0, d_0) \neq (0,0)$ such that (c,d) satisfies $(*)$ if and only if $(c,d) = \lambda(c_0, d_0)$ for some $\lambda \in k$. It follows that the pairs (c,d) yielding $g'(0) = 0$ all parametrize the same line. This shows that there is a unique line which intersects C at P with multiplicity ≥ 2, and hence (i) is proved. □

1.2. Definition (i) If $\nabla f(P) \neq (0,0)$, then the *tangent line* of C at P is the unique line through P intersecting C with multiplicity ≥ 2 at P. In this case, we say that P is a *nonsingular point* (or a *simple point*) of C.
(ii) If $\nabla f(P) = (0,0)$, then we say that P is a *singular point* of C.

Obviously, this definition of singularity at a point on a curve is consistent with the definition given in the last section.

Let $P = (a,b) \in C = \mathbf{V}(f)$. In the foregoing section we defined the tangent space of C at P as

$$T_P V = \mathbf{V}\left(\frac{\partial f}{\partial x}(P)x + \frac{\partial f}{\partial y}(P)y\right).$$

It is clear that if P is a simple point, then $T_P V$ is nothing but the tangent line at C in P which intersects C at P with multiplicity ≥ 2; if P is sigular, then $T_P V = \mathbf{A}^2$ and every line through $P = (a,b)$ also intersects C with multiplicity ≥ 2. This means that $T_P V$ cannot provide more information about the varying degrees of the singularity at a point.
Since an affine change of coordinates does not change the singularity of a point (if we restrict to an irreducible component of C containing P, CH.VI §5), we may now assume that $P = (0,0)$. Thus, f has a unique decomposition into homogeneous components:

$$(**) \qquad f = F_m + F_{m+1} + \cdots + F_n$$

where $n = \deg(f)$, each F_p is a homogeneous polynomial of degree p, and $F_m \neq 0$. Since k is algebraically closed, F_m factors into a product of linear factors (CH.III §5 Corollary 5.3):

$$(***) \qquad F_m = \prod_{i=1}^{s}(a_i x + b_i y)^{r_i}, \quad r_1 + \cdots + r_s = m.$$

Write L_i for the line $a_i x + b_i y = 0$, $i = 1, ..., s$, and put

$$\mathcal{L} = L_1 \cup \cdots \cup L_s.$$

We call $\mathcal{L}$ the *tangent space* of C at $P = (0,0)$, and each L_i a *tangent line* to C at P in the following sense:

- If $m = 1$, then P is a simple point and F_m is the tangent line of C at P which intersects C at P with multiplicity $\geq 2 = m + 1$.
- If $m > 1$, then $\nabla f(0,0) = (0,0)$, i.e., P is a singular point, and every line $L_i \in \mathcal{L}$ intersects C at P with multiplicity $\geq m + 1$.
- Any line L which intersects C at P with multiplicity $\geq m + 1$ coincides with some $L_i \in \mathcal{L}$.

1.3. Definition (i) The number m appearing in (**) is called the *multiplicity* of C at $P = (0,0)$, and is denoted $m_P(C)$. If $m = 1$, P is called a *simple* point as before, if $m = 2$, a *double point*, if $m = 3$, a *triple point*, etc.
(ii) In the decomposition (* * *) of F_m, L_i is called *simple* (resp. *double*, etc.) tangent if $r_i = 1$ (resp. $r_i = 2$, etc.). If C has m distinct (simple) tangents L_i at P, we say that $P = (0,0)$ is an *ordinary multiple point* of C.

Example (i) From the definition we see that if $P = (0,0)$ is an ordinary double point then it is just a *node* with two distinct tangents, e.g., the curve $\mathbf{V}(y^2 - x^3 - x^2)$; and if $m = 2$ but $P = (0,0)$ is not an ordinary double point, then P is a *cusp* with one tangent line which passes through P twice, e.g., look at the curve $\mathbf{V}(y^2 - x^3)$. A similar interpretation may be made for $m \geq 3$. (Suggestion: Use any available computer algebra system to plot the curves considered in this example by using the parametrizations given in CH.I §3.)

Let $f = \prod f_i^{e_i}$ be the factorization of f into irreducible components, $C_i = \mathbf{V}(f_i)$, $P = (0,0) \in C = \mathbf{V}(f)$. Then it is easy to see that

a. $m_P(C) = \sum e_i m_P(C_i)$;
b. P is a simple point of C if and only if P belongs to just one component C_i of C, $e_i = 1$ and P is a simple point of C_i (see CH.II §2).

So, from now on, we always assume that $C = \mathbf{V}(f)$ is *irreducible*, i.e., f is an irreducible polynomial in $k[x,y]$, and $\mathbf{I}(C) = \langle f \rangle$ (CH.II §2). Let $\mathcal{O}_P$ be the local ring of $P = (0,0)$ with maximal ideal $\mathcal{M}_P$. Then $k \cong \mathcal{O}_P/\mathcal{M}_P$ (CH.II §6), and hence $\mathcal{M}_P^n/\mathcal{M}_P^{n+1}$ may be regarded as a k-space for every $n \geq 0$.

1.4. Theorem Let $P = (0,0) \in C$. Then

$$m_P(C) = \dim_k \left(\mathcal{M}_P^n/\mathcal{M}_P^{n+1}\right) \quad \text{for } n \gg 0.$$

Proof Write $\mathcal{O}$, $\mathcal{M}$ for $\mathcal{O}_P$, $\mathcal{M}_P$ respectively. From the isomorphism

$$\left(\mathcal{O}/\mathcal{M}^{n+1}\right) / \left(\mathcal{M}^n/\mathcal{M}^{n+1}\right) \cong \mathcal{O}/\mathcal{M}^n$$

it follows that it is enough to prove that

$$\dim_k \left(\mathcal{O}/\mathcal{M}^n\right) = n \cdot m_P(C) + s$$

for some constant s and all $n \geq m_P(C)$. Since $P = (0,0)$, we have $\mathcal{M}^n = \Omega^n\mathcal{O}$, where $\Omega = \langle x, y \rangle \subset k[x,y]$. Note that $\mathbf{V}(\Omega^n) = \{P\} = \mathbf{V}(\Omega^n, f)$, it follows from CH.II §6 Exercise 6 that

$$\frac{k[x,y]}{\langle \Omega^n, f \rangle} \cong \frac{O_P}{\langle \Omega^n, f \rangle O_P} \cong \frac{\mathcal{O}}{\Omega^n \mathcal{O}} = \frac{\mathcal{O}}{\mathcal{M}^n}$$

where O_P is the local ring of P in $k(\mathbf{A}^2) = k(x,y)$. So we reduce the problem to the calculation of the dimension of $k[x,y]/\langle \Omega^n, f \rangle$. Let $m = m_P(C)$. Then $fg \in \Omega^n$ whenever $g \in \Omega^{n-m}$. There is a natural ring homomorphism φ: $k[x,y]/\Omega^n \to k[x,y]/\langle \Omega^n, f \rangle$, and a k-linear map ψ: $k[x,y]/\Omega^{n-m} \to k[x,y]/\Omega^n$ defined by $\psi([g]) = [fg]$. It is easy to verify that $\varphi \circ \psi$ induces the isomorphism of k-spaces:

$$\left(\frac{k[x,y]}{\Omega^n}\right) \Big/ \left(\frac{\langle f \rangle + \Omega^n}{\Omega^n}\right) \cong \frac{k[x,y]}{\langle \Omega^n, f \rangle}.$$

It follows from CH.I §1 Exercise 3(b) that

$$\dim_k \left(\frac{k[x,y]}{\langle \Omega^n, f \rangle}\right) = nm - \frac{m(m-1)}{2}, \quad \text{for all } n \geq m,$$

as desired. □

Remark (i) From Theorem 1.4 it is clear that $m_P(C)$ only depends on the local ring of $P = (0,0)$, and hence the multiplicity of C at a point P is an intrinsic property. Considering any point $P = (a,b) \in C$, let α be the translation which takes $(0,0)$ to P, i.e., $\alpha(x,y) = (x+a, y+b)$. Then $f^\alpha = f(x+a, y+b)$. Put $C' = \mathbf{V}(f^\alpha)$. Define $m_P(C)$ to be $m_{(0,0)}(C')$, then this definition only depends on the local ring $\mathcal{O}_{P,C}$ of P. Accordingly, the *tangent space* of C at P is defined to be the tangent space of C' at $(0,0)$.

(ii) The function $\chi(n) = \dim_k(\mathcal{O}_P/\mathcal{M}_P^n)$ is a polynomial in n (for large n) which is called the Hilbert-Samuel polynomial of the local ring $\mathcal{O}_P$.

Another important application of Theorem 1.4 is the following

1.5. Theorem Let $P = (a,b) \in C = \mathbf{V}(f)$. Then P is a simple point of the plane curve C if and only if the local ring $\mathcal{O}_P$ of P is a discrete valuation ring in the sense of App.III. In this case, if $L = ax + by + c$ is any line through P which is not tangent to C at P, then the image of L in $\mathcal{O}_P$ generates the maximal ideal $\mathcal{M}_P$ of $\mathcal{O}_P$.

Proof Since L is not the tangent line of C at P, after an appropriate affine change of coordinates we may assume that $P = (0,0)$, that y is the tangent line, and that $L = x$. Then it it sufficient to show that $\mathcal{M}_P$ is generated by the image of x in $\mathcal{O}_P$.
As usual we write $[x]$, $[y]$ for the image of x, y in $k[C]$. Then $P = (0,0)$ means $\mathcal{M}_P = \langle [x], [y] \rangle \subset \mathcal{O}_P$, whether P is simple or not. Moreover, we have

$$f = y + \text{higher terms.}$$

Grouping together terms involving y, we write

$$f = yg - x^2h, \quad \text{where } g = 1 + \text{higher terms, and } h \in k[x].$$

Thus, $[yg] = [x^2h] \in k[C]$, so $[y] = [x^2][h][g]^{-1}$, since $g(P) \neq 0$. It follows that $\mathcal{M}_P = \langle [x], [y] \rangle = \langle [x] \rangle$, as desired.
Conversely, if $\mathcal{O}_P$ is a discrete valuation ring with maximal ideal $\mathcal{M}_P$ then the latter is principal. Theorem 1.4 implies that $m_P(C) = 1$, so P is simple. This completes the proof of the theorem. □

So far, everything is valid for plane curves in $\mathbf{A}^2$. For a projective plane curve $C = \mathbf{V}(F) \subset \mathbf{P}^2$ defined by a homogeneous polynomial $F \in k[x_1, x_2, x_3]$, of course we may define the singularity of a point $P \in C$ by passing to the affine case, i.e., since $C = V_1 \cup V_2 \cup V_3$ with $V_i = C \cap U_i$, $U_i = \mathbf{P}^2 - \mathbf{V}(x_i)$, $i = 1, 2, 3$, if $P \in V_i$, we may define the *multiplicity* of C at P to be

$$m_P(C) = m_P(\mathbf{V}(F_*))$$

where F_* is the dehomogenization of F with respect to x_i. Obviously, m_P is independent of the choice of V_i, and it is locally invariant. So, we say that P is a *nonsingular* (or *simple*) point if $m_P = 1$.
It is convenient (e.g. see CH.VII, CH.VIII) to have a similar description of the singularity of points on a projective curve as we did for an affine curve (Definition 1.2).

Let $C = \mathbf{V}(F) \subset \mathbf{P}^2$ be a projective curve given by a homogeneous polynomial of degree n. Suppose $P = (a_1, a_2, a_3)$ is a point on C and L is a line passing through P and another point $(b_1, b_2, b_3) \in \mathbf{P}^2$ which is distinct from P. Then all points on L are given by

$$(sa_1 + tb_1, sa_2 + tb_2, sa_3 + tb_3), \quad s,\ t \in k.$$

The intersections of L and C are obtained by finding the values of s and t such that

$$(1) \qquad F(sa_1 + tb_1, sa_2 + tb_2, sa_3 + tb_3) = 0.$$

The intersection point P corresponding to the case $s = 1$, $t = 0$. If we consider the left-hand side of (1) as a function of t alone and denote it by $g(t)$, we have its formal MacLaurin expansion

$$g(t) = g(0) + \frac{g'(0)}{1!}t + \frac{g''(0)}{2!}t^2 + \cdots + \frac{g^{(n)}(0)}{n!}t^n.$$

As in the affine case, we say that *L intersects C at P with multiplicity m* if $g^{(m)}(0) \neq 0$, and $g^{(l)}(0) = 0$ for $l < m$. It is easy to show, as in the affine case, that this definition depends on the line L and not on any particular parametrization of L.
Now suppose L intersects C at P with multiplicity $m \geq 2$. This means $g'(0) = 0$. Going back to (1), we see that this is equivalent to the condition

$$(2) \qquad \frac{\partial F}{\partial x_1}(P)b_1 + \frac{\partial F}{\partial x_2}(P)b_2 + \frac{\partial F}{\partial x_3}(P)b_3 = 0.$$

If $\nabla F = \left(\frac{\partial F}{\partial x_1}, \frac{\partial F}{\partial x_2}, \frac{\partial F}{\partial x_3}\right) = (0,0,0)$, then P is called a *singular* point; otherwise it is a *simple* point.
Suppose P is simple. Then (2) means that the point (b_1, b_2, b_3) must lie on the line $\sum_{i=1}^{3} \frac{\partial F}{\partial x_i}(P)x_i = 0$. By using Euler's theorem (CH.I §1 Exercise 6) it is verified immediately that P lies on this line. Thus we see that if P is a simple point, there is a unique line intersecting C at P with multiplicity $m \geq 2$ and its equation is

$$\frac{\partial F}{\partial x_1}(P)x_1 + \frac{\partial F}{\partial x_2}(P)x_2 + \frac{\partial F}{\partial x_3}(P)x_3 = 0. \tag{3}$$

We call this line the *tangent* at P. Indeed, we define tangents in terms of multiplicities for the projective case in exactly the same way as we did for the affine case.
At this point the natural question arises as to whether what we have done so far is consistent with the results obtained for the affine case. It is. To see this, let us start with a curve $f(x_1, x_2) = 0$ of degree n in $\mathbf{A}^2$ and suppose $P = (x_0, y_0)$ is a point on the curve. Homogenizing $f(x_1, x_2)$ by setting $f^* = F(x_1, x_2, x_3)$ and writing $Q = (x_0, y_0, 1)$, one checks immediately that

$$\frac{\partial F}{\partial x_1}(Q) = \frac{\partial f}{\partial x_1}(P), \quad \frac{\partial F}{\partial x_2}(Q) = \frac{\partial f}{\partial x_2}(P).$$

Also, applying Euler's theorem to $F(x_1, x_2, x_3)$, we see that

$$\frac{\partial F}{\partial x_3}(Q) \cdot 1 = -\frac{\partial F}{\partial x_1}(Q) \cdot x_0 - \frac{\partial F}{\partial x_2}(Q) \cdot y_0.$$

Using these results, we see that P is singular in the affine sense if and only if it is singular in the projective sense. If P is simple, the equation (3) reduces to the equation for the tangent in the affine case.

We now deal with curves in a more general setting.
First, we refine some results of CH.VI §5 for curves. Recall that for any variety X, the dimension of X, denoted $\dim X$, is defined to be $\dim X = \text{tr.deg}_k \mathcal{R}(X)$.

1.6. Lemma (i) A variety X has dimension equal to 0 if and only if X contains only one point.

(ii) Let X be a variety of dimension one, and V a proper closed subvariety of X (CH.VI Definition 1.5). Then V is a point.
(iii) A closed subvariety V of $\mathbf{A}^2$ (resp. $\mathbf{P}^2$) has dimension one if and only if it is an affine (resp. projective) plane curve.

Proof (i) This follows from CH.V Proposition 5.3.
(ii) This follows from CH.V Proposition 5.10 and above (i).
(iii) Assume $V = \mathbf{V}(f) \subset \mathbf{A}^2$, where f is an irreducible polynomial in $k[x, y]$. Then by CH.V Proposition 5.2 $\dim V = 1$. Conversely, suppose $V \subset \mathbf{A}^2$ is a closed subvariety of dimension one. By Ch.II §2 Corollary 2.5, V is of the form $\mathbf{V}(f)$ with f irreducible in $k[x, y]$. Since any projective variety V is the closure of some affine variety W (CH.III §6), and $\dim V = \dim W$ (CH.V Theorem 4.7), the projective case is clear because of CH.III §6. □

1.7. Corollary Every variety X of dimension one is birationally equivalent to a plane curve.

Proof Since $\dim X = 1$, $\mathcal{R}(X) = k(u,, v)$ for some $u, v \in \mathcal{R}(X)$ (see App.I Proposition 4.5). Let I be the kernel of the natural homomorphism $k[t_1, t_2] \to k[u, v] \subset \mathcal{R}(X)$. Then I is prime, so $V = \mathbf{V}(I) \subset \mathbf{A}^2$ is a variety. Since $k[t_1, t_2]/I \cong k[u, v]$, it follows that $k(V) \cong k(u, v) = \mathcal{R}(X)$. So $\dim V = 1$, and V is a plane curve by the lemma above. □

The foregoing result and Theorem 1.5 inspire the following definition.

1.8. Definition A variety X of dimension one is called a *curve.* If a curve X has its function field $k(X)$ which is isomorphic to the rational function field $k(t)$ in one variable t (i.e., the function field of $\mathbf{A}^1$, or equivalently, of $\mathbf{P}^1$), then X is called a *rational curve.*
Let X be a curve. A point $P \in X$ is called a *nonsingular point* (or a *simple point*) if its local ring $\mathcal{O}_{P,X}$ is a discrete valuation ring in the sense of App.III.

Note that a curve X is rational if and only if it can be parametrized in the form $x_i = f_i(t)$, $i = 1, ..., n$ ($i = 1, ..., n+1$ in case $X \subset \mathbf{P}^n$), where the $f_i(t)$ are rational functions in one variable t (CH.VI Corollary 3.10). Recall from the end of CH.I §3 that *not every plane curve is rational.*

Let K be a field extension of k. We say that a local ring A is a *local ring of* K if A is a subring of K containing k, and K is the quotient field of A. For example, if X is any variety, $P \in V$, then $\mathcal{O}_{P,X}$ is a local ring of of $K(V)$. Similarly, a *discrete valuation ring of* K is a DVR which is a local ring of K.

1.9. Theorem Let C be a curve in $\mathbf{P}^n$, $K = \mathcal{R}(C)$. Let L be a field extension of K, and R a DVR of L with maximal ideal $\mathfrak{m}$. Assume that $R \not\supset K$. Then there is a unique point $P \in C$ such that $\mathfrak{m} \supset \mathcal{M}_{P,C}$.

Proof Let $o(l)$ be the order function $L \to \mathbb{Z} \cup \{\infty\}$ determined by R (App.III). For any $P \in C$, write $\mathcal{O}_P$ for the local ring of P, and write $\mathcal{M}_P$ for the maximal ideal of $\mathcal{O}_P$.

Uniqueness: If $\mathfrak{m} \supset \mathcal{M}_P, \mathcal{M}_Q$ for $P, Q \in C$, choose $f \in \mathcal{M}_P$, $\frac{1}{f} \in \mathcal{O}_Q$ (CH.II Theorem 4.3 and Lemma 6.13). Then $o(f) > 0$ and $o\left(\frac{1}{f}\right) \geq 0$, which is a contradiction by App.III §2.

Existence: we may assume that C is a closed subvariety of $\mathbf{P}^n$, and that $C \cap U_i \neq \emptyset$, $i = 1, \ldots, n+1$ (§2 Exercise 5). Then $k^g[C] = k[x_1, \ldots, x_{n+1}]/\mathbf{I}(C)$ with each $[x_i] \neq 0$. Let

$$N = \max\left\{ o\left(\frac{[x_i]}{[x_j]}\right) \,\middle|\, i, j = 1, \ldots, n+1 \right\}.$$

Assume that $o\left(\frac{[x_j]}{[x_{n+1}]}\right) = N$ for some j (change coordinates if necessary). Then for all i

$$o\left(\frac{[x_i]}{[x_{n+1}]}\right) = o\left(\left(\frac{[x_j]}{[x_{n+1}]}\right) \cdot \left(\frac{[x_i]}{[x_j]}\right)\right) = N - o\left(\frac{[x_j]}{[x_i]}\right) \geq 0.$$

If C_* is the affine curve corresponding to $C \cap U_{n+1}$, then $k[C_*]$ may be identified with

$$k^g[C][x_{n+1}^{-1}]_{(0)} = \left\{ \frac{[F]}{[x_{n+1}^n]} \,\middle|\, F \text{ is a homogeneous of degree } n \geq 0 \right\}$$

$$\subset k(C).$$

Thus, $o(a) \geq 0$ for any $a \in k[C_*]$, and hence $R \supset k[C_*]$. Let $p = \mathfrak{m} \cap k[C_*]$. Then p is a prime ideal of $k[C_*]$, and so, p corresponds to a closed subvariety W of C_* (CH.II Theorem 4.3). If $W = C_*$, then $p = 0$ and every nonzero element of $k[C_*]$ is a unit in R (see App.III); but then $K \subset R$, which is contrary to our assumption. So $W = \{P\}$ is a point by Lemma 1.6(ii). It is now easy to check that $\mathcal{M}_{P,C} = \mathcal{M}_{P,C_*} = \mathcal{M}_{P,W} \subset \mathfrak{m}$, finishing the proof. □

1.10. Corollary Let C be a non-singular curve in $\mathbf{P}^n$, $K = \mathcal{R}(C)$. Then there is a natural bijective correspondence

$$\{\text{points of } C\} \longleftrightarrow \{\text{DVR of } K\}.$$

Proof Each $\mathcal{O}_P$ is clearly a DVR of K (Theorem 1.5), and the direction "$\longrightarrow$" is one-to-one by CH.II §6. If R is any DVR of K with maximal ideal $\mathfrak{m}$, then by the theorem there is a unique $P \in C$ such that $\mathfrak{m} \supset \mathcal{M}_P$. Since R and $\mathcal{O}_P$ are both DVR's of K, it follows from App.III that $R = \mathcal{O}_P$. This proves that the direction "$\longrightarrow$" is also onto. □

Let C, K be as in Corollary 1.10. Let $\mathcal{X}$ be the set of all DVR's of K over k. Put a topology on $\mathcal{X}$ as follows: a nonempty set U of $\mathcal{X}$ is open if $\mathcal{X} - U$ is finite. Then the correspondence

$$\begin{array}{ccc} C & \longrightarrow & \mathcal{X} \\ P & \mapsto & \mathcal{O}_P \end{array}$$

is a homeomorphism. And if U is open in C, $\mathcal{O}(U) = \cap_{P \in U} \mathcal{O}_P$, so all the rings of functions on C may be recorvered from $\mathcal{X}$. Since $\mathcal{X}$ is determined by K alone, this means that C is determined up to isomorphism by K alone (Exercise 6).

This topological space $\mathcal{X}$ is called the Riemann surface of K/k. For any field extension K/k of transcendence degree one the corresponding Riemann surface is an abstract nonsingular model of a curve. The geometry of a variety is therefore closely linked to an arithmetical theory of the corresponding function field. The theory of algebraic function fields presents another approach to algebraic geometry; this was exploited in a systematic way a. o. in the work of C. Chevalley:

Introduction to the Theory of Algebraic Functions in One variable, Amer. Math. Soc. New York, 1951.

Exercises for §1

1. With notation of Theorem 1.4, and $\mathcal{M} = \mathcal{M}_{P,C}$, show that $\dim_k (\mathcal{M}^n/\mathcal{M}^{n+1}) = n+1$ for $0 \leq n < m_P(C)$. In particular, P is a simple point if and only if $\dim_k (\mathcal{M}/\mathcal{M}^2) = 1$; otherwise $\dim_k (\mathcal{M}/\mathcal{M}^2) = 2$.
2. Let $V = \mathbf{V}(x^2 - y^3, y^2 - z^3) \subset \mathbf{A}^3$, $P = (0,0,0)$, $\mathcal{M} = \mathcal{M}_{P,V}$. Show that $\dim_k (\mathcal{M}/\mathcal{M}^2) = 3$.
3. Let $\mathcal{O} = \mathcal{O}_{P,\mathbf{A}^2}$ for some $P \in \mathbf{A}^2$, $\mathcal{M} = \mathcal{M}_{P,\mathbf{A}^2}$. Calculate $\chi(n) = \dim_k (\mathcal{O}/\mathcal{M}^n)$.
4. Let $f \in k[x_1, ..., x_r]$ define a hypersurface $V = \mathbf{V}(f) \subset \mathbf{A}^r$. Write $f = F_m + F_{m+1} + \cdots$, and let $m = m_P(V)$ where $P = (0, ..., 0)$. Suppose f is irreducible, and let $\mathcal{O} = \mathcal{O}_{P,V}$, $\mathcal{M}$ its maximal ideal. Show that $\chi(n) = \dim_k (\mathcal{O}/\mathcal{M}^n)$ is a polynomial of degree $r-1$ for $n \gg 0$, and that the leading coefficient of $\chi(n)$ is $m_P(V)/(r-1)!$.
5. Show that any curve has only a finite number of multiple points.
6. Let φ be a rational map from a curve C' to a projective curve C. Let $U = \cup U_\psi$, where (U_ψ, ψ) runs over all representative of φ. Show that U contains every simple point of C'. If C' is nonsingular, φ is a morphism.
7. Show that two nonsingular projective curves are isomorphic if and only if their function fields are isomorphic.
8. Show that the elliptic curve $y^2 = (x-a_1)(x-a_2)(x-a_3)$ in $\mathbf{A}^2_{\mathbb{C}}$, where a_1, a_2, a_3 are distinct complex numbers, has no singularities; similar for the corresponding curve in $\mathbf{P}^2_{\mathbb{C}}$.
9. Show that intersection multiplicities for lines and curves, and the property of tangency, are preserved by projective change of coordinates (or projective transformation) of $\mathbf{P}^2_k$.
10. Show that
    a. all nonsingular conics are rational curves, and hence are birationally equivalent to a line (indeed to all rational curves) (hint: use the example given at the end of CH.VI §5.);
    b. up to projective equivalence, there is only one irreducible conic in $\mathbf{P}^2_{\mathbb{C}}$. Any irreducible conic in $\mathbf{P}^2_{\mathbb{C}}$ is nonsingular.

11. Let $V = \mathbf{V}(F) \subset \mathbf{P}^n$ be a hypersurface defined by a homogeneous polynomial in $k[x_1, ..., x_n, x_{n+1}]$. Suppose $P \in V$ is simple (see CH.VI §5 Exercise 4). Show that, as in the case of $n = 2$, there is a unique tangent hyperplane passing through P given by the equation

$$\sum_{i=1}^{n+1} \left(\frac{\partial F}{\partial x_i}\right)(P) = 0.$$

12. Let $V = \mathbf{P}^1_k$, $k^g[V] = k[x, y]$. Let $t = \frac{x}{y} \in k(V)$, and show $k(V) = k(t)$. Show that there is a natural one-to-one correspondence between the points of $\mathbf{P}^1_k$ and the DVR's of $k(V)$; which DVR corresponds to the point at infinity? (See Apendix.II §2 Example (i).)
13. Let $P_1, ..., P_n \in C \subset \mathbf{P}^2_{\mathbf{C}}$ where C is a curve defined by a homogeneous polynomial. Show that there are an infinite number of lines passing through P_1, but not through $P_2, ..., P_n$. If P_1 is a simple point on C, we may take these lines transversal to C at P_1. (Hint: Use CH.I §2 Exercise 2.)
14. Let C be an irreducible curve in $\mathbf{P}^2_{\mathbf{C}}$, $P_1, ..., P_n$ simple points on C, $m_1, ..., m_n$ integers. Show that there is a $\frac{\phi}{\psi} \in k(C)$ with $v_{P_i}\left(\frac{\phi}{\psi}\right) = m_i$ for $i = 1, ..., n$, where v_{P_i} denotes the discrete valuation function given by the local ring $\mathcal{O}_{P_i,C}$. (Hint: Take lines L_i as in above exercise for P_i, and a line L_0 not through any P_j, and let $\frac{\phi}{\psi} = \prod L_i^{m_i} L_0^{-\sum m_i}$.)

§2. Intersection of Curves

Let k be a field of Char$k = 0$, and let $\mathbf{A}^2 = \mathbf{A}^2_k$, $\mathbf{P}^2 = \mathbf{P}^2_k$.

Example (i) Let C be the parabola $y = x^2$. Consider the line $y = tx$ through $(0, 0)$ with variable slope t. In all cases when t varies, except for $L =$ the x-axis and $L =$ the y-axis, we see that the intersection of L and C has two points $\{P, (0, 0)\}$:

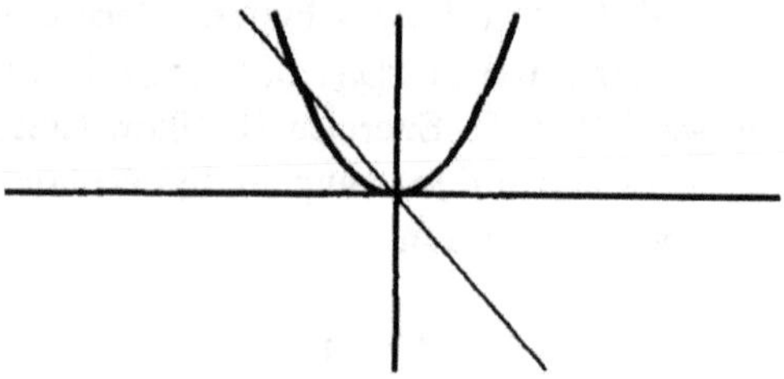

Figure 5

If t reaches 0, we see that the two points of intersection coincide at $(0,0)$ and that L coincides with the x-axis; we also see that L and C have a tangent line in common at $(0,0)$ (which is L itself). So in this case it is reasonable to say that C and L have an "intersection number" at $(0,0)$ which is equal to 2. When L reaches the y-axis, again we only see one point of intersection, namely $(0,0)$, while the other point of intersection just "moves" to the point at ∞. The missing point P of intersection may be recovered by extending C to the the quadratic curve $\mathcal{C}$: $YZ = X^2$ in $\mathbf{P}^2$ as in CH.III. Now the y-axis corresponds to the line $\mathcal{L}$ at ∞: $X = 0$ in $\mathbf{P}^2$. We see that $\mathcal{C}$ intersects $\mathcal{L}$ at two points $\{(0,0,1),(0,1,0)\}$.

The procedure described above holds for any line in $\mathbf{A}^2$. Consequently, if we want "intersection numbers" for the number of points in the intersection of C and any line, then we always obtain 2 when we work in $\mathbf{P}^2$.

(ii) In this example we consider the parabola C: $y = x^2$ and the ellipse D: $x^2 + 4(y-\lambda)^2 = 4$, where λ is a parameter. For example, when $\lambda = 2$ or 0 and $k = I\!R$ we obtain different numbers of points in the intersection of C and D, i.e., 4 and 2 respectively. And it is also clear that there are values of λ for which there are no points of intersection (e.g. when $\lambda < -1$).

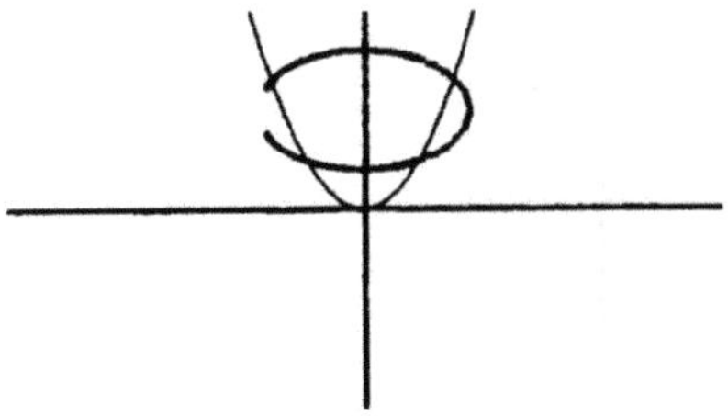

$\lambda = 2$

Figure 6

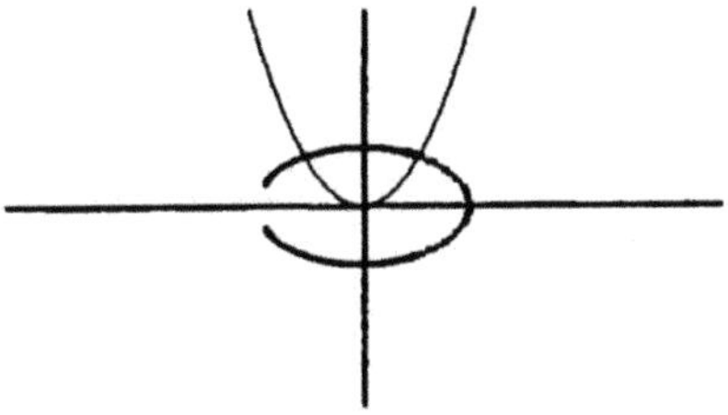

$\lambda = 0$

Figure 7

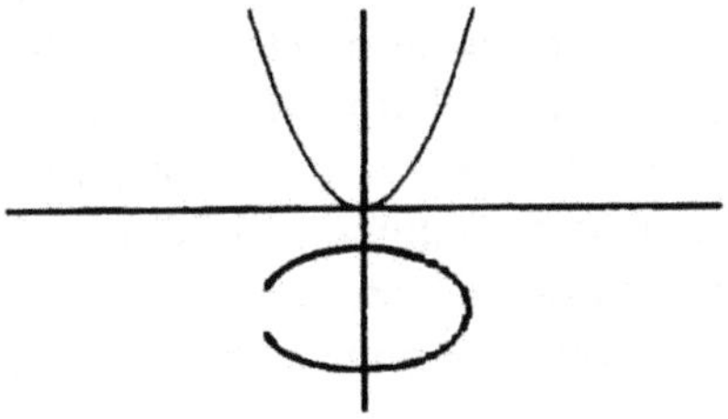

$\lambda < -1$

Figure 8

However, over $\mathbb{C}$ we have four points of intersection in both of the foregoing cases. Hence, over $\mathbb{C}$ the number of points of intersection seems to be more predictable. The reader may check that in the cases where there are no points of intersection over $\mathbb{R}$, we still get four points over $\mathbb{C}$.

Similarly, even over $\mathbb{C}$, as in Example (i) above, we meet the problem how to "count" the "intersection number" at a point of intersection. For instance, if we let λ vary from 2 to 1, from the picture

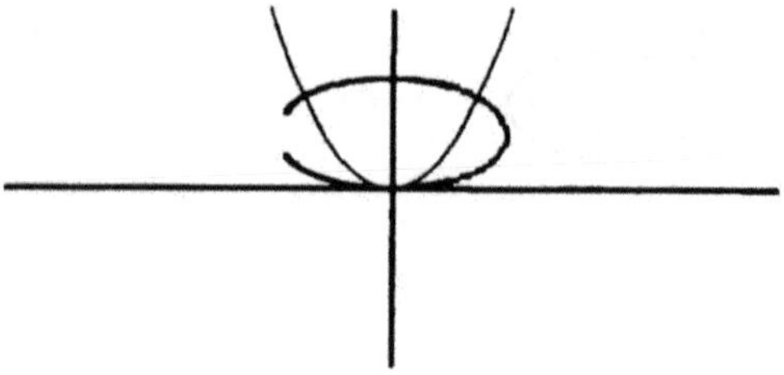

$\lambda = 1$
Figure 9

we see that when λ reaches 1, two intersection points coincide at the origin, i.e., over $\mathbb{C}$ there are only three different points of intersection left. In this case $(0,0)$ is a simple point on both curves, and we see that the curves have a tangent line in common at the origin $(0,0)$. It is plausable to say that the "intersection number" of C and D at $(0,0)$ is 2. If we take "intersection numbers" into account instead of just counting points then we still obtain the number four in total.

Summarizing:

- For giving a correct interpretation of the intersection of two curves it is necessary to work in $\mathbf{P}^2_{\mathbb{C}}$ and take the "intersection number" at every point of intersection into account.

Thus, from now on we will always work in $\mathbf{P}^2 = \mathbf{P}^2_{\mathbb{C}}$ or in $\mathbf{A}^2 = \mathbf{A}^2_{\mathbb{C}}$.

Let $\mathcal{C} = \mathbf{V}(F)$ and $\mathcal{D} = \mathbf{V}(G)$ be two curves in $\mathbf{P}^2$ defined by two nonconstant homogeneous polynomials F and G in $k[x_1, x_2, x_3]$. Then $\mathcal{C} \cap \mathcal{D} = \mathbf{V}(F, G)$. Geometrically, however, we can only "see" $\mathcal{C} \cap \mathcal{D}$ by looking at its affine parts, as in CH.III §6. We have

$$\mathbf{V}(F,G) = V_1 \cup V_2 \cup V_3,$$
$$\text{where } V_i = \mathbf{V}(F,G) \cap U_i,\ U_i = \mathbf{P}^2 - V(x_i), i = 1,2,3.$$

If $V_i \neq \emptyset$, then by CH.III §6 we have $V_i \cong \mathbf{V}(F_*, G_*) \subset \mathbf{A}^2$, where F_* and G_* are the dehomogenizations of F and G with respect to x_i. If $P \in \mathcal{C} \cap \mathcal{D}$ and $P \in V_i$, then we may say that $P \in \mathbf{V}(F_*) \cap \mathbf{V}(G_*)$. So, in view of the foregoing examples, counting the "number" of points of $\mathcal{C} \cap \mathcal{D}$ by adding up the "intersection number" at every point $P \in \mathcal{C} \cap \mathcal{D}$ is equivalent to counting the number of points of each $V_i = \mathbf{V}(F_*, G_*)$ by adding up the "intersection number" at each $P \in \mathbf{V}(F_*) \cap \mathbf{V}(G_*)$. But then, hoping to avoid problems arising from such definition, we should have that $\mathcal{C} \cap \mathcal{D}$ is finite and the "intersection number" at a point of intersection should be well-defined.

From CH.V Proposition 5.3 it follows that $\mathbf{V}(F, G)$ is finite if and only if $\dim \mathbf{V}(F, G) = 0$, and this dimension can be computed by using F and G. However, we also have another useful criterion which is given in the following lemma.

2.1. Lemma With notation as above, $\mathcal{C} \cap \mathcal{D}$ is finite if and only if F and G have no common factors if and only if $\mathcal{C}$ and $\mathcal{D}$ do not have common components.

Proof Since we are working over $\mathbb{C}$, this follows from CH.I §2 Exercise 2, CH.II Proposition 2.3 and CH.III §5 Exercise 5. □

We now define the "intersection number" at a point of intersection in the affine case. There are many ways to do this, but we adopt the local method given in [Ful] which, at the same time, points a very easy algorithmic way to compute the "intersection number".

Let $C = \mathbf{V}(f)$ and $D = \mathbf{V}(g)$ be two plane curves in $\mathbf{A}^2_{\mathbb{C}}$ defined by $f, g \in \mathbb{C}[x, y]$ respectively. We assume that $C \cap D$ is finite. Let $P \in C \cap D$. Assume for the moment that $P = (0,0)$, and let $m_P(C) = m$, $m_P(D) = n$, which is the multiplicity of C at P resp. of D at P. By looking at the foregoing examples we intuitively conclude that

a. if C and D do not have tangent lines in common at P, then C and D meet at P with multiplicity mn because P is now an m-fold point on C resp. an n-fold point on C;
b. if C and D have tangent lines in common at P, then certainly C and D meet at P with multiplicity at least mn.

Also recall that the multiplicity of a curve at a point P is a local property. So, we want to have a proper definition of the "intersection number" which fits the geometric intuition and depends only on the local ring of P. The following proposition deals with the algebraic properties corresponding to a. and b. (**Warning**: Since C and D are not necessarily irreducible, one has to be careful about the local rings to be used hereafter!)

2.2. Proposition With notation as above, let $\Omega = \langle x, y\rangle \subset \mathbb{C}[x,y]$ be the ideal of $P = (0,0)$, $\mathcal{O} = \mathcal{O}_{P,\mathbf{A}^2}$ the local ring of P, and $m = m_P(C)$, $n = m_P(D)$.
(i) If C and D do not have tangent lines in common at P, then $\Omega^d \subset \langle f, g\rangle\mathcal{O}$ for $d \geq m+n-1$.
(ii) C and D do not have tangent lines in common at P if and only if the linear map

$$\psi: \frac{\mathbb{C}[x,y]}{\Omega^n} \bigoplus \frac{\mathbb{C}[x,y]}{\Omega^m} \longrightarrow \frac{\mathbb{C}[x,y]}{\Omega^{m+n}}$$

$$([h],[s]) \mapsto [hf+sg]$$

is injective.

Proof (i) Let $L_1, \ldots, L_m$ be the tangents to C at P, $M_1, \ldots, M_n$ the tangents to D at P. Let $L_i = L_m$ if $i > m$, $M_j = M_n$ if $j > n$, and let

$$F_{ij} = L_1 \cdots L_i M_1 \cdots M_j \text{ for all } i, j \geq 0 \ (F_{00} = 1).$$

Then $\{F_{ij} \mid i+j = d\}$ forms a basis for the k-subspace $\mathbb{C}[x,y]_d$ of all homogeneous polynomials of degree d in $\mathbb{C}[x,y]$ (CH.III §5 Exercise 7).
To prove (i), it therefore suffices to show that $F_{ij} \in \langle f, g\rangle\mathcal{O}$ for all $i+j \geq m+n-1$. But $i+j \geq m+n-1$ implies that either $i \geq m$ or $j \geq n$. Say $i \geq m$, so $F_{ij} = F_{m0}B$, where B is a homogeneous of degree $i+j-m$. Note that f is now the form $f = F_{m0} + f'$ where all terms of f' are of degree $\geq m+1$, we have $Bf = BF_{m0} + Bf' = F_{ij} + Bf'$. This implies $F_{ij} = Bf - Bf'$ where each term of Bf' has degree $\geq (i+j-m) + (m+1) = i+j+1$. The proof of (i) will be finished if we can show that $\Omega^t \subset \langle f, g\rangle\mathcal{O}$ for $t \gg 0$ (why?)
This fact is a consequence of the Nullstellensatz: let $\mathbf{V}(f,g) = \{P, Q_1, \ldots, Q_s\}$, and choose a polynomial h so that $h(Q_i) = 0$, $h(P) \neq 0$ (CH.I §4 Exercise 3). Since $hx, hy \in \mathbf{I}(\mathbf{V}(f,g))$, so $(hx)^N, (hy)^N \in \langle f, g\rangle$ for some N. But h^N is a unit in $\mathcal{O}$, so $x^N, y^N \in \langle f, g\rangle\mathcal{O}$, and therefore $\Omega^{2N} \subset \langle f, g\rangle\mathcal{O}$, as desired.

(ii) Suppose that the tangents are distinct, and that $\psi([h],[s]) = [hf+sg] = 0$ for some nonzero $([h],[s]) \in \mathbb{C}[x,y]/\Omega^n \oplus \mathbb{C}[x,y]/\Omega^m$. Then $hf+sg$ consists entirely of terms of degree $\geq m+n$. Write

$$\begin{aligned} h &= H_r + \text{ higher terms} \\ s &= S_l + \text{ higher terms} \\ f &= F_m + \text{ higher terms} \\ g &= G_n + \text{ higher terms.} \end{aligned}$$

So

$$hf + sg = H_r F_m + S_l G_n + \text{ higher terms.}$$

Since not both $[h]$ and $[s]$ are zero, we must have

$$r+m = l+n \text{ and } H_r F_m = -S_l G_n.$$

But F_m and G_n have no common factors, so F_m divides S_l, and G_n divides H_r. Therefore $s \geq m$, $r \geq n$, and hence $([h],[s]) = ([0],[0])$, a contradiction. This shows that ψ is injective.
Conversely, if L were a common tangent to C and D at P, then $F_m = LF'_{m-1}$, $G_n = LG'_{n-1}$. But then $\psi\left([G'_{n-1}],[-F'_{m-1}]\right) = 0$, i.e., ψ is not injective. □

2.3. Corollary Let $P = (0,0) \in C \cap D$, Ω, $\mathcal{O}$, $m = m_P(C)$, $n = m_P(D)$ be as in the proposition. Then

$$\dim_{\mathbb{C}}\left(\frac{\mathcal{O}}{\langle f,g\rangle\mathcal{O}}\right) \geq mn,$$

and the equality holds if and only if C and D do not have tangent lines in common at P.

Proof First, by CH.II §6 Exercise 6(c) we have

$$(*)\quad \left\{\begin{aligned} \dim_{\mathbb{C}}\left(\frac{\mathcal{O}}{\langle f,g\rangle\mathcal{O}}\right) &\geq \dim_{\mathbb{C}}\left(\frac{\mathcal{O}}{\langle \Omega^{m+n},f,g\rangle\mathcal{O}}\right) \\ &= \dim_{\mathbb{C}}\left(\frac{\mathbb{C}[x,y]}{\langle \Omega^{m+n},f,g\rangle}\right). \end{aligned}\right.$$

Secondly, if we consider the natural k-linear map

$$\varphi: \quad \frac{\mathbb{C}[x,y]}{\Omega^{n+m}} \longrightarrow \frac{\mathbb{C}[x,y]}{\langle \Omega^{m+n},f,g\rangle}$$

and the k-linear map ψ in the proposition, then it is not hard to see that $\mathrm{Im}\psi = \mathrm{Ker}\varphi = \langle \Omega^{m+n}, f, g\rangle/\Omega^{m+n}$. Hence by CH.I §1 Exercise 3 we obtain

$$(**)\quad \begin{cases} \dim_{\mathbb{C}}\left(\dfrac{\mathbb{C}[x,y]}{\langle\Omega^{m+n},f,g\rangle}\right) & \geq \dim_{\mathbb{C}}\left(\dfrac{\mathbb{C}[x,y]}{\langle\Omega^{m+n}\rangle}\right) - \\ & \quad -\dim_{\mathbb{C}}\left(\dfrac{\mathbb{C}[x,y]}{\langle\Omega^{m}\rangle}\right) - \\ & \quad -\dim_{\mathbb{C}}\left(\dfrac{\mathbb{C}[x,y]}{\langle\Omega^{n}\rangle}\right) \\ & = mn. \end{cases}$$

This shows that $\dim_{\mathbb{C}}(\mathcal{O}/\langle f,g\rangle\mathcal{O}) \geq mn$, and that the equality holds if and only if both inequalities in (*) and (**) are equalities. The inequality in (*) is an equality if $\Omega^{m+n} \subset \langle f,g\rangle\mathcal{O}$; while the second is an equality if and only if ψ is injective. But it follows from the proposition that this is equivalent to say that C and D do not have tangent lines in common at P. □

2.4. Proposition Let I be an ideal in $\mathbb{C}[x_1,...,x_n]$, and suppose $\mathbf{V}(I) = \{P_1,...,P_N\}$ is finite. Let $\mathcal{O}_i = \mathcal{O}_{P_i,\mathbf{A}^n}$, the local ring of P_i. Then there is a natural k-vector space isomorphism of $\mathbb{C}[x_1,...,x_n]/I$ with $\oplus_{i=1}^{N}(\mathcal{O}_i/I\mathcal{O}_i)$, where the latter is the direct sum of k-vector spaces.

Proof Let $\Omega_i = \mathbf{I}(P_i) \subset \mathbb{C}[x_1,...,x_n]$ be the distinct maximal ideals corresponding to the P_i which contain I. Let $R = \mathbb{C}[x_1,...,x_n]/I$, $R_i = \mathcal{O}_i/I\mathcal{O}_i$. The natural ring homomorphisms (which are also k-linear maps) φ_i: $R \to R_i$ induce a k-linear map:

$$\begin{array}{rcl} \varphi:\ R & \longrightarrow & \displaystyle\bigoplus_{i=1}^{N} R_i \\ {[f]} & \mapsto & \displaystyle\sum_{i=1}^{N} \varphi_i([f]) \end{array}$$

By the Nullstellensatz and CH.II Corollary 1.9, $\sqrt{I} = \mathbf{I}(\{P_1,...,P_N\}) = \cap_{i=1}^{N}\Omega_i$, so $\left(\cap_{i=1}^{N}\Omega_i\right)^d \subset I$ for some d (CH.I §6 Exercise 5). Since

$\cap_{j \neq i} \Omega_j$ and I_i are comaximal (see CH.I §6 Exercise 6), it follows from CH.I §6 Exercise 6 that $\cap_{i=1}^N \Omega_i^d = (\Omega_1 \cdots \Omega_N)^d = \left(\cap_{i=1}^N \Omega_i\right)^d \subset I$. Now choose $f_i \in \mathbb{C}[x_1, ..., x_n]$ such that $f_i(P_j) = 0$ if $i \neq j$, $f_i(P_i) = 1$ (CH.I §4 Exercise 3). Let

$$e_i = 1 - \left(1 - f_i^d\right)^d, \quad i = 1, ..., N.$$

Note that $e_i = f_i^d D_i$ for some $D_i \in \mathbb{C}[x_1, ..., x_n]$, hence

$$(*) \qquad \begin{cases} e_i \in \Omega_j^d \quad \text{if } i \neq j, \text{ and} \\ \\ 1 - \sum_i e_i = (1 - e_j) - \sum_{i \neq j} e_i \in \bigcap_{i=1}^N \Omega_i^d \subset I. \end{cases}$$

If we let $[e_i]$ be the class of e_i in R, we get

$$(**) \qquad \begin{cases} [e_i^2] = [e_i] \quad (\text{note that } e_i^2 - e \in \mathbf{I}(\{P_1, ..., P_N\})), \\ \\ [e_i][e_j] = 0 \text{ if } i \neq j, \text{ and} \\ \\ \sum_{i=1}^N [e_i] = 1. \end{cases}$$

Claim: If $g \in \mathbb{C}[x_1, ..., x_n]$, and $g(P_i) \neq 0$, then there is an $[h] \in R$ such that $[hg] = [e_i]$. To prove this, we may assume that $g(P_i) = 1$, and let $h = 1 - g$. Then $h \in \Omega_i$ and hence $h^d e_i \in I$. It follows from the equality

$$(1 - h)(e_i + he_i + \cdots + h^{d-1}e_i) = e_i - h^d e_i$$

that $\left[g\left(e_i + he_i + \cdots + h^{d-1}e_i\right)\right] = [e_i]$, as claimed.

Now we are ready to conclude that φ is an isomorphism:

φ is injective: If $\varphi([f]) = 0$, then for each i there is a g_i with $g_i(P_i) \neq 0$ and $g_i f \in I$. Let $[h_i g_i] = [e_i]$ as we claimed above. Then by the foregoing $(**)$,

$$[f] = \left(\sum_i [e_i]\right)[f] = \sum_i [h_i g_i f] = 0.$$

φ **is surjective:** Since $e_i(P_i) = 1$, $\varphi_i([e_i])$ is a unit in $R_i = \mathcal{O}_i/I\mathcal{O}_i$; since $\varphi_i([e_i])\varphi_i([e_j]) = \varphi_i([e_ie_j]) = 0$, if $i \neq j$, then $\varphi_i([e_j]) = 0$ by the foregoing (*). Therefore, $\varphi_i([e_i]) = \varphi_i\left(\sum_{j=1}^N e_j\right) = \varphi_i(1) = 1$. Now, suppose $z = \sum_{i=1}^N \left[\frac{a_i}{s_i}\right] \in \oplus_{i=1}^N R_i$, where $\frac{a_i}{s_i} \in \mathcal{O}_i$. As established in the claim, we may set $[h_is_i] = [e_i]$ in R; then $\left[\frac{a_i}{s_i}\right] = [a_i][h_i]$ in R_i, and $\varphi\left(\sum_{j=1}^N [h_ja_je_j]\right) = z$. This completes the proof of the proposition. □

2.5. Corollary It follows immediately from CH.V Theorem 2.4 and the above proposition that

$$\dim_{\mathbb{C}}\left(\frac{\mathbb{C}[x_1,\ldots,x_n]}{I}\right) = \sum_{i=1}^N \dim_{\mathbb{C}}\left(\frac{\mathcal{O}_i}{I\mathcal{O}_i}\right).$$

In particular, every $\mathcal{O}_i/I\mathcal{O}_i$ is a finite dimensional k-vector space.

□

Our next goal is to cumpute $\dim_{\mathbb{C}}(\mathcal{O}/\langle f,g\rangle\mathcal{O})$ for $P = (0,0) \in C \cap D$, where $\mathcal{O} = \mathcal{O}_{P,\mathbf{A}^2}$ is the local ring of P.

First a lemma.

2.6. Lemma let $q, h \in \mathbb{C}[x,y]$. Suppose that qh and g have no common factors. then

$$\dim_{\mathbb{C}}\left(\frac{\mathcal{O}}{\langle qh,g\rangle\mathcal{O}}\right) = \dim_{\mathbb{C}}\left(\frac{\mathcal{O}}{\langle q,g\rangle\mathcal{O}}\right) + \dim_{\mathbb{C}}\left(\frac{\mathcal{O}}{\langle h,g\rangle\mathcal{O}}\right).$$

Proof Let Ω be the maximal ideal of P. If $q, h \notin \Omega$ (hence $qh \notin \Omega$), then the lemma is clear. Suppose $q \in \Omega$. then we have the natural k-linear map

$$\varphi: \quad \frac{\mathcal{O}}{\langle qh,g\rangle\mathcal{O}} \longrightarrow \frac{\mathcal{O}}{\langle h,g\rangle\mathcal{O}}$$

and the k-linear map

$$\begin{aligned}\psi: \quad \frac{\mathcal{O}}{\langle q,g\rangle\mathcal{O}} &\longrightarrow \frac{\mathcal{O}}{\langle qh,g\rangle\mathcal{O}}\\ [s] &\mapsto [hs]\end{aligned}$$

We claim that ψ is injective. For if $[hs] = 0$, then $hs = uqh + vg$ for some $u, v \in \mathcal{O}$. Clearing the denominators (if necessary) we may assume $s, u, v \in \mathbb{C}[x, y]$. Thus, $(s - uq)h = vg$. But there are no common factors for h and g, so g divides $s - uq$, i.e., $s - uq = gw$ with $w \in \mathbb{C}[x, y]$. Hence $s = gw + uq$, $[s] = 0$. Moreover, it is clear that $\mathrm{Im}\psi = \mathrm{Ker}\varphi$, and hence the lemma is proved. □

2.7. Theorem With notation as before, for $P = (0,0) \in C \cap D$, $\dim_{\mathbb{C}} (\mathcal{O}/\langle f, g\rangle\mathcal{O})$ may be calculated in a finite number of steps.

Proof In view Corollary 2.5 we may assume $\dim_{\mathbb{C}} (\mathcal{O}/\langle f, g\rangle\mathcal{O}) = n < \infty$. Since $\langle f, g\rangle = \langle f, g + hf\rangle$ for any $h \in \mathbb{C}[x, y]$, this will make the calculation much easier.

Let $f(x, 0)$ $g(x, 0) \in \mathbb{C}[x]$ be of degrees r, s respectively. We may assume $r \leq s$.

Case 1: $r = 0$. Then y divides f, $f = yh$ for some $h \in \mathbb{C}[x, y]$, but y does not divide g because there are no common factors for f and g. From Lemma 2.6 we retain:

$$(*) \qquad \dim_{\mathbb{C}} \left(\frac{\mathcal{O}}{\langle f, g\rangle\mathcal{O}}\right) = \dim_{\mathbb{C}} \left(\frac{\mathcal{O}}{\langle y, g\rangle\mathcal{O}}\right) + \dim_{\mathbb{C}} \left(\frac{\mathcal{O}}{\langle h, g\rangle\mathcal{O}}\right).$$

If $g(x, 0) = x^m(a_0 + a_1x + \cdots)$, $a_0 \neq 0$, then since $P \in \mathbf{V}(y) \cap \mathbf{V}(g)$, the finiteness of $\mathbf{V}(y) \cap \mathbf{V}(g)$ yields that g can be written as $g = g(x, 0) + ys$ with $s \in \mathbb{C}[x, y]$. Thus we obtain

$$\begin{aligned}
\dim_{\mathbb{C}} \left(\frac{\mathcal{O}}{\langle y, g\rangle\mathcal{O}}\right) &= \dim_{\mathbb{C}} \left(\frac{\mathcal{O}}{\langle y, g(x, 0) + ys\rangle\mathcal{O}}\right) \\
&= \dim_{\mathbb{C}} \left(\frac{\mathcal{O}}{\langle y, g(x, 0)\rangle\mathcal{O}}\right) \\
&= m \text{ (Corollary 2.3).}
\end{aligned}$$

If $h(P) \neq 0$, then $\dim_{\mathbb{C}} (\mathcal{O}/\langle h, g\rangle\mathcal{O}) = 0$; otherwise from (*) we derive that

$$\dim_{\mathbb{C}} \left(\frac{\mathcal{O}}{\langle h, g\rangle\mathcal{O}}\right) < \dim_{\mathbb{C}} \left(\frac{\mathcal{O}}{\langle f, g\rangle\mathcal{O}}\right),$$

so induction on n finishes the proof.

Case 2: $r > 0$. We may multiply f and g by constants to make f and g monic. Let $q = g - x^{s-r}f$. Then

$$\begin{aligned}\dim_{\mathbb{C}}\left(\frac{\mathcal{O}}{\langle f, g(x,0)\rangle\mathcal{O}}\right) &= \dim_{\mathbb{C}}\left(\frac{\mathcal{O}}{\langle f, q + x^{s-r}f\rangle\mathcal{O}}\right)\\ &= \dim_{\mathbb{C}}\left(\frac{\mathcal{O}}{\langle f, q\rangle\mathcal{O}}\right),\end{aligned}$$

and $\deg(q(x,0)) = t < s$. Note that since $P \in \mathbf{V}(f) \cap \mathbf{V}(q)$ and there are no common factors for f and q, repeating this process (interchanging the order of f and q if $t < r$) a finite number of times we eventually reach a pair of curves u, v which fit Case 1, and such that

$$\dim_{\mathbb{C}}\left(\frac{\mathcal{O}}{\langle f, g\rangle\mathcal{O}}\right) = \dim_{\mathbb{C}}\left(\frac{\mathcal{O}}{\langle u, v\rangle\mathcal{O}}\right).$$

This finishes the proof. □

Now, let $P \in C \cap D$ be an arbitrary point. If α is an affine change of coordinates on $\mathbf{A}^2$, and $\alpha(Q) = P$ for some $Q \in \mathbf{A}^2$. Then it follows from CH.II §5 that

$$\begin{aligned}Q = \alpha^{-1}(P) \in \alpha^{-1}(C) \cap \alpha^{-1}(D) &= C^{\alpha} \cap D^{\alpha}\\ &= \mathbf{V}(\alpha_*(f)) \cap \mathbf{V}(\alpha_*(g)),\end{aligned}$$

where α_* is the ring homomorphism $\mathbb{C}[x,y] \to \mathbb{C}[x,y]$ induced by α. Moreover, α_* induces the k-linear ring isomorphism

$$\frac{\mathcal{O}}{\langle f, g\rangle\mathcal{O}} \xrightarrow{\cong} \frac{\mathcal{O}_{Q,\mathbf{A}^2}}{\langle \alpha_*(f), \alpha_*(g)\rangle\mathcal{O}_{Q,\mathbf{A}^2}}$$

2.8. Definiton Let $P \in C \cap D$ be an arbitrary point, and $\mathcal{O} = \mathcal{O}_{P,\mathbf{A}^2}$ the local ring of P. The *intersection number* of C and D at P, denoted $\mathbf{I}(P, C \cap D)$, is defined to be the unique number:

$$\dim_{\mathbb{C}}\left(\frac{\mathcal{O}}{\langle f, g\rangle\mathcal{O}}\right).$$

Obviously, if $P \in C$, $P \notin D$, i.e., C and D do not intersect in P, then it follows from CH.II §6 Exercise 6(c) that $\dim_k(\mathcal{O}/\langle f, g\rangle\mathcal{O}) = 0$, which fits with our intuitive observation that C and D have intersection number 0 at P.

In conclusion, we may count the "number" of points in $C \cap D$ by adding up the intersection number at every $P \in C \cap D$ as follows.

2.9. Theorem Let $C = \mathbf{V}(f)$ and $D = \mathbf{V}(g)$ be two curves in $\mathbf{A}^2$. If f and g have no common factors, then $C \cap D$ is finite and

$$\sum_{P \in C \cap D} \mathbf{I}(P, C \cap D) = \dim_{\mathbb{C}} \left(\frac{\mathbb{C}[x,y]}{\langle f, g \rangle} \right).$$

Proof This follows from CH.II Proposition 2.3 and the previous Proposition 2.4. □

Exercises for §2

1. Find the intersection numbers of various pairs of the following given curves at the point $P = (0,0)$.
   a. $C = \mathbf{V}(y^2 - x^3)$.
   b. $D = \mathbf{V}(y^2 - x^3 - x^2)$.
   c. $E = \mathbf{V}((x^2 + y^2)^2 + 3x^2y - y^3)$.
   d. $F = \mathbf{V}((x^2 + y^2)^3 - 4x^2y^2)$.
2. A line L is tangent to a curve $C = \mathbf{V}(f)$ at a point P if and only if $\mathbf{I}(P, L \cap C) > m_P(C)$.
3. Suppose P is a double point on a curve $C = \mathbf{V}(f)$, and suppose C has only one tangent line L at P. Show that $\mathbf{I}(P, C \cap L) \geq 3$.

§3. Bezout's Theorem

Let $\mathcal{C} = \mathbf{V}(F)$ and $\mathcal{D} = \mathbf{V}(G)$ be two projective plane curves in $\mathbf{P}^2 = \mathbf{P}^2_{\mathbb{C}}$ defined by homogeneous polynomials $F, G \in \mathbb{C}[x,y,z]$. In the foregoing section, we have reduced the study of the intersection number of $\mathcal{C}$ and $\mathcal{D}$ at a point $P \in \mathcal{C} \cap \mathcal{D}$ to the study of the intersection number of the affine plane curves $C = \mathbf{V}(F_*)$ and $D = \mathbf{V}(G_*)$ at $P \in C \cap D \subset \mathbf{A}^2 = \mathbf{A}^2_{\mathbb{C}}$, and we have also seen how to calculate the intersection number. In this section, we will prove Bezout's

Theorem which tells us more about the intersection number of two projective plane curves at a point of intersection.

Let $P \in \mathcal{C} \cap \mathcal{D}$. Suppose that $\mathcal{C} \cap \mathcal{D}$ is finite and that P is in some affine part of $\mathcal{C} \cap \mathcal{D}$, say $P \in \mathbf{P}^2 - \mathbf{V}(z)$. Then the discussion in the foregoing seciton prompts us to define the *intersection number* of $\mathcal{C}$ and $\mathcal{D}$ at P, denoted $\mathbf{I}(P, \mathcal{C} \cap \mathcal{D})$, as

$$\mathbf{I}(P, \mathcal{C} \cap \mathcal{D}) = \mathbf{I}(P, \mathbf{V}(F_*) \cap \mathbf{V}(G_*)),$$

where F_* and G_* are the dehomogenizations of F and G with respect to z respectively.

From CH.III Proposition 6.9 and CH.III Proposition 5.9 we obtain the following fact.

3.1. Lemma If none of the points in $\mathcal{C} \cap \mathcal{D}$ is on the line $z = 0$, then $\mathcal{C} \cap \mathcal{D}$ is the projective closure of $C \cap D$ in $\mathbf{P}^2$, where $C = \mathbf{V}(F_*)$, $D = \mathbf{V}(G_*)$. Moreover, the ideal $\langle F, G \rangle$ is $(\phi_*)^*$-closed with respect to z (in the sense of CH.II §5).

□

3.2. Corollary (i) Let $\mathcal{C}$ and $\mathcal{D}$ be as in the Lemma. There is a k-linear isomorphism

$$\frac{\mathbb{C}[x,y,z]_d + \langle F, G \rangle}{\langle F, G \rangle} \xrightarrow{\cong} \frac{\mathbb{C}[x,y]_{\leq d} + \langle F_*, G_* \rangle}{\langle F_*, G_* \rangle}$$

where $\mathbb{C}[x,y,z]_d$ is the k-vector space of all homogeneous of degree d in $\mathbb{C}[x,y,z]$, $\mathbb{C}[x,y]_{\leq d}$ is the k-vector space of all polynomials of degree $\leq d$.

(ii) If furthermore $\mathcal{C} \cap \mathcal{D}$ is finite, then

$$\dim_{\mathbb{C}}\left(\frac{\mathbb{C}[x,y]}{\langle F_*, G_* \rangle}\right) = \dim_{\mathbb{C}}\left(\frac{\mathbb{C}[x,y,z]_d + \langle F, G \rangle}{\langle F, G \rangle}\right) = mn$$

for $d \geq m + n$, where $m = \deg(F)$, $n = \deg(G)$.

Proof (i) this follows from CH.V Proposition 4.6(ii) because $\langle F, G \rangle$ is $(\phi_*)^*$-closed.

(ii) For $d \geq m + n$, consider the k-linear map

$$\begin{array}{rccc} \psi: & \mathbb{C}[x,y,z]_{d-m-n} & \longrightarrow & \mathbb{C}[x,y,z]_{d-m} \oplus \mathbb{C}[x,y,z]_{d-n} \\ & H & \mapsto & (GH, -FH) \end{array}$$

the k-linear map

$$\begin{array}{rccc} \varphi: & \mathbb{C}[x,y,z]_{d-m} \oplus \mathbb{C}[x,y,z]_{d-n} & \longrightarrow & \mathbb{C}[x,y,z]_d \\ & (A,B) & \mapsto & AF+BG \end{array}$$

and the natural k-linear map

$$\pi: \quad \mathbb{C}[x,y,z]_d \longrightarrow \frac{\mathbb{C}[x,y,z]_d + \langle F,G\rangle}{\langle F,G\rangle}$$

then since F and G have no common factors, it may be verified directly that ψ is injective, $\mathrm{Im}\psi = \mathrm{Ker}\varphi$, and $\mathrm{Im}\varphi = \mathrm{Ker}\pi$. Hence,

$$\dim_{\mathbb{C}}\left(\frac{\mathbb{C}[x,y,z]_d + \langle F,G\rangle}{\langle F,G\rangle}\right) = \dim_{\mathbb{C}}(\mathrm{Im}\pi)$$

and

$$\begin{aligned} \dim_{\mathbb{C}}(\mathrm{Im}\pi) &= \dim_{\mathbb{C}}\mathbb{C}[x,y,z]_d - \dim_{\mathbb{C}}\mathbb{C}[x,y,z]_{d-m} - \\ &\quad -\dim_{\mathbb{C}}\mathbb{C}[x,y,z]_{d-n} + \dim_{\mathbb{C}}\mathbb{C}[x,y,z]_{d-m-n}. \end{aligned}$$

Since $\dim_{\mathbb{C}}\mathbb{C}[x,y,z]_d = \dfrac{(d+1)(d+2)}{2}$, it follows from a simple calculation that

$$\dim_{\mathbb{C}}\left(\frac{\mathbb{C}[x,y,z]_d + \langle F,G\rangle}{\langle F,G\rangle}\right) = mn$$

if $d \geq m+n$. □

3.3. Theorem (Bezout's Theorem) Let $\mathcal{C} = \mathbf{V}(F)$ and $\mathcal{D} = \mathbf{V}(G)$ be projective plane curves in $\mathbf{P}^2$ with $\deg(F) = m$ and $\deg(G) = n$ respectively. Assume F and G have no common factors. Then

$$\sum_{P \in \mathcal{C}\cap\mathcal{D}} \mathbf{I}(P, \mathcal{C}\cap\mathcal{D}) = mn.$$

Proof Since $\mathcal{C}\cap\mathcal{D}$ is finite by Lemma 2.1, we may assume, by a projective change of coordinates if necessary, that none of the points in $\mathcal{C}\cap\mathcal{D}$ is on the line at infinity $z = 0$. Let $C = \mathbf{V}(F_*)$, $D = \mathbf{V}(G_*)$, where F_* and G_* are the dehomogenizations of F and G with respect to z respectively. Then it follows from Theorem 2.9 and Corollary 3.2 that

$$\begin{aligned}\sum_{P\in\mathcal{C}\cap\mathcal{D}} \mathbf{I}(P,\mathcal{C}\cap\mathcal{D}) &= \sum_{P\in C\cap D} \mathbf{I}(P,C\cap D)\\ &= \dim_{\mathbb{C}}\left(\frac{\mathbb{C}[x,y]}{\langle F_*,G_*\rangle}\right)\\ &= mn.\end{aligned}$$

□

Combining Corollary 2.3 with Bezout's Theorem we deduce

3.4. Corollary If two plane curves $\mathcal{C} = \mathbf{V}(F)$, $\mathcal{D} = \mathbf{V}(G)$ in $\mathbf{P}^2$ have no common components, then

$$\sum_{P\in\mathcal{C}\cap\mathcal{D}} m_P(\mathcal{C})m_P(\mathcal{D}) \leq \deg(F)\cdot\deg(G).$$

□

3.5. Corollary (i) If two plane curves $\mathcal{C} = \mathbf{V}(F)$, $\mathcal{D} = \mathbf{V}(G)$ in $\mathbf{P}^2$ meet in mn distinct points, $m = \deg(F)$, $n = \deg(G)$, then these points are all simple on F and on G.
(ii) if $\mathcal{C}$ and $\mathcal{D}$, as in (i), have more than mn points in common, then they have a common component.

□

As an application of Bezout's Theorem, we will prove the following result due to Pascal. Suppose we have six distinct points $P_1,\ldots,P_6$ on an *irreducible* conic in $\mathbf{P}^2$ (we may assume that the six points entirely lie in some affine part of P^2 so that we can "see" them in $\mathbf{A}^2$). By Bezout's Theorem, a line meets the conic in at most 2 points. Hence, we get six distinct lines by connecting P_1 to P_2, P_2 to P_3, ..., and P_6 to P_1. If we label these lines $L_1,\ldots,L_6$, then we get the following picture:

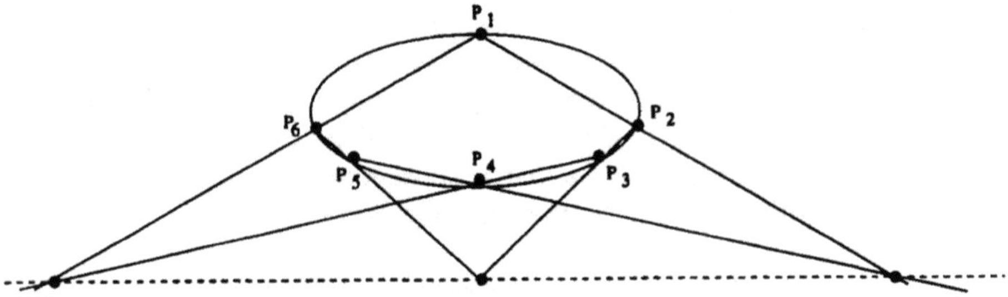

Figure 10

We say that lines L_1, L_4 are *opposite*, and similarly the pairs L_2, L_5 and L_3, L_6 are opposite. The portions of the lines lying inside the conic form a hexagon, and opposite lines correspond to opposite sides of the hexagon.
In the above picture, the intersections of the opposite pairs of lines appear to lie on the same line. The following theorem reveals that this is not an accident.

3.6. Theorem (Pascal's Mystic Hexagon) Given six points on an irreducible conic, connected by six lines as above, the points of intersection of the three pairs of opposite lines are collinear.

Proof Let the conic be C. As above, we have six points $P_1, ..., P_6$ and three pairs of opposite lines $\{L_1, L_4\}$, $\{L_2, L_5\}$, and $\{L_3, L_6\}$. Now consider the curves $C_1 = L_1 \cup L_2 \cup L_3$ and $C_2 = L_4 \cup L_5 \cup L_6$. These curves are defined by cubic equations, so that by Bezout's Theorem, the number of points in $C_1 \cap C_2$ is 9 (counted with intersection numbers). However, note that $C_1 \cap C_2$ contains the six original points $P_1, ..., P_6$ and the three points of intersection of opposite pairs of lines (check it!). Thus, these are all of the points of intersection, and all of the intersection numbers are 1.
Suppose that $C = \mathbf{V}(F)$, $C_1 = \mathbf{V}(G_1)$, and $C_2 = \mathbf{V}(G_2)$, where F has total degree 2 and G_1 and G_2 have total degree 3. Now pick a point $P \in C$ distinct from $P_1, ..., P_6$. Thus, $G_1(P)$ and $G_2(P)$ are nonzero (why?), so that $G = G_2(P)G_1 - G_1(P)G_2$ is a cubic polynomial which vanishes at $P, P_1, ..., P_6$. Furthermore, G is nonzero since otherwise G_1 would be a multiple of G_2 (or vice versa). Hence, the cubic $\mathbf{V}(G)$ meets the conic C in at least seven points, so that the hypotheses for Bezout's Theorem are not satisfied. Thus, by Corollary 3.5(ii), C

and $\mathbf{V}(G)$ have a common component. But C is irreducible, which implies that $C = \mathbf{V}(F)$ is a component of $\mathbf{V}(G)$, and consequently F must divide G. Therefore, we get a factorization $G = FL$, where L has total degree 1. Since G vanishes where the opposite lines meet and F doesn't, it follows that L vanishes at these points. Since $\mathbf{V}(L)$ is a projective line, the theorem is proved. □

Bezout's Theorem serves as a nice introduction to the study of curves in $\mathbf{P}^2$. This part of algebraic geometry is traditionally called *algebraic curves* and includes many interesting topics we have omitted (inflection points, dual curves, etc.).

Exercises for §3

1. This exercise is concerned with the parabola $y = x^2$ and the ellipse $x^2 + 4(y - \lambda)^2 = 4$ from Example (ii) of this section.
   a. Show that these curves have empty intersection over $\mathbb{R}$ when $\lambda < -1$. Illustrate the cases $\lambda < -1$ and $\lambda = -1$ with a picture.
   b. Find the smallest positive real number λ_0 such that the intersection over $\mathbb{R}$ is empty when $\lambda > \lambda_0$. Illustrate the cases $\lambda > \lambda_0$ and $\lambda = \lambda_0$ with a picture.
   c. When $-1 < \lambda < \lambda_0$, describe the possible types of intersections that can occur over $\mathbb{R}$ and illustrate each case with a picture.
   d. In the pictures for parts a, b, and c, use the intuitive idea of intersection number described in Example (ii) to determine which ones represent intersections with intersection number > 1.
   e. Without using Bezout's Theorem, explain why (over $\mathbb{C}$), the number of points of intersection, counted with intersection number, adds up to 4 when λ is real.
2. Let C and D be curves in $\mathbf{P}^2_{\mathbb{C}}$.
   a. (Comparing with CH.V Proposition 5.7) Prove that $C \cap D \neq \emptyset$.
   b. Suppose that C is nonsingular, show that C is irreducible. (Compare this with §1 Exercise 10(b).) Is this true for affine curves?
3. Let C be an irreducible conic in $\mathbf{P}^2_{\mathbb{C}}$. Use Bezout's Theorem

to explain why a line L meets in at most two points. What happens when C is reducible? What about when C is a curve defined by an irreducible polynomial of total degree n?

4. In the picture for Pascal's Mystic Hexagon, the six points are ordered clockwise around the conic. if we change the order of the points, we can still form a "hexagon", though opposite lines might intersect inside the conic. For example, the picture could be as follows:

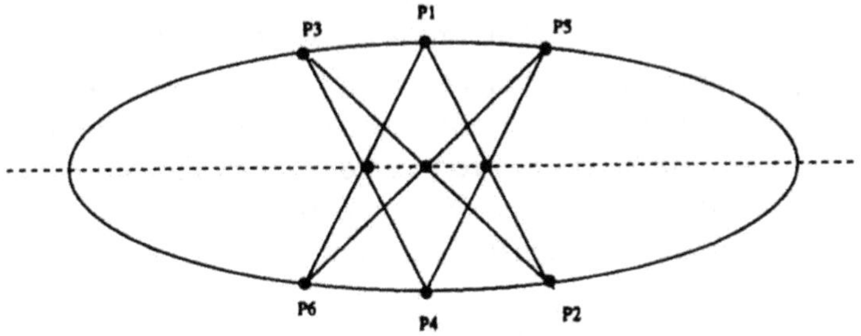

Figure 10

Investigate if the theorem remains true in this case and explain.

5. In Pascal's Mystic Hexagon, suppose that the conic is a circle and the six lines come from a regular hexagon inscribed inside the circle. Where do the opposite lines meet and on what line do their intersections lie?

§4. Divisors on Curves

In previous sections the singularity of points on a plane algebraic curve and the intersection number of curves have been defined and described in $\mathbf{P}_k^2$ by using local rings of regular (rational) functions associated to points. From CH.II and CH.III we retain that every rational function is an "analytic-like" function if we are working over $\mathbb{C}$. However, so far we have not seen, viewing each rational function $\frac{\phi}{\psi} \in k(C)$ as a function $C \to \mathbb{C} \cup \{\infty\}$, what information about this function may be obtained by looking at its zeros and poles.

In complex function theory, rational functions on $\mathbf{P}_{\mathbb{C}}^1$ (the Riemann sphere) are just meromorphic functions, and several results are obtained to relate meromorphic functions on $\mathbf{P}^1_{\mathbb{C}}$ to their zeros and poles. Here are few of them:

(1) Each nonzero meromorphic function on $\mathbf{P}^1_{\mathbb{C}}$ has only finitely many zeros and poles. That is, any meromorphic function f on $\mathbf{P}^1_{\mathbb{C}}$ defines a divisor $\mathbf{div}(f) = \sum_i m_i P_i$, where $|m_i|$ is the multiplicity of the zero ($m_i > 0$) or pole ($m_i < 0$).

(2) Each nonzero meromorphic function on $\mathbf{P}^1_{\mathbb{C}}$ has as many zeros as poles. More precisely, $\deg(\mathbf{div}(f)) = \sum_i m_i = 0$.

(3) A meromorphic function on $\mathbf{P}^1_{\mathbb{C}}$ has no zeros and poles if and only if it is a nonzero constant. (This implies that the only functions everywhere holomorphic on $\mathbf{P}^1_{\mathbb{C}}$ are the constant functions. It is also implies that multiplying a function by a nonzero constant does not affect its zero or poles.)

(4) For each divisor D, which is a formal sum $D = \sum_{P \in \mathbf{P}^1_{\mathbb{C}}} n_P P$ with only finitely many nonzero $n_P \in \mathbb{Z}$, if $\deg D = \sum n_P = 0$ there is a meromorphic function f on $\mathbf{P}^1_{\mathbb{C}}$ with $D = \mathbf{div}(f)$; f is unique up to a nonzero constant factor.

It is natural to ask:

- What part of complex function theory on $\mathbf{P}^1_{\mathbb{C}}$ carries over to an arbitrary projective curve C?

We start to introduce such "elementary function theory" on plane algebraic curves by extending "divisors" to nonsingular projective curves, and finish (in §6) with the Riemann-Roch theorem, which is one of the most fundamental results in the algebraic geometry of curves. One of its most important applications allows to describe the functions on a curve having prescribed zeros and poles, i.e., an analogue of the statement in (4) above.

Let C be an *irreducible* projective plane curve in $\mathbf{P}^2 = \mathbf{P}^2_k$, where $k = \mathbb{C}$. From CH.II §2 Exercise 6 it follows that $C = \mathbf{V}(F)$ where F is an irreducible homogeneous polynomial in $k[x, y, z]$, say $\deg F = n$. Let G be a homogeneous polynomial of degree m in $k[x, y, z]$ and $\mathcal{D} = \mathbf{V}(G)$ the curve determined by G. Suppose that $C \not\subset \mathcal{D}$. Then it follows from Lemma 2.1 that $C \cap \mathcal{D}$ is a finite set of points, and it follows from Bezout's theorem that

$$\sum_{P \in \mathcal{C} \cap \mathcal{D}} \mathbf{I}(P, \mathcal{C} \cap \mathcal{D}) = mn.$$

Let $P \in \mathcal{C} \cap \mathcal{D}$, say $P = (a, b, 1)$, and let $f = F_*$, $g = G_*$ be the dehomogenizations of F and G with respect to z. If we write $\mathcal{C}_*$ for the affine curve $\mathbf{V}(f) \subset \mathbf{A}^2 = \mathbf{A}_k^2$, $\mathcal{O}_{P,\mathbf{A}^2}$ for the local ring of P on $\mathbf{A}^2$ and $\mathcal{O}_{P,\mathcal{C}_*}$ for the local ring of P on $\mathcal{C}_*$ as in CH.II, then by CH.II §6 Exercise 6(c) we have

$$\begin{cases} \dfrac{\mathcal{O}_{P,\mathbf{A}^2}}{\langle f \rangle \mathcal{O}_{P,\mathbf{A}^2}} \cong \mathcal{O}_{P,\mathcal{C}_*}, \\[2ex] \dfrac{\mathcal{O}_{P,\mathbf{A}^2}}{\langle f, g \rangle \mathcal{O}_{P,\mathbf{A}^2}} \cong \dfrac{\mathcal{O}_{P,\mathcal{C}_*}}{\langle [f], [g] \rangle \mathcal{O}_{P,\mathcal{C}_*}} = \dfrac{\mathcal{O}_{P,\mathcal{C}_*}}{\langle [g] \rangle \mathcal{O}_{P,\mathcal{C}_*}}, \end{cases}$$

where $[f]$, $[g]$ are the classes of f and g in the coordinate ring $k[\mathcal{C}_*]$ of $\mathcal{C}_*$. Thus, the definition of intersection number of curves and Theorem 2.7 together yield the following *computable* number:

$$\begin{aligned} \mathbf{I}(P, \mathcal{C} \cap \mathcal{D}) &= \mathbf{I}(P, \mathbf{V}(f) \cap \mathbf{V}(g)) \\ &= \dim_k \left(\frac{\mathcal{O}_{P,\mathcal{C}_*}}{\langle [g] \rangle \mathcal{O}_{P,\mathcal{C}_*}} \right). \end{aligned}$$

If P is a *simple* point on $\mathcal{C}$, then $\mathcal{O}_{P,\mathcal{C}_*}$ is a discrete valuation ring in $k(\mathcal{C})$ (Theorem 1.5). Let t be a uniformizing parameter of $\mathcal{O}_{P,\mathcal{C}_*}$, i.e., the maximal ideal of $\mathcal{O}_{P,\mathcal{C}_*}$ is a principle ideal generated by t. Then $\langle [g] \rangle = t^n \mathcal{O}_{P,\mathcal{C}_*}$ for some $n \geq 1$ (why?), and $[g] = t^n u$ for some unit u in $\mathcal{O}_{P,\mathcal{C}_*}$. Writing v_P for the valuation function on $k(\mathcal{C})$ determined by t, (by Appendix.II §2 Exercise 3) we obtain

$$\text{(D1)} \qquad v_P([g]) = n = \dim_k \left(\frac{\mathcal{O}_{P,\mathcal{C}_*}}{\langle [g] \rangle \mathcal{O}_{P,\mathcal{C}_*}} \right) = \mathbf{I}(P, \mathcal{C} \cap \mathcal{D}).$$

Now let us look at an arbitrary *nonzero* function $\frac{\phi}{\psi} \in k(\mathcal{C})$. Suppose that $\frac{\phi}{\psi}$ is represented by $\frac{G}{H}$, where G and H are homogeneous polynomials of the same degree m, say. $\frac{\phi}{\psi}$ is associated with two subsets of $\mathcal{C}$, i.e., the set of *zeros* and the set of *poles* (see CH.II §6 and CH.III §3):

$$\mathcal{Z}\left(\frac{\phi}{\psi}\right) = \left\{P \in \mathcal{C} \,\middle|\, \frac{\phi}{\psi}(P) = 0\right\},$$

$$\mathcal{P}\left(\frac{\phi}{\psi}\right) = \left\{P \in \mathcal{C} \,\middle|\, \frac{\phi}{\psi} \text{ is not defined at } P\right\}.$$

Since $\frac{\phi}{\psi} \neq 0$ and $\mathcal{Z}\left(\frac{\phi}{\psi}\right) \subset \mathbf{V}(G) \cap \mathcal{C}$, $\mathcal{P}\left(\frac{\phi}{\psi}\right) \subset \mathbf{V}(H) \cap \mathcal{C}$, we conclude that

(D2) Both $\mathcal{Z}\left(\frac{\phi}{\psi}\right)$ and $\mathcal{P}\left(\frac{\phi}{\psi}\right)$ are finite sets of points.

Again suppose that P is a simple point on $\mathcal{C}$. From CH.II §6 and APP.II §2 it follow sthat

$$\text{(D3)} \quad \begin{cases} P \in \mathcal{Z}\left(\frac{\phi}{\psi}\right) & \text{if and only if } v_P\left(\frac{\phi}{\psi}\right) > 0, \\ P \in \mathcal{P}\left(\frac{\phi}{\psi}\right) & \text{if and only if } v_P\left(\frac{\phi}{\psi}\right) < 0, \\ P \notin \mathcal{Z}\left(\frac{\phi}{\psi}\right),\ P \notin \mathcal{P}\left(\frac{\phi}{\psi}\right) & \text{if and only if } v_P\left(\frac{\phi}{\psi}\right) = 0. \end{cases}$$

Moreover, write $[g]$ and $[h]$ for the classes of G_* and H_* in $k[\mathcal{C}_*]$. By (D1) we have

$$\begin{aligned} v_P\left(\frac{\phi}{\psi}\right) &= v_P([g]) - v_P([h]) \\ &= \mathbf{I}(P, \mathcal{C} \cap \mathbf{V}(G)) - \mathbf{I}(P, \mathcal{C} \cap \mathbf{V}(H)). \end{aligned}$$

If $\mathcal{C}$ is a nonsingular curve, then since $v_P\left(\frac{\phi}{\psi}\right)$ is *independent* of the choice of its representative $\frac{G}{H}$ we have by Bezout's theorem that

$$\text{(D4)} \quad \begin{cases} \displaystyle\sum_{P \in \mathcal{C}} v_P\left(\frac{\phi}{\psi}\right) &= \displaystyle\sum_{P \in \mathcal{C}} \mathbf{I}(P, \mathcal{C} \cap \mathbf{V}(G)) - \\ & \qquad\qquad - \displaystyle\sum_{P \in \mathcal{C}} \mathbf{I}(P, \mathcal{C} \cap \mathbf{V}(H)) \\ &= mn - mn \\ &= 0. \end{cases}$$

From (D3) it follows that

(D5) $\frac{\phi}{\psi}$ has as many zeros as it does poles on C.

Again let C be nonsingular, and $\frac{\phi}{\psi} \in k(C)$. Suppose that $\mathcal{Z}\left(\frac{\phi}{\psi}\right) = \mathcal{P}\left(\frac{\phi}{\psi}\right) = \emptyset$, and that $\frac{\phi}{\psi}$ is nonconstant. For any $P_0 \in C$, $\frac{\phi}{\psi}(P_0) = c \in \mathbb{C} - \{0\}$. Then $\frac{\phi}{\psi} - c$ has a zero at P_0, but still nonconstant. Since $\frac{\phi}{\psi}$ has no poles, so does $\frac{\phi}{\psi} - c$. This contradicts (D5) above. Hence,

(D6)

$\frac{\phi}{\psi}$ has no zeros and no poles if and only if it is a nonzero constant.

So, we have seen that any nonzero rational function on a nonsingular projective curve $C \subset \mathbf{P}^2_{\mathbb{C}}$ has the properties (1)–(3) that a nonzero meromorphic function on $\mathbf{P}^1_{\mathbb{C}}$ does. In order to extend the foregoing (4) to a projective curve in $\mathbf{P}^2_{\mathbb{C}}$, we need first to extend the notion of a "divisor" to a curve.

In what follows $C = V(F)$ always denotes an *irreducible nonsingular* curve in $\mathbf{P}^2$.

4.1. Definition A *divisor* on C is a formal sum $\sum_{P \in C} n_P P$, where $n_i \in \mathbb{Z}$ and all but a finite number of the integers n_P are zero. The *support* of a divisor $D = \sum_{P \in C} n_P P$, denoted Supp(D), is defined to be

$$\operatorname{Supp}(D) = \{P \in C \mid n_P \neq 0\}.$$

Divisors on a curve C form a set and this set becomes an abelian group with respect to the following operation: for $D = \sum_i n_i P_i$, $D' = \sum_i n'_i P_i$, we put

$$D + D' = \sum_i (n_i + n'_i) P_i.$$

We write $\mathbf{Div}(C)$ for the group of divisors on C. We say that a divisor D is *positive*, denoted by $D \succ 0$, if $D = \sum n_P P$ with $n_P \geq 0$ for all $P \in C$; in particular the zero divisor is considered to be positive. Thus, we say $D \succ D'$ if $D - D'$ is positive.

Example (i) For curves C and C' we define the *intersection divisor* $C \circ C'$ as

$$C \circ C' = \sum_{P \in C \cap C'} \mathbf{I}(P, C \cap C')P,$$

If we look at curves C, C', C'', then $C \circ (C' \cup C'') = C \circ C' + C \circ C''$ because $\mathbf{I}(P, C \cap (C' \cup C'')) = \mathbf{I}(P, C \cap C') + \mathbf{I}(P, C \cap C'')$.

(ii) Using the notation as before, take $\frac{\phi}{\psi} \in k(C)$, $\frac{\phi}{\psi} \neq 0$. We define the *divisor* of $\frac{\phi}{\psi}$, $\operatorname{div}\left(\frac{\phi}{\psi}\right)$, by putting

$$\operatorname{div}\left(\frac{\phi}{\psi}\right) = \sum_{P \in C} v_P\left(\frac{\phi}{\psi}\right)P.$$

Since $\frac{\phi}{\psi}$ has only a finite number of zeros and poles (see (D2) above), $\operatorname{div}\left(\frac{\phi}{\psi}\right)$ is a well-defined divisor. If we let

$$\left(\frac{\phi}{\psi}\right)_0 = \sum_{v_P\left(\frac{\phi}{\psi}\right)>0} v_P\left(\frac{\phi}{\psi}\right)P, \text{ the } \textit{divisor of zeros} \text{ of } \frac{\phi}{\psi},$$

$$\left(\frac{\phi}{\psi}\right)_\infty = -\sum_{v_P\left(\frac{\phi}{\psi}\right)<0} v_P\left(\frac{\phi}{\psi}\right)P, \text{ the } \textit{divisor of poles} \text{ of } \frac{\phi}{\psi},$$

then it follows from (D3) that

$$\operatorname{div}\left(\frac{\phi}{\psi}\right) = \sum_{P \in C} v_P\left(\frac{\phi}{\psi}\right)P = \left(\frac{\phi}{\psi}\right)_0 - \left(\frac{\phi}{\psi}\right)_\infty.$$

If $\frac{\phi}{\psi}$ and $\frac{\eta}{\xi}$ are nonzero functions in $k(C)$ then $\operatorname{div}\left(\frac{\phi}{\psi} \cdot \frac{\eta}{\xi}\right) = \operatorname{div}\left(\frac{\phi}{\psi}\right) + \operatorname{div}\left(\frac{\eta}{\xi}\right)$, and $\operatorname{div}\left(\left(\frac{\phi}{\psi}\right)^{-1}\right) = -\operatorname{div}\left(\frac{\phi}{\psi}\right)$.

If $D = \sum n_P P$ is a divisor on C then the *degree* of D, denoted $\deg D$, is defined to be $\sum n_P$ (note that this is a finite sum).
A function $\frac{\phi}{\psi} \in k(C)$ is said to have *valence* r if it takes every value $c \in k$ exactly r-times.

4.2. Proposition (i) If $\frac{\phi}{\psi} \neq 0$ in $k(C)$ then $\deg\left(\frac{\phi}{\psi}\right)_0 = \deg\left(\frac{\phi}{\psi}\right)_\infty$, i.e., the number of zeros equals the number of poles of $\frac{\phi}{\psi}$ (if the multiplicity of each point is counted).

(ii) If $\frac{\phi}{\psi}$ is not a constant in $k(C)$ then $\frac{\phi}{\psi}$ has valence $\deg\left(\frac{\phi}{\psi}\right)_\infty$.
(iii) If $\frac{\phi}{\psi}, \frac{\eta}{\xi} \neq 0$ in $k(C)$ then $\mathbf{div}\left(\frac{\phi}{\psi} \cdot \frac{\eta}{\xi}\right) = \mathbf{div}\left(\frac{\phi}{\psi}\right) + \mathbf{div}\left(\frac{\eta}{\xi}\right)$, $\mathbf{div}\left(\left(\frac{\phi}{\psi}\right)^{-1}\right) = -\mathbf{div}\left(\frac{\phi}{\psi}\right)$; hence $\mathbf{div}\left(\frac{\phi}{\psi}/\frac{\eta}{\xi}\right) = \mathbf{div}\left(\frac{\phi}{\psi}\right) - \mathbf{div}\left(\frac{\eta}{\xi}\right)$.

Proof (i) This follows from the foregoing (D5).
(ii) If $\frac{\phi}{\psi}$ is not a constant and $c \in k$, look at $\frac{\phi}{\psi} - c$. It is clear that $\left(\frac{\phi}{\psi} - c\right)_\infty = \left(\frac{\phi}{\psi}\right)_\infty$. Hence $\frac{\phi}{\psi} - c$ has $\deg\left(\frac{\phi}{\psi}\right)_\infty$ zeros. The latter just expresses that $\frac{\phi}{\psi}$ takes the value $c \in k$ exactly $\deg\left(\frac{\phi}{\psi}\right)_\infty$ times, and hence $\frac{\phi}{\psi}$ has valence $\deg\left(\frac{\phi}{\psi}\right)_\infty$.
(iii) Exercise. □

4.3. Observation (i) The set of divisors on C consisting of the divisors $\mathbf{div}\left(\frac{\phi}{\psi}\right)$ for $\frac{\phi}{\psi} \neq 0$ in $k(C)$ is a subgroup of $\mathbf{Div}(C)$, denoted by $\mathbf{Div}_v(C)$. The set of divisors on C consisting of the divisors of degree zero is also a subgroup denoted by $\mathbf{Div}^0(C)$, and we have $\mathbf{Div}_v(C) \subset \mathbf{Div}^0(C) \subset \mathbf{Div}(C)$. Write

$$\mathbf{Pic}(C) = \frac{\mathbf{Div}(C)}{\mathbf{Div}_v(C)}.$$

We call divisors D_1 and D_2 in $\mathbf{Div}(C)$ *equivalent*, denoted $D_1 \sim D_2$, if equivalently: $D_1 = D_2 + \mathbf{div}\left(\frac{\phi}{\psi}\right)$ for some $\frac{\phi}{\psi} \in k(C)$, or $D_1 - D_2 \in \mathbf{Div}_v(C)$. An equivalence class is called a *divisor class*. The group of divisor classes, $\mathbf{Pic}(C)$, is called the *divisor class group*, or the *Picard group* of C.
(ii) If $D_1 \sim D_2$, i.e., $D_1 - D_2 \in \mathbf{Div}_v(C)$, then $\deg D_1 = \deg D_2$, and hence we may view deg as being defined on $\mathbf{Pic}(C)$, $\deg\colon \mathbf{Pic}(C) \to \mathbb{Z}$, and $\ker(\deg) = \mathbf{Div}^0(C)/\mathbf{Div}_v(C)$. One may hope that the latter is zero, we return to this problem later. In fact if C is a line L then the equality between $\mathbf{Div}^0(L)$ and $\mathbf{Div}_v(L)$ may be checked easily by using coordinates on the line (exercise).

Exercises for §4

1. Prove Proposition 4.2(iii).
2. let $C = \mathbf{P}^1_{\mathbb{C}}$, $\mathbb{C}(C) = \mathbb{C}(t)$, where $t = \frac{x}{y}$, x, y homogeneous coordinates in $\mathbf{P}^1_{\mathbb{C}}$.

a. calculate $\mathbf{div}(t)$.
b. Calculate $\mathbf{div}\left(\frac{f}{g}\right)$, f, g relatively prime in $k[t]$.

3. Let $C = \mathbf{V}\left(Y^2Z - X(X-Z)(X-\lambda Z)\right) \subset \mathbf{P}^2_{\mathbb{C}}$, $\lambda \in k$, $\lambda \neq 0, 1$ (see the example given in the end of CH.I §3). let $x = \frac{X}{Z}$, $y = \frac{Y}{Z} \in k(x,y)$. Calculate $\mathbf{div}(x)$, $\mathbf{div}(y)$.

§5. The Linear Space $\mathcal{L}(D)$

Let $C \subset \mathbf{P}^2 = \mathbf{P}^2_k$ be an irreducible nonsingular curve over $k = \mathbb{C}$. When we try to extend property (4) (see beginning of foregoing section) we have to determine when there exists a rational function $\frac{\phi}{\psi}$ such that $\mathbf{div}\left(\frac{\phi}{\psi}\right) = D$ for some given divisor $D = \sum n_P P$ that picks out finitely many points and designates numbers to them. More modestly, we look for $\frac{\phi}{\psi}$ having poles only at the chosen points and with poles no "worse" than order n_P at P; if so, how many such functions are there?

We maintain the notation of last section. To a divisor D on C we associate the *linear space* of D, denoted by $\mathcal{L}(D)$, as follows:

$$\mathcal{L}(D) = \left\{ \frac{\phi}{\psi} \in k(C) \;\middle|\; D + \mathbf{div}\left(\frac{\phi}{\psi}\right) \succ 0 \right\}$$

Let $D = \sum n_P P$. If $\frac{\phi}{\psi} \in \mathcal{L}(D)$, then $D + \mathbf{div}\left(\frac{\phi}{\psi}\right)$ is a positive divisor, or in other words, $v_P\left(\frac{\phi}{\psi}\right) \geq -n_P$ for all n_P appearing in D. Such a function $\frac{\phi}{\psi}$ has, in a given set of points, zeros of an order larger than a prescribed number, and in another given set of points, poles of order smaller than a prescribed number.

5.1. Lemma For $D = \sum_i n_i P_i \in \mathbf{Div}(C)$, $\mathcal{L}(D)$ is a k-vector space.

Proof The zero function is in $\mathcal{L}(D)$ because $v_P(0) = \infty$ for every $P \in C$. For $c \in k$ we have $\mathbf{div}\left(c \cdot \frac{\phi}{\psi}\right) = \mathbf{div}\left(\frac{\phi}{\psi}\right)$ for $\frac{\phi}{\psi} \in k(C)$. Hence, if $\frac{\phi}{\psi} \in \mathcal{L}(D)$ then $c \cdot \frac{\phi}{\psi} \in \mathcal{L}(D)$. If $\frac{\phi}{\psi}, \frac{\eta}{\xi} \in \mathcal{L}(D)$, then since

$$v_P\left(\frac{\phi}{\psi} + \frac{\eta}{\xi}\right) \geq \min\left(v_P\left(\frac{\phi}{\psi}\right), v_P\left(\frac{\eta}{\xi}\right)\right) \geq -n_i,$$

we have $\frac{\phi}{\psi} + \frac{\eta}{\xi} \in \mathcal{L}(D)$. □

One of the main problems is to calculate the dimension $\ell(D)$ of $\mathcal{L}(D)$ over k, after establishing that $\ell(D)$ is finite. This is actually the topic of the Riemann-Roch theorem.
First let us observe that $\ell(D)$ depends only on the divisor class of D and not on the representative D; this follows from the following proposition.

5.2. Proposition If $D \sim D_1$ then $\mathcal{L}(D) \cong \mathcal{L}(D_1)$ as k-spaces.

Proof From $D \sim D_1$ we obtain $D - D_1 = \text{div}\left(\frac{\phi}{\psi}\right)$ for some $\frac{\phi}{\psi} \neq 0$ in $k(C)$. Define

$$\begin{array}{rccc} \mu: & \mathcal{L}(D) & \longrightarrow & \mathcal{L}(D_1) \\ & \dfrac{\eta}{\xi} & \mapsto & \dfrac{\eta}{\xi} \cdot \dfrac{\phi}{\psi} \end{array}$$

This is obviously well-defined and injective. Note that since $\text{div}\left(\frac{\eta}{\xi}\right) + D \succ 0$, it follows that $\text{div}\left(\frac{\eta}{\xi}\right) + D_1 + \text{div}\left(\frac{\phi}{\psi}\right) \succ 0$ and hence $\frac{\eta}{\xi} \cdot \frac{\phi}{\psi} \in \mathcal{L}(D_1)$. If $\frac{\tau}{\chi} \in \mathcal{L}(D_1)$ then $\left(\frac{\phi}{\psi}\right)^{-1} \cdot \frac{\tau}{\chi}$ satisfies $\text{div}\left(\left(\frac{\phi}{\psi}\right)^{-1} \cdot \frac{\tau}{\chi}\right) + \text{div}\left(\frac{\phi}{\psi}\right) + D_1 \succ 0$, hence $\text{div}\left(\left(\frac{\phi}{\psi}\right)^{-1} \cdot \frac{\tau}{\chi}\right) + D \succ 0$, or $\left(\frac{\phi}{\psi}\right)^{-1} \cdot \frac{\tau}{\chi} \in \mathcal{L}(D)$. It follows that μ is a k-linear isomorphism.
□

The next proposition establishes that $\ell(D)$ is finite.

5.3. Proposition (i) If $D \prec D'$, then $\mathcal{L}(D) \subset \mathcal{L}(D')$, and

$$\dim_k \frac{\mathcal{L}(D')}{\mathcal{L}(D)} = \ell(D') - \ell(D) \leq \deg(D' - D).$$

(ii) $\mathcal{L}(0) = k$; $\mathcal{L}(D) = \{0\}$ if $\deg D < 0$.
(iii) $\mathcal{L}(D)$ is finite dimensional for all D. If $\deg D \geq 0$, then $\ell(D) \leq \deg D + 1$.

Proof (i) Since $D' = D + P_1 + \cdots + P_s$, and

$$\mathcal{L}(D) \subset \mathcal{L}(D + P_1) \subset \cdots \subset \mathcal{L}(D + P_1 + \cdots + P_s),$$

the assertion will follow inductively if we can show

$$\dim_k(\mathcal{L}(D+P)/\mathcal{L}(D)) \leq 1, \text{ for any } P \in C.$$

To prove this, let t be a uniformizing parameter of $\mathcal{O}_{P,C}$, and let $r = n_P$ be the coefficient of P in D. Define

$$\begin{array}{rccc} \varphi: & \mathcal{L}(D+P) & \longrightarrow & k \\ & \dfrac{\phi}{\psi} & \mapsto & \left(t^{r+1}\cdot\dfrac{\phi}{\psi}\right)(P) \end{array}$$

Since $v_P\left(\frac{\phi}{\psi}\right) \geq -r-1$, this is well defined. φ is a linear map, and $\mathrm{Ker}\varphi = \mathcal{L}(D)$ (check this!), hence φ induces a one-to-one linear map $\overline{\varphi}: \quad \mathcal{L}(D+P)/\mathcal{L}(D) \longrightarrow k$ which gives the desired result.
(ii) $\mathcal{L}(0) = k$ follows from Proposition 4.2. Note that $\deg D = \deg\left(\mathrm{div}\left(\frac{\phi}{\psi}\right) + \deg D\right) = \deg\left(D + \mathrm{div}\left(\frac{\phi}{\psi}\right)\right)$. Now, $0 \neq \frac{\phi}{\psi} \in \mathcal{L}(D)$ is equivalent to $D + \mathrm{div}\left(\frac{\phi}{\psi}\right) \succ 0$, but then $\deg\left(D + \mathrm{div}\left(\frac{\phi}{\psi}\right)\right)$ cannot be negative. It follows that $\mathcal{L}(D) = \{0\}$ or $\ell(D) = 0$.
(iii) If $\deg D = n \geq 0$, choose $P \in C$, and let $D' = D - (n+1)P$. Then $\mathcal{L}(D') = \{0\}$ by (ii), and by (i), $\dim_k(\mathcal{L}(D)/\mathcal{L}D')) \leq n+1$. Hence $\ell(D) \leq n+1 = \deg D + 1$. □

5.4. Example (i) The linear space $\mathcal{L}(D)$ for a line L.
Up to a coordinate transform we may obtain D in some affine $\mathbf{A}^2$, $D = \sum_{i=1}^{r} n_i P_i - \sum_{j=1}^{s} m_j Q_j$, and we may also assume that L is transformed to the line given by $y = 0$.
That $\ell(D) = 0$ if $\deg D < 0$ has been observed above. If $\deg D \geq 0$ then $\ell(D) = 1+ \deg D$ holds in this case. This is proved as follows. On the line L, $y = 0$, let the P_i correspond to the coordinate a_i, the Q_j to the coordinate b_j. Consider the polynomials

$$f = \prod_{j=1}^{s}(x-b_j)^{m_j}\prod_{i=1}^{r}(x-a_i)^{n_i},$$

$$g = f\cdot T(x) \text{ with } \deg T(x) \leq \sum_i n_i - \sum_j m_j = \deg D.$$

Only g of this form can be in $\mathcal{L}(D)$ because we cannot have ∞ as a pole. Thus, we arrive at the following k-basis for the space $\mathcal{L}(D)$:

$$\left\{f, xf, ..., x^d f\right\}, \text{ where } d = \deg D.$$

Hence, $\dim_k \mathcal{L}(D) = 1 + d$, as claimed.

If two nonsingular curves C and C' are birationally equivalent, then we have seen in CH.VI that points corresponding under the equivalence have isomorphic local rings, the isomorphisms of the corresponding local rings being induced by an isomorphism of the function fields $k(C)$ and $k(C')$ that we may fix (or use an identification). If points P_i' in C' correspond to points P_i in C, then a divisor $D = \sum n_i P_i$ is transformed to $D' = \sum n_i P_i'$.

5.5. Corollary With notation as above, if D' corresponds to D under a birational equivalence between the curves C and C', then $\mathcal{L}(D) \cong \mathcal{L}(D')$, hence $\ell(D)$ is a birational invariant.

□

5.6. Corollary Let C be a nonsingular conic (or in fact any rational curve) and D a divisor on C. Then $\ell(D) = 0$ if $\deg D < 0$ and $\ell(D) = 1 + \deg D$ otherwise.

Proof Using §1 Exercise 10, this may be obtained directly from Example 5.4 and the foregoing corollary. □

More generally, for any subset S of C, and any divisor $D = \sum n_P P$, define

$$\deg^S D = \sum_{P \in S} n_P,$$

$$\mathcal{L}^S(D) = \left\{ \frac{\phi}{\psi} \in k(C) \;\middle|\; v_P\left(\frac{\phi}{\psi}\right) \geq -n_P \text{ for all } P \in S \right\}.$$

A similar argument as for $\mathcal{L}(D)$ shows that $\mathcal{L}^S(D)$ is a k-linear space.

5.7. Lemma If $D \prec D'$, then $\mathcal{L}^S(D) \subset \mathcal{L}^S(D')$. If S is finite, then

$$\dim_k \left(\frac{\mathcal{L}^S(D')}{\mathcal{L}^S(D)} \right) = \deg^S(D' - D).$$

Proof As in the proof of Proposition 5.3 we may assume $D' = D + P$, and define $\varphi\colon \mathcal{L}^S(D + P) \to k$ the same way. This time we have to show that $\varphi \neq 0$ in order to derive that $\overline{\varphi}$ is an isomorphism. Thus we need to find some $\frac{\eta}{\xi} \in k(C)$ with $v_P\left(\frac{\eta}{\xi}\right) = -r-1$, and with $v_Q\left(\frac{\eta}{\xi}\right) \geq -n_Q$ for all $Q \in S$. Since S is finite, this can be done easily by using §1 Exercise 14. □

The proof of the next proposition depends only on field theoretic properties of algebraic function fields in one variable (APP.I §4).

5.8. Proposition Let $z = \frac{\phi}{\psi} \in k(C) - k$ and let $n = [k(C) : k(z)]$. Then
(i) $D_0 = \left(\frac{\phi}{\psi}\right)_0$ is a positive divisor of degree n.
(ii) There is a constant c such that

$$\ell(m \cdot D_0) \geq mn - c \text{ for all } m \geq 0.$$

Proof Put $D_0 = \sum n_P P$, $r = \deg D_0$. We show first $n \geq r$.
Let $S = \{P \in C \mid n_P > 0\}$. By Lemma 5.7, we may choose $z_1, ..., z_r \in \mathcal{L}^S(0)$ so that the classes $\overline{z_1}, ..., \overline{z_r} \in \mathcal{L}^S(0)/\mathcal{L}^S(-D_0)$ form a k-basis for this vector space. We claim that $z_1, ..., z_r$ are linearly independent over $k(z)$. If not (by clearing denominators and multiplying by a power of z), there would be polynomial $g_i = \lambda_i + zh_i \in k[z]$ with $\lambda_i \in k$ and $\sum g_i z_i = 0$, not all $\lambda_i = 0$. But then

$$\sum \lambda_i z_i = -z \sum h_i z_i \in \mathcal{L}^S(-D_0),$$

i.e., $\sum \lambda_i \overline{z_i} = 0$, a contradiction. Hence

$$r \leq n. \tag{1}$$

Next we prove (ii). Let $w_1, ..., w_n$ be a basis of $k(C)$ over $k(z)$ (APP.I §4). We may assume that each w_i satisfies an equation

$$w_i^{n_i} + a_{i1} w_i^{n_i - 1} + \cdots = 0, \quad a_{ij} \in k[z^{-1}].$$

Then $v_P(a_{ij}) \geq 0$ if $P \neq\in S$. If $v_P(w_i) < 0$, $P \notin S$, then

$$v_P\left(w_i^{n_i}\right) < v_P\left(a_{ij} w_i^{n_i - j}\right),$$

which is impossible (APP.II §2 Exercise 1). It follows that for some $h > 0$,

$$\begin{aligned} &\mathbf{div}(w_i) + hD_0 \prec 0, \ i = 1, ..., n \text{ and then} \\ &w_i z^{-j} \in \mathcal{L}((m+h) \cdot D_0), \ i = 1, ..., n, \ j = 0, 1, ..., m. \end{aligned}$$

Since the w_i are independent over $k(z)$, and $1, z^{-1}, ..., z^{-m}$ are independent over k, $\{w_i z^{-j} \mid i = 1, ..., n,\ j = 0, 1, ..., m\}$ are independent over k. So

$$\ell((m+h)\cdot D_0) \geq n(m+1). \tag{2}$$

But

$$\begin{aligned} \ell((m+h)\cdot D_0) &= \ell(m\cdot D_0) + \dim_k\left(\frac{\mathcal{L}((m+h)\cdot D_0)}{\mathcal{L}(m\cdot D_0)}\right) \\ &\leq \ell(m\cdot D_0) + hr \text{ (Proposition 5.3(i))}. \end{aligned} \tag{3}$$

Therefore, a combination of (2) and (3) yields

$$\ell(m\cdot D_0) \geq n(m+1) - hr = mn - c,$$

as desired.
Finally, since $mn - c \leq \ell(m\cdot D_0) \leq mr + 1$ by Proposition 5.3(iii), if we let m get large enough, we see that $n \leq r$, and this finishes the proof of (i) if the foregoing (1) is combined. □

5.9. Corollary The following are equivalent for a nonsingular C:
(i) C is a rational curve;
(ii) C is isomorphic to $\mathbf{P}^1$;
(iii) There exists one $\frac{\phi}{\psi} \in k(C)$ such that $\deg\left(\frac{\phi}{\psi}\right)_0 = 1$;
(iv) There exists a $P \in C$ such that $\ell(P) > 1$.

Proof From (iv) it follows that there is an $\frac{\phi}{\psi} \in k(C) - k$ such that $\left(\frac{\phi}{\psi}\right)_\infty = P$. Thus, $\deg\left(\frac{\phi}{\psi}\right)_0 = \deg\left(\frac{\phi}{\psi}\right)_\infty = 1$. Consequently $\left[k(C) : k\left(\frac{\phi}{\psi}\right)\right] = 1$, or $k(C) = k\left(\frac{\phi}{\psi}\right)$ and C is rational. The other implications are clear enough (see §4 Exercise 2). □

5.10. Observation (i) if $D \succ D'$ then we have

$$\ell(D') - \ell(D) \leq \deg(D' - D),$$

hence $\deg D - \ell(D) \leq \deg D' - \ell(D')$.
(ii) For any divisor D, $\ell(D) > 0$ and $\deg D = 0$ if and only if D is in $\mathrm{Div}_v(C)$, i.e., $D \sim 0$. In fact, $\ell(D) > 0$ implies that D is equivalent

to a positive divisor (exercise) and then $\deg D = 0$ forces D to be equivalent to the trivial divisor.

□

One may wonder what the algebraic meaning of the divisor class group can be. The divisor group is the free abelian group generated by the points of the nonsingular curve C; if we look at the projective model (see §1) these points correspond to the discrete k-valuation rings of $k(C)$. An affine part of C, say C^a, corresponds to some coordinate ring $k[C^a]$ that is a Dedekind domain (see Appendix II §3) and points of C^a correspond exactly to the discrete k-valuation rings of $k(C)$ containing $k[C^a]$. Every such k-valuation ring is given by a maximal ideal $\mathfrak{m}_v$ of $k[C^a]$ and the localization of $k[C^a]$ at $\mathfrak{m}$ (see Appendix II §1), denoted $\mathcal{O}_v$, is the discrete valuation ring under consideration. Every ideal I of $k[C^a]$ is in an essentially unique way a product of prime ideals (Appendix II §3, Theorem 3), but we need some kind of inverses in order to have the possibility to let the maximal ideals "generate" a free group. This may be obtained by using the so called *fractional ideals* $J \subset k(C)$ introduced in Appendix II §3, i.e., the $k[C^a]$-submodules of $k[C]$ such that $bJ \subset k[C^a]$ for some $b \in k[C^a]$, or equivalently, the finitely generated $k[C^a]$-modules in $k(C)$. A fractional ideal J has an inverse

$$J^{-1} = \left\{ b \in k(C) \;\middle|\; bJ \subset k[C^a] \right\}$$

which is again a fractional ideal and indeed it also satisfies

$$JJ^{-1} = J^{-1}J = k[C^a].$$

The group of fractional ideals is freely generated by the maximal ideals of $k[C^a]$ (and their inverses) and is denoted by $\mathbf{Div}k[C]$. The set of principal fractional ideals, i.e., $J = k[C^a]b$ for some $b \in k(C)$, is a subgroup (of principal divisors). These principal fractional ideals are exactly the free fractional ideals (the others are always projective but not free, this follows from elementary theory about modules over Dedekind domains). We write $\mathbf{Prin}k[C]$ for that group, and refer to any book on Commutative Algebra for the proof of the following proposition.

5.11. Proposition With notation as above, we obtain

(i) $\mathbf{Div}(C) \cong \mathbf{Div}k[C]$;
(ii) $\mathbf{Pic}(C) \cong \mathbf{Pic}k[C] = \mathbf{Div}k[C]/\mathbf{Prin}k[C]$.

□

For curves the Picard group and the class group mean the same thing; we point out that the Picard group defined in terms of invertible ideals (or more generally invertible bimodules) is generally smaller than the class group defined for Noetherian integrally closed rings of higher dimension in terms of divisorial ideals (or divisorial lattices in general).

Exercises for §5

1. Prove Observation 5.10.
2. let $C = \mathbf{P}^1_C$, t as in §4 Exercise 2. Calculate $\mathcal{L}(m\cdot(t)_0)$ explicitly, and show that $\ell(m\cdot(t)_0) = m+1$, where $m \geq 0$.
3. Let C, x, y be as in §4 Exercise 3. Let $z = x^{-1}$. Show that $\mathcal{L}(m\cdot(z)_0) \subset k[x,y]$, and show that $\ell(m\cdot(z)_0) = 2m$ if $m > 0$.
4. Suppose $\ell(D) > 0$, and let $\frac{\phi}{\psi} \neq 0$, $\frac{\phi}{\psi} \in \mathcal{L}(D)$. Show that $\frac{\phi}{\psi} \not\subset \mathcal{L}(D-P)$ for all but a finite number of P. Hence, $\ell(D-P) = \ell(D)-1$ for all but a finite number of P.

§6. The Riemann-Roch Theorem

Let C be a nonsingular plane curve in $\mathbf{P}^2 = \mathbf{P}^2_k$ with $k = \mathbb{C}$, and we maintain notation as before. The problem we focus on in this section is the explicite calculation of $\ell(D)$. As a first step we introduce the genus of a curve C.

6.1. Theorem (Riemann's Theorem) For a nonsingular plane curve C there is a constant g such that $\ell(D) \geq \deg D + 1 - g$ for every divisor D on C, where $g \in \mathbb{N}$. The smallest such g is called the *genus* of C.

Proof For $D = 0$ we have $\deg 0 + 1 - \ell(0) = 0$, so $g \geq 0$ if it exists. From Proposition 5.2 we see that $\deg D + 1 - \ell(D)$ is invariant under equivalence and clearly if $D \prec D_1$ then

$$\deg D + 1 - \ell(D) \leq \deg D_1 + 1 - \ell(D_1).$$

For $z = \frac{\phi}{\psi} \in k(C) - k$, let c be the smallest integer for which the claim of Proposition 5.8 holds. Since for all m we have

$$\deg(m \cdot (z)_0) + 1 - \ell(m \cdot (z)_0) \leq c + 1,$$

and $m \cdot (z)_0 \prec (m+1) \cdot (z)_0$, it follows that for large enough m we must have

$$\deg(m \cdot (z)_0) + 1 - \ell(m \cdot (z)_0) = c + 1.$$

Put $g = c + 1$. It is now enough to establish that for a given divisor D there is an equivalent divisor D_1 and an $m \in \mathbb{N}$ such that $D_1 \prec m \cdot (z)_0$. Let us write

$$(z)_0 = \sum n_P P, \qquad D = \sum d_P P,$$

where we have almost all n_P and d_P equal to zero. We look for a function $w = \frac{\eta}{\xi} \in k(C)$ such that $D_1 = D - \mathrm{div}(w) \prec m \cdot (z)_0$ for some $m \in \mathbb{N}$. This means

$$d_P - v_P(w) \leq m \cdot n_P \text{ for all } P \in C.$$

Put

$$E = \left\{ P \in C \;\middle|\; d_P > 0 \text{ and } v_P\left(z^{-1}\right) \geq 0 \right\}$$

and take

$$w = \prod_{P \in E} \left(z^{-1} - z^{-1}(P)\right)^{d_P}.$$

Thus, if $v_P\left(z^{-1}\right) \geq 0$ then $d_P - v_P(w) \leq 0$ and on the other hand, when $v_P\left(z^{-1}\right) < 0$ then $n_P > 0$. By taking m large enough we arrive at the desired w and D_1. □

6.2. Corollary (i) If $\ell(D) = \deg D + 1 - g$, then for every $D' \sim D_1$ with $D' \succ D$ we have $\ell(D') = \deg D' + 1 - g$.
(ii) For any $z = \frac{\phi}{\psi} \in k(C) - k$, $g = \deg(m \cdot (z)_0) - \ell(m \cdot (z)_0) + 1$ for m large enough.
(iii) Combining both (i) and (ii) we obtain that there exists an $N \in \mathbb{N}$ such that for all divisors D with $\deg D > N$ the equality

$$\ell(D) = \deg D + 1 - g$$

holds.

Proof Exercise. □

Example (i) $g = 0$ if and only if C is rational.
Indeed, if C is rational then $g = 0$ follows from the foregoing because $\deg(m \cdot (z)_0) - \ell(m \cdot (z)_0) + 1 = 0$ for m large enough. Now the converse is true too because if $g = 0$ then $\ell(P) > 1$ for any $P \in C$ where P is viewed as the divisor $1 \cdot P$, then apply Corollary 5.9.

Remark The notion of genus is one of the most important (topological, geometrical and algebraic) invariants for a curve. In a first course of Algebraic Geometry, we are not so ambitious to include a deep theory concerning the genus of a curve. We even do not intend to include a detailed proof of the following important result which associates a plane curve to a nonsingular curve so that the notion of genus also make sense for general curves, because we do not intend to include the theory of "resolution of singularities" in a Primer of Algebraic Geometry. (The interested reader can connect without problem to more specialized literature, e.g. [Ful], [Har].)

Theorem (cf. [Ful] P.179) Let $\mathcal{D}$ be an irreducible projective curve in $\mathbf{P}_k^2$. Then there is a nonsingular projective curve $C \subset \mathbf{P}_k^2$ and a rational map φ from C *onto* $\mathcal{D}$; up to isomorphism, such C is unique.

Using the birational morphism in the above theorem, one may define the genus of the curve $\mathcal{D}$ as that of C; and the genus calculation for curves with only ordinary multiple points (see §1 for the definition) will finish the preparation for the Riemann-Roch theorem.

6.3. Lemma If C is a plane curve with only ordinary multiple points, say P with multiplicity $m_P(C) = m_P$, then we have

$$g = \frac{(n-1)(n-2)}{2} - \sum_{P \in C} \frac{m_P(m_P - 1)}{2}$$

where n is the degree of the curve C.

□

6.4. Corollary A nonsingular curve of degree n has genus

$$\frac{(n-1)(n-2)}{2}.$$

□

Example (ii) Lines and conics are rational. Nonsingular cubics have genus one. Singular cubics are rational. No nonsingular plane curve has genus 2, hence not every curve is birationally equivalent to a nonsingular plane curve (compare to the theorem in the foregoing remark), for example, $y^2xz = x^4 + z^4$ has genus 2.

We do not intend to provide full detail relating to the Riemann-Roch theorem but just indicate how "differentials" do enter the theory. Let A be a k-algebra, M an A-module. A *k-derivation* δ: $A \to M$ is a k-linear map satisfying

$$\delta(ab) = a\delta(b) + \delta(a)b, \quad a, b \in A.$$

If A is a domain with quotient field K then any k-derivation δ: $A \to M$ extends in a unique way to a k-derivation δ^e: $K \to M$ (exercise).
For $a \in A$, let $[a]$ be a symbol and consider the free A-module generated by $\{[a] \mid a \in A\}$, say $[A]$. Consider the A-submodule $\mathcal{D} \subset [A]$ generated by the following set of elements:

$$\left\{[a+b]-[a]-[b] \,\middle|\, a,b \in A\right\}, \quad \left\{[\lambda a]-\lambda[a] \,\middle|\, \lambda \in k,\ a \in A\right\},$$

$$\left\{[ab]-a[b]-[a]b \,\middle|\, a,b \in A\right\},$$

and consider

$$\Omega_k(A) = [A]/\mathcal{D}.$$

We have a map

$$\begin{array}{rccc} d: & A & \longrightarrow & \Omega_k(A) \\ & a & \mapsto & [a] \bmod \mathcal{D} \end{array}$$

which is indeed a derivation. For $a \in A$ we write da for the image of a under d. The A-module $\Omega_k(A)$ is called the *module of* (k-)

differentials of A. It is clear that the construction of $\Omega_k(A)$ in this abstract way has a certain "universal" quality, this is an important advantage highlighted in the following.

6.5. Proposition For any derivation δ: $A \to M$ there is a unique A-module morphism $\mathcal{E}_\delta$: $\Omega_k(A) \to M$ such that for all $a \in A$, $\delta(a) = \mathcal{E}_\delta(da)$, i.e., the following diagram is commutative

$$\begin{array}{ccc} A & \xrightarrow{\ d\ } & \Omega_k(A) \\ \delta\downarrow & \swarrow \mathcal{E}_\delta & \\ M & & \end{array}$$

Proof Define

$$\begin{array}{cccc} \mathcal{E}_1 : & [A] & \longrightarrow & M \\ & \sum b_i[a_i] & \mapsto & \sum b_i\delta(a_i) \end{array}$$

Clearly, $\mathcal{E}_1(\mathcal{D}) = 0$, so $\mathcal{E}_1$ induces $\mathcal{E}_\delta$: $\Omega_k(A) \to M$. □

If we look at a polynomial $f \in k[x_1, ..., x_n]$ then for $a_1, ..., a_n \in A$ we have

$$d(f(a_1, ..., a_n)) = \sum \frac{\partial f}{\partial x_i}(a_1, ..., a_n)da_i.$$

For $A = k[x_1, ..., x_n]$ it is clear that the module of differentials $\Omega_k(A)$ is generated as a $k[x_1, ..., x_n]$-module by $dx_1, ..., dx_n$. For funciton fieldds we have the following proposition.

6.6. Proposition Let C be an irreducible plane curve in $\mathbf{A}_k^2$, and $K = k(C)$ the function field of C. Then $\Omega_k(K)$ is one-dimensional over K. If Char$k \neq 0$, then $\Omega_k(K) = K \cdot dz$ for any $z \in k(C) - k$.

Proof Put $k[C] = k[x, y]/\langle f\rangle = A = k[a, b]$, where a and b are the classes of x and y in A respectively. Since f is irreducible we may assume $\frac{\partial f}{\partial y} \neq 0$. So f cannot divide $\frac{\partial f}{\partial y}$, hence $\frac{\partial f}{\partial y}(a, b) \neq 0$. It is clear from the foregoing argumentation that da and db generate

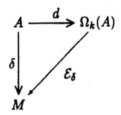

$\Omega_k(K)$ over K. From $df(a,b) = 0$ we obtain

$$db = \left(\frac{\partial f}{\partial x}(a,b) \Big/ \frac{\partial f}{\partial y}(a,b)\right) da.$$

Hence $\dim_K \Omega_k(K) \leq 1$.
In view of Proposition 6.5 it will follow that $\Omega_k(K) \neq 0$ if we find any nonzero derivation δ: $A \to M$ for some K-vector space M. Take $M = K$ and $g(a,b) \in A$, $g \in k[x,y]$, we define

$$\delta\,(g(a,b)) = \frac{\partial g}{\partial x}(a,b) - n\frac{\partial g}{\partial y}(a,b).$$

Since $\delta(a) = 1$, this is a nontrivial derivation (verification is left as an exercise). □

In the case char$k = 0$, for any $a, b \in K$ and $\delta \notin k$, we have $da = udb$ for some $u \in k$. It is customary to write

$$u = \frac{da}{db}$$

and to call it the *derivative of a with respect to b.*
Since a uniformizing parameter in a local ring of a nonsingular point P on C may be used for expressing functions in $k(C)$ as a series expansion, the derivative with respect to a uniformizing parameter may be an interesting "local tool" for dealing with differentials.

6.7. Proposition Let C be as in Proposition 6.6. Put $K = k(C)$. For a nonsingular point P on C, we write $\mathcal{O}_P$ for the local ring of P, which is a discrete valuation ring of K with uniformizing parameter ω, say. If $z = \frac{\phi}{\psi} \in \mathcal{O}_P$, then $\frac{dz}{d\omega} \in \mathcal{O}_P$.

Proof By an affine change of coordinate, harmless to all statements in the proposition, we may assume that P is the origin. As before, put $k[C] = k[a,b]$; write $a' = \frac{da}{d\omega}$, $b' = \frac{db}{d\omega}$. Choose N large enough such that $v_P(a') \geq -N$, $v_P(b') \geq -N$. For $z = [f] \in k[C]$ with $f \in k[x,y]$ we have

$$z' = \frac{d\phi}{d\omega} = \frac{\partial f}{\partial x}(a,b)a' + \frac{\partial f}{\partial y}(a,b)b',$$

hence $v_P([f]') \geq -N$. If $z \in \mathcal{O}_P$, say, $z = \frac{[g]}{[h]}$ with $h(P) \neq 0$, where $g, h \in k[x, y]$ and $[g]$ and $[h]$ are the classes of g and h in $k[C]$ respectively. We conclude

$$(*) \qquad z' = \frac{[h][g]' - [g][h]'}{[h]^2}, \text{ and hence } v_P(z') \geq -N.$$

Since z has a series expansion in ω with highest power of ω appearing at most equal to N (and coefficients in $\mathcal{O}_P$), the highest power of ω in the expansion for z' is again N, the coefficientof this ω^N being the derivative of the coefficient of ω^N in z. Hence, in view of (*) above, this coefficient is in $\mathcal{O}_P$. The other coefficients of lower ω powers are certainly in $\mathcal{O}_P$ too, whence the statement! (Note: for any $\frac{\phi}{\psi} \in \mathcal{O}_P$ there are unique $\lambda_1, ..., \lambda_{N-1} \in k$ and $\left(\frac{\phi}{\psi}\right)_N \in \mathcal{O}_P$ such that $\frac{\phi}{\psi} = \lambda_0 + \lambda_1\omega + \cdots + \lambda_{N-1}\omega^{N-1} + \left(\frac{\phi}{\psi}\right)_N \omega^N$.) □

Considering a nonsingular projective plane curve C and any $P \in C$, we may identify $k(C)$ with $k(C)$ where C is any affine part of C containing P. So we may just write K for the function field of C. Put $\Omega = \Omega_k(K)$. For $\omega \neq 0$ in Ω, define $v_P(\omega)$ by choosing a uniformizing parameter t in $\mathcal{O}_P$ and expressing $\omega = zdt$ for some unique $z = \frac{\phi}{\psi} \in K$ by Proposition 6.6, then put $v_P(\omega) = v_P(z)$ (Check that this is well-defined). The divisor of ω is defined to be

$$\mathbf{div}(\omega) = \sum_{P \in C} v_P(\omega)P.$$

This divisor is called a *canonical divisor*. If $\omega' \in \Omega$ is nonzero, then $\mathbf{div}(\omega') = \mathbf{div}(z') + \mathbf{div}(\omega)$ for some $z' = \frac{\eta}{\xi} \in K$ such that $\omega' = z'\omega$. Hence $\mathbf{div}(\omega') \sim \mathbf{div}(\omega)$. It is the clear that the canonical divisors form just one equivalence class and so they all have the same degree. It is possible to prove that $\mathbf{div}(\omega)$ is indeed a divisor (i.e., almost all $v_P(\omega)$ are zero) and

6.8. Proposition Let $W = \mathbf{div}(\omega)$ be a canonical divisor. Then $\deg W = 2g - 2$ and $\ell(W) \geq g$.

□

This leads to the Riemann-Roch theorem:

6.9. Theorem Let W be a canonical divisor on C, then for any divisor D we have

$$\ell(D) = \deg D + 1 - g + \ell(W - D).$$

□

6.10. Corollary (i) $\ell(W) = g$.
(ii) If $\deg D \geq 2g - 1$, then $\ell(D) = \deg D + 1 - g$.
(iii) If $\deg D \geq 2g$, then $\ell(D - P) = \ell(D) - 1$ for all $P \in C$.
(iv) If $\ell(D) > 0$ and $\ell(W - D) > 0$, then $\ell(D) \leq \frac{1}{2}\deg D + 1$.
□For the proof of 6.9 and 6.10 we refer to the literature, e.g., [Ful], [Ken].

For a divisor D on C we may define a k-subspace $\Omega(D)$ of Ω as follows.

$$\Omega(D) = \left\{\omega \in \Omega \;\middle|\; \mathrm{div}(\omega) > D\right\}.$$

Put $\delta(D) = \dim_k \Omega(D)$. We call $\delta(D)$ the *index* of D. The differentials in $\Omega(0)$ are said to be of the *first kind*; in case $k\mathbb{C}$ they are also known as the *holomorphic differentials*.

6.11. Proposition With notation as before:
(i) $\delta(D) = \ell(W - D)$.
(ii) There are g k-independent differentials of the first kind on C.
(iii) $\ell(D) = \deg D + 1 - g + \delta(D)$.

Proof Put $N = \mathbf{div}(\omega)$ for some canonical differential. Define

$$\begin{array}{cccc} \gamma: & \mathcal{L}(W - D) & \longrightarrow & \Omega(D) \\ & \dfrac{\phi}{\psi} & \mapsto & \dfrac{\phi}{\psi}\cdot\omega \end{array}$$

One easily verifies that γ is an isomorphism. Hence (i) has been established. (iii) follows directly from (i) and the Riemann-Roch theorem. (ii) follows from Corollary 6.10(i). □

We have included the results in this section because they are fundamental in the further theory of curves (and generalizations of it to higher dimensional varieties) but also because the calculations with divisors provide other effective calculation methods to deal with geometric properties. The methods used in the proofs, also those we did not include here, are not deeper than linear algebra, applied in an ingenious way however.

Exercise for §6

1. Show that the curve $y^2xz = x^4 + z^4$ has one node, and so is of genus 2.
2. Let $\mathbf{Pic}(C)$ be the divisor class group of a nonsingular curve $C \subset \mathbf{P}^2_k$. Show that $\mathbf{Pic}(C) = \{0\}$ if and only if C is rational.
3. Prove Corollary 6.2. (Hint: Take a D_0 such that $\ell(D_0) = \deg D_0 + 1 - g$ and put $N = \deg D_0 + g$; then apply Theorem 6.1 to D with $\deg D \geq N$, find $\ell(D - D_0) > 0$ and conclude $D \sim D + \mathbf{div}\left(\frac{\eta}{\xi}\right) > D_0$ for an $\frac{\eta}{\xi}$ such that $D - D_0 + \mathbf{div}\left(\frac{\eta}{\xi}\right) > 0$.)
4. Let A be a k-algebra and M an A-module. Show that if A is a domain and $\delta\colon A \to M$ is a k-derivartion, then δ extends in a unique way to a k-derivation $\delta^e\colon K \to M$ by putting
$$\delta^e\left(\frac{a}{b}\right) = \frac{b\delta(a) - a\delta(b)}{b^2},$$
where K is the quotient field of A.
5. Check that the δ defined in the proof of Proposition 6.6 is a derivation.

CHAPTER VIII
Elliptic Curves

A lot of modern geometry has been developed in order to deal with problems of a more arithmetical nature that are usually easily formulated in terms of numbers or solutions of polynomial equations in integers or rational numbers. Diophantes of Alexandria ($\pm$ 250AD) gathered his mathematical ideas in the thirteen volumes of the "Arithmetica", seven of which have been rediscovered. The basic problem of Diophantine geometry, as we now call it and as we have mentioned in CH.I §2, deals with a polynomial $f \in \mathbb{Z}[x_1, ..., x_n]$; the question is to find a method of deciding whether the equation $f = 0$ has a solution in $\mathbb{Z}$, or in $\mathbb{Q}$, to decide whether there are infinitely or only finitely many solutions and to describe all of these. Even when some equations define "relatively easy" geometric objects, e.g., rational surfaces, the Diophantine problems cannot easily be solved. A classical example is obtained by looking at

$$ax^p + by^q + cz^r = d,$$

where a, b, c, d are nonzero in $\mathbb{Q}$ and p, q, r are positive integers such that $\frac{1}{p} + \frac{1}{q} + \frac{1}{r} \geq 1$ (note: there are only finitely many such p, q, r). The case $p = q = r = 2$ may be treated by classical methods. Even in the case $p = q = r = 3$, e.g., $x^3 + y^3 + z^3 = d$ one needs heavy modern mathematical technology (K-theory, Chow groups,...) in order to understand the diophantine problems here, not withstanding the fact that Ryley (± 1825) already knew that there is a "simple" solution

$$d = \left(\frac{d^3 - 3^6}{3^2d^2 + 3^4d + 3^6}\right)^3 + \left(\frac{-d^3 - 3^5d + 3^6}{3^2d^2 + 3^4d + 3^6}\right)^3 + \left(\frac{3^3d^2 + 3^5d}{3^2d^2 + 3^4d + 3^6}\right)^3$$

The affine-projective transfer (CH.III) obtains a new aspect in the framework of the diophantine theory. Indeed, looking for rational solutions (i.e. in $\mathbb{Q}$) of e.g. $x^n + y^n = 1$ is equivalent to finding integral (i.e. in $\mathbb{Z}$) solutions of $x^n + y^n = z^n$. For curves of degree 1 it is clear that there are infinitely many rational points. For irreducible curves of degree 2 having at least one rational point one may construct infinitely many rational points by Euclid's method that shows (in our language) that a curve of degree 2 is birationally equivalent to the projective line over $\mathbb{C}$, moreover if the curve has at least one rational point over $\mathbb{Q}$ then the curve is birationally equivalent to the projective line over $\mathbb{Q}$. Now, two curves are birationally equivalent if and only if they have the same genus. So we arrive at the following conclusion concerning the diophantine problems: for a curve of genus 0 we can solve the diophantine problems. In the Arithmetica Diophantes already looked at some cubic curves of genus 1, this is exactly the class of curves we want to study in this chapter.

§1. Standard Equation for a Nonsingular Cubic Curve

In this section we assume $k = \mathbb{C}$ and let C be an irreducible curve of degree 3 in $\mathbf{P}^2_{\mathbb{C}}$. Such a curve can have at most one multiple point and its multiplicity is at most 2; this follows from the fact that a line through the multiple point intersects the curve in 3 points counting multiplicity (Bezout's Theorem).

First assume that there is a *double* point P on C. In homogeneous coordinates we may assume that $P = (0,0,1)$ and C has homogeneous equation:

$$F(x_1, x_2, x_3) = x_3Q(x_1x_2) + F_1(x_1, x_2)$$

where Q and F_1 are homogeneous of degree 2, 3 respectively. We may view $Q = L_1L_2$ with L_1 and L_2 linear since we are working over $\mathbb{C}$.

Case 1: $L_1 \neq L_2$.
By a coordinate transformation we may assume $L_1 = x_1$, $L_2 = x_2$. Then

$$F(x_1, x_2, x_3) = x_1x_2 + \alpha x_1^3 + \beta x_1^2x_2 + \gamma x_1x_2^2 + \delta x_2^3,$$

for some $\alpha, \beta, \gamma, \delta \in \mathbb{C}$. Apply the following coordinate transformation:

$$x_1 \mapsto \alpha^{\frac{1}{3}}x_1, \quad x_2 \mapsto \delta^{\frac{1}{3}}x_2, \quad x_3 \mapsto (\alpha\delta)^{\frac{-1}{3}}x_3.$$

We obtain a new equation, for certain new $\beta_1, \gamma_1 \in \mathbb{C}$:

$$F(x_1, x_2, x_3) = x_1x_2x_3 + x_1^3 + \beta_1x_1^2x_2 + \gamma_1x_1x_2^2 + x_2^3.$$

After another coordinate transformation:

$$x_1 \mapsto x_1, \quad x_2 \mapsto x_2, \quad x_3 \mapsto -x_3 - \beta_1x_1 - \gamma_1x_2,$$

we obtain a rather simple equation expressing $F(x_1, x_2, x_3) = 0$, i.e.,

$$x_1x_2x_3 = x_1^3 + x_2^3.$$

We may obtain a rational parametrization by intersecting the curve and $x_2 = tx_1$:

$$\begin{cases} x_1 &= \dfrac{t}{1+t^3}, \\[2ex] x_2 &= \dfrac{t^2}{1+t^3}. \end{cases}$$

Conclusion in case 1, the curve C is birationally equivalent to a line.

Case 2: $L_1 = L_2 = L$, $Q = L^2$.
We may assume, up to a coordinate transformation, that $L = x_2$. Hence,

$$F(x_1, x_2, x_3) = x_2^2x_3 + \alpha x_1^3 + \beta x_1^2x_2 + \gamma x_1x_2^2 + \delta x_2^3.$$

Since F is irreducible α must be nonzero; then we may assume $\alpha = 1$ by rescaling x_1. Again perform a coordinate transformation

$$x_1 \mapsto x_1 + \frac{\beta}{3\alpha}x_2, \quad x_2 \mapsto x_2, \quad x_3 \mapsto x_3.$$

We then arrive at an equation of the form

$$F(x_1, x_2, x_3) = x_2^2 x_3 + \alpha x_1^3 + \beta x_1 x_2^2 + \delta x_2^3$$

with a new $\alpha \neq 0$. Next perform the coordinate transform

$$x_1 \mapsto \alpha^{\frac{1}{3}} x_1, \quad x_2 \mapsto x_2, \quad x_3 \mapsto -x_3 - \beta\alpha^{\frac{-1}{3}} x_1 - \delta x_2.$$

Then we obtain $F(x_1, x_2, x_3) = -x_2^2 x_3 + x_1^3$, i.e., in affine form

$$x_2^2 = x_1^3.$$

Again, by intersecting with $x_2 = tx_1$, we obtain a rational parametrization

$$\begin{cases} x_1 &= t^2, \\ x_2 &= t^3. \end{cases}$$

Note that in this case, the singular point $(0, 0, 1)$ does correspond with only one value of t, nevertheless the local rings on the curve and the rational parametrization are different!

Conclusion: A curve C of degree 3 with a double point P is birationally equivalent to a line.

Now the problem is reduced to looking at a curve C of degree 3 without multiple points!
Let Γ be any curve of degree n, given by an homogeneous equation $F(x_1, x_2, x_3) = 0$. Let P be a simple point of Γ. Let L be the tangent to Γ at P, it is given by the equation

$$\frac{\partial F}{\partial x_1} x_1 + \frac{\partial F}{\partial x_2} x_2 + \frac{\partial F}{\partial x_3} x_3 = 0.$$

We say that P is a *flex* of Γ if the intersection number $\mathbf{I}(P, \Gamma \cap L) > 2$. Put $P = (p_1, p_2, p_3)$. Take $Q = (q_1, q_2, q_3) \in L$. The points on L correspond to $\lambda P + \mu Q$, i.e., points with coordinates $(\lambda p_1 + \mu q_1, \lambda p_2 + \mu q_2, \lambda p_3 + \mu q_3)$. Calculating the value of F at such a point we obtain

$$\begin{aligned} F(\lambda P + \mu Q) &= \lambda^n F(P) + \lambda^{n-1}\mu \sum_i \frac{\partial F}{\partial x_i}(P) q_i + \\ &\quad + \lambda^{n-2}\mu^2 \sum_{i,j} \frac{\partial^2 F}{\partial x_i \partial x_j}(P) q_i q_j + \cdots \end{aligned}$$

Here $F(P) = 0$ because P is on Γ, $\sum \frac{\partial F}{\partial x_i}(P) = 0$ because (q_1, q_2, q_3) is on L.

Consequently, P is a flex on γ if and only if we have

$$\sum_{i,j} \frac{\partial^2 F}{\partial x_i \partial x_j}(P) = 0,$$

or in other words, if and only if Q lies on the conic

$$\mathcal{E}: \quad \sum_{i,j} \frac{\partial^2 F}{\partial x_i \partial x_j}(P) x_i x_j = 0.$$

1.1. Lemma Let $P \in \Gamma$ and $\mathcal{E}$ the corresponding conic. With notation as above,

(i) $\mathcal{E}$ is singular if and only if P is singular or P is a flex of Γ.

(ii) If $\mathcal{E}$ is singular and P is a flex of Γ, then $\mathcal{E}$ is the product of two lines one of which is the tangent L in P at Γ.

Proof We remind the reader to go back to CH.VI §6 Example (iii) for some discussion about the singularity of a conic.

(i) If P is singular then $\frac{\partial F}{\partial x_i}(P) = 0$ for $i = 1, 2, 3$. Hence $\mathbf{I}(P, \Gamma \cap PQ) > 2$ if Q is on $\mathcal{E}$; but then the line PQ is on $\mathcal{E}$ or $\mathcal{E}$ is singular. If P is a flex then the same conclusion follows from foregoing remarks. Conversely, if $\mathcal{E}$ is singular, then $\mathcal{E}$ is the product of lines L_1, L_2. From Euler's theorem it follows that $\sum \frac{\partial^2 F}{\partial x_i \partial x_j}(P) p_i p_j = 0$, hence P is on $\mathcal{E}$.

(ii) Suppose that P is on L_1 and take $Q \in L_1$. Then

$$\sum_{i,j} \frac{\partial^2 F}{\partial x_i \partial x_j}(P)(\lambda p_i + \mu q_i)(\lambda p_j + \mu q_j) = 0,$$

or

$$\lambda^2 \left(\sum_{i,j} \frac{\partial^2 F}{\partial x_i \partial x_j}(P) p_i p_j \right) + 2\lambda\mu \sum_{i,j} \frac{\partial^2 F}{\partial x_i \partial x_j}(P) p_i q_j + \\ + \mu^2 \sum_{i,j} \frac{\partial^2 F}{\partial x_i \partial x_j}(P) q_i q_j = 0.$$

It follows that $\sum_{i,j} \frac{\partial^2 F}{\partial x_i \partial x_j}(P) p_i q_j = 0$. Hence

$$\sum_j \sum_i \frac{\partial}{\partial x_i}\left(\frac{\partial F}{\partial x_j}\right)(P) = (n-1)\sum_{j=1}^{3} \frac{\partial F}{\partial x_j}(P) q_j = 0,$$

or $\sum_j \frac{\partial F}{\partial x_j}(P) = 0$.

So either P is singular, i.e., all $\frac{\partial F}{\partial x_j}(P) = 0$, or Q is on the tangent in P at Γ. In the latter case $L_1 = L$ and $\mathbf{I}(P, \Gamma \cap L) > 2$, hence P is flex; thus (ii) also follows from this. □

1.2. Corollary (i) $P \in \Gamma$ is either singular or a flex if P is on

$$(*) \qquad \det\left(\frac{\partial^2 F}{\partial x_i \partial x_j}(x_1, x_2, x_3)\right) = 0 \quad \text{(the Hesse determinant)}.$$

Note that $(*)$ defines a curve of degree $3(n-2)$, so if the degree of Γ is also 3 then there are exactly 9 points of intersection (counting multiplicities). If all multiplicities are one then we find exactly 9 different flexes!

(ii) A nonsingular cubic curve (i.e. nonsingular curve of degree 3) has at least one flex. In fact it is not hard to see (cf. foregoing) that there are indeed 9 flexes in that case.

Let us return to our cubic curve $C = \mathbf{V}(F)$ without multiple points. Select a flex P of C (this is possible because of Corollary 1.2(ii)), and apply a coordinate transform such that $P = (0,1,0)$ and $x_3 = 0$ is the tangent in P at C. We arrive at a homogeneous equation for C

$$\begin{aligned} 0 &= F(x_1, x_2, x_3) \\ &= x_3 x_2^2 + (\alpha x_1^2 + \beta x_1 x_3 + \gamma x_3^2) x_2 + \lambda x_1^3 + \mu x_1^2 x_3 + \nu x_1 x_3^2 + \rho x_3^3. \end{aligned}$$

We consider the corresponding affine equation in x_1, x_3. The line $x_3 = 0$ is tangent in P if and only if $F(x_1, x_2, 0) = 0$ has at least a triple solution $x_1 = 0$. So α must be zero! Irreducibility of F then yields $\lambda \neq 0$. Hence

$$0 = F(x_1, x_2, x_3) = x_3 x_2^2 + (\beta x_1 x_3 + \gamma x_3^2) x_2 + \lambda x_1^3 + \mu x_1^2 x_3 + \nu x_1 x_3^2 + \rho x_3^3.$$

Apply a coordinate transformation

$$x_2 \mapsto x_2 + \frac{1}{2}\beta x_1 + \frac{1}{2}\gamma x_3, \quad x_1 \mapsto x_1, \quad x_3 \mapsto x_3.$$

We obtain a new equation

$$0 = F(x_1, x_2, x_3) = x_2^2 x_3 + \lambda x_1^3 + \mu x_1^2 x_3 + \nu x_1 x_3^2 + \rho x_3^3.$$

A new transformation

$$x_1 \mapsto x_1 + \frac{\mu}{3\lambda} x_3, \quad x_2 \mapsto x_2, \quad x_3 \mapsto -x_3$$

yields an equation (with new coefficients):

$$0 = F(x_1, x_2, x_3) = -x_2^2 x_3 + \lambda x_1^3 + \gamma x_1 x_3^2 + \rho x_3^3.$$

The corresponding affine equation may thus be rewritten as:

$$x_2^2 = \lambda x_1^3 + \nu x_1 + \rho.$$

1.3. Proposition A (nonsingular) cubic curve C may be given by an equation in normal form:

$$y^2 = 4x^3 - g_2 x - g_3$$

(Weierstrass normal form).

Proof As above with $x = x_1 c$, $y = x_2$ and c chosen suitably to arrive at $\lambda = 4$. Of course one may reduce to an equation with $\lambda = 1$ but in view of the analytic function theory connected to these cubic curves, the equation given has a few technical advantages. □

1.4. Corollary A cubic curve $C = \mathbf{V}(F)$ given by its normal form equation $y^2 = 4x^3 - g_2 x - g_3$ is nonsingular if and only if

$$\Delta = g_2^3 - 27 g_3^2 \neq 0.$$

Proof Here we give a sketch of the proof and leave the details to the reader as an exercise.
$P \in C$ is singular if and only if

$$\frac{\partial F}{\partial x_1} = \frac{\partial F}{\partial x_2} = \frac{\partial F}{\partial x_3} = 0;$$

express this for $F = x_2^2 - \psi(x_1, x_3)$ where $\psi(x_1, x_3) = 4x_1^3 - g_2x_1x_3^2 - g_3x_3^3$. Put $P = (p_1, p_2, 1)$ (note first that $x_3 = 0$ only leads to the point $(0,1,0)$ and $\frac{\partial F}{\partial x_3}(0,1,0) = -1$, so that point is simple on C). Then $\frac{\partial F}{\partial x_2}(P) = -2p_2 = 0$ if and only if $p_2 = 0$, i.e., $\psi(p_1) = 0$. Observe that P is singular if and only if $\psi(p_1) = 0$, $\frac{\partial \psi}{\partial x_1}(p_1) = 0$; i.e., if and only if p_1 is a multiple solution of $\psi(x_1) = 0$ (affine version). Express $\psi(x) = 4\prod_{i=1}^3(x - \xi_i)$, the ξ_i being the roots of $\psi(x) = 0$ and $\Delta = 4^4\prod_{i\neq j=1}^3(\xi_i - \xi_j)^2$. Verify that nonsingularity of C comes down to $\Delta \neq 0$ and calculate $\Delta = g_2^3 - 27g_3^2$. □

1.5. Example and Observation Let Γ be the Fermat curve

$$X^3 + Y^3 = 1.$$

If we apply the substitution

$$x = \frac{6}{X} + \frac{Y}{6X}, \quad y = \frac{6}{X} - \frac{Y}{6X},$$

then we obtain a curve with equation $y^2 = x^3 - 432$ and, if we put $Y = 2y$, $X = x$, then $Y^2 = 4X^3 - 1728$. The latter is a cubic curve given by its normal equation. We have proved that the Fermat curve $X^3 + Y^3 = 1$ is birationally equivalent to the elliptic curve $Y^2 = 4X^3 - 1728$ (note that $1728 = 4^3 3^3 = (2^3)^2 3^3$).

1.6. Remark The curve given by $y^2 = \psi(x)$ is nonsingular if and only if it has genus 1. This follows directly from the formula calculating the genus (CH.VII Corollary 6.4). In fact it is not difficult (using the ingredients of the Riemann-Roch theorem ...) to prove that any curve (even over fields $k \neq \mathbb{C}$ but with char$k \neq 2,3$) has genus 1 exactly when it is birationally equivalent to a plane nonsingular cubic curve.

Moreover, a result of Mordell establishes that for a quartic curve Γ given by $y^2 = \phi(x)$ where $\phi(x)$ is a polynomial of degree 4, the existence of a rational point forces Γ to be birationally equivalent to a curve given by an equation $y^2 = f(x)$ with $\deg_x f \leq 3$.

§2. Addition on an Elliptic Curve

From the foregoing section we retain that any curve of genus 1, defined over Q is birationally equivalent over Q to a nonsingular curve with normal equation

$$y^2 = 4x^3 - g_2x - g_3$$

with $g_2^3 - 27g_3^2 \neq 0$. Such a curve is called an *elliptic curve* over Q; the terminology may be extended to any field k with char$k \neq 2, 3$. In this section we establish that there exists a natural group structure on the points of an elliptic curve. The rational points form a subgroup and one may hope to use this extra structure in the solution of the Diophantine problems. As an introductary motivation let us mention a few highlights of the theory, which we shall not treat in detail in these notes. First we should mention a famous theorem due to Mordell (1922).

2.1. Theorem (Mordell) The group $E(Q)$ of rational points on an elliptic curve E, is finitely generated.

There will be elliptic curves having a finite number of rational points as well as curves having infinitely many rational points. The problem of deciding which case holds for a given curve is related to finding a description of the group $E(Q)$, e.g., is it a torsion group or is there a torsionfree (hence infinite) part?
Of course one may wonder whether it is a reasonable use of energy to concentrate so deeply on curves of genus 1 when there are so much more curves of higher genus. However, Mordell conjectured that any rational curve of genus higher than one could have only finitely many rational points. It is one of the most famous results of 20-th century mathematics that this is indeed a fact!

2.2. Theorem (G. Faltings 1983) Rational curves of genus > 1 have finitely many rational points.

Of course this has importance for the haunting problem of proving Fermat's last "theorem". G. Faltings established that the Fermat curve

$$x^n + y^n = z^n$$

with $n > 2$, could have only finitely many solutions and in fact he obtained the corrolary that for every prime number p and for almost all multiples $n = \gamma p$, $x^n + y^n = z^n$ has only trivial solutions in Q. After that many mathematicians predicted a solution to the long standing Fermat problem and indeed we now know that Fermat's theorem is correct by the celebrated work of Wiles.

For us it is important to retain that in the case $g = 0$ and $g > 1$ the Diophantine problems are essentially solved and all the remaining problems are concentrated in the case $g = 1$! This shows the importance of elliptic curves from the algorithmic (or number theoretic) point of view.

We consider curves in $\mathbf{P}^2_{\mathbf{C}}$, but any algebraically closed field k such that char$k \neq 2, 3$ may be substituted without harming the validity of all statements. If one does not use the normal form of the equation of an elliptic curve then even the assumption char$k \neq 2, 3$ may be droped!

Let E be a nonsingular cubic curve in $\mathbf{P}^2$ and fix a point O on E *arbitrarily.*

2.3. Definition We define a binary operation $\odot$ on E as follows: for $P, Q \in E$ we write $\overline{PQ}$ for the line determined by P and Q, and let S' be the intersection of $\overline{PQ}$ and E (there is a third point which may or may not coincide with P or Q), and let S be the third intersection point of the line $\overline{S'O}$ and E, and we write

$$S = P \odot Q.$$

This is illustrated in the following sketch (just for visualization, we do not pretend to be drawing a real elliptic curve).

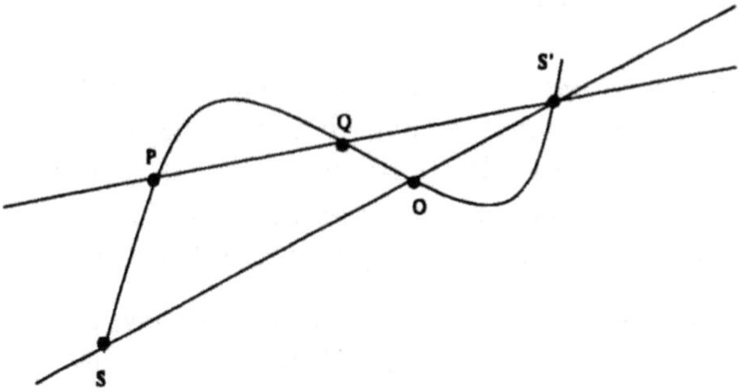

Figure 12

2.4. Lemma The operation $\odot$ satisfies the following properties
(i) $P \odot Q = Q \odot P$, for $P, Q \in E$.
(ii) $P \odot O = O \odot P$, for $P \in E$.
(iii) For every $P \in E$ there is a $\overline{P} \in E$ such that $P \odot \overline{P} = \overline{P} \odot P = O$.

Proof (i) The line $\overline{PQ}$ does not depend on the order of P and Q.
(ii) Obvious from the construction.
(iii) Let L be the tangent in O at E and let O' be the other intersection point of L and E (note $O \neq O'$ because there is no triple point on E). The intersection of $\overline{PO'}$ and E is $\overline{P}$. It is clear from the construction that $P \odot \overline{P} = O$.

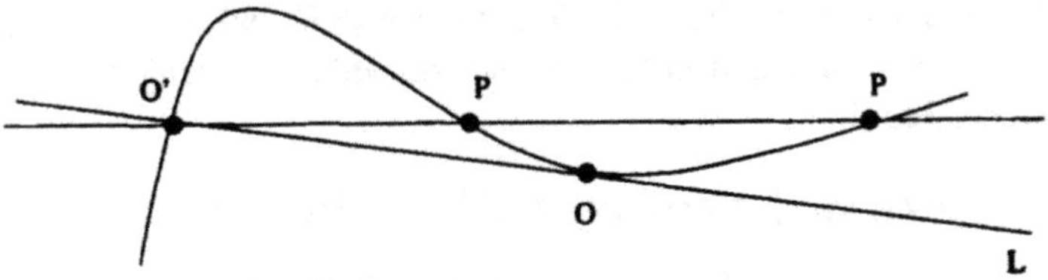

Figure 13

□

In order to prove that $(E, \odot)$ is an abelian group we have to establish associativity for $\odot$. This can be done by a tedious calculation with coordinates but there is a rather nice conceptual proof based on Max Noether's theorem. In fact we need a weaker version of the latter theorem, so we only include a proof for the weaker version.

2.5. Lemma Consider curves $\mathcal{C} = \mathbf{V}(F)$, $\mathcal{D} = \mathbf{V}(G)$ in $\mathbf{P}^2$ defined by the homogeneous polynomials F and G, respectively, and a line L, where F is irreducible. Assume that $P \in \mathcal{C} \cap \mathcal{D} \cap L$, assume moreover that P is a simple point on $\mathcal{C}$, L not a component of $\mathcal{D}$. If $\mathbf{I}(P, \mathcal{C} \cap L) = r$ and $\mathbf{I}(P, \mathcal{C} \cap \mathcal{D}) \geq r$, then $\mathbf{I}(P, \mathcal{D} \cap L) \geq r$.

Proof After a coordinate transformation we may assume that L is the line $x_2 = 0$ and $P = (0,0,1)$. There is nothing to prove if $r = 1$ so let us assume $r > 1$. now $r = \mathbf{I}(P, \mathcal{C} \cap L)$ is the multiplicity of the solution $x_3 = 0$ of $F(x_1, 0, x_3)$.
We have

$$F(x_1, x_2, 1) = x_2 F_1(x_1, x_2, 1) + x_1^r F_2(x_1, 1),$$

where $F_2(0,1) \neq 0$, $F_1(0,0,1) = c \neq 0$ because P is simple on $F = 0$. Let ξ, η represent x_1, x_2 in the coordinate ring of $F = 0$, then

$$0 \neq F(\xi, \eta) = \eta F_1(\xi, \eta) + \xi^r F_2(\xi),$$

or

$$\eta = \frac{-\xi^r F_2(\xi)}{F_1(\xi, \eta)} \in \mathcal{O}_{P,F},$$

where $\mathcal{O}_{P,\mathcal{C}}$ is the local ring of P on $F = 0$. Since P has multiplicity 1 on $\mathcal{C}$, $\mathcal{O}_{P,\mathcal{C}}$ is a discrete valuation ring and ξ is a uniformizing parameter of $\mathcal{O}_{P,\mathcal{C}}$. Consequently, $v_P(\eta) = r$ (note that $F_1(\xi, \eta)$ is a unit in $\mathcal{O}_{P,\mathcal{C}}$ because $F_1(P) = F_1(0,0,1) = c \neq 0$), where v_P is the valuation function associated to the discrete valuation ring $\mathcal{O}_{P,\mathcal{C}}$. On the other hand,

$$G(x_1, x_2, 1) = x_2 G_1(x_1, x_2, 1) + x_1^s G_2(x_1, 1),$$

and $\mathbf{I}(P, \mathcal{D} \cap L) = s$. It is given that $\mathbf{I}(P, \mathcal{D} \cap \mathcal{C}) \geq r$, hence $v_P(G(\xi, \eta)) \geq r$. From

$$G(\xi, \eta) = \eta G_1(\xi, \eta) + \xi^s G_2(\xi)$$

we get

$$\xi^s G_2(\xi) = G(\xi, \eta) - \eta G_1(\xi, \eta).$$

Consequently,

$$s = v_P\left(\xi^s G_2(\xi)\right) > \min\left\{v_P(G(\xi,\eta)) \;\middle|\; v_P(\eta G_1(\xi,\eta))\right\} \geq r.$$

□

2.6. Proposition (Special case of M. Noether's theorem) Consider curves $C = \mathbf{V}(F)$ and $\mathbf{V}(G)$ of degree n resp. m in $\mathbf{P}^2$. Suppose that C is a nonsingular irreducible curve and let L be a line. If the intersection divisors satisfy $C \circ \mathcal{D} - C \circ L \succ 0$, then there exists a curve $\mathcal{H}$ of degree $m-1$ in $\mathbf{P}^2_{\mathbf{C}}$ such that $C \circ \mathcal{D} - C \circ L = C \circ \mathcal{H}$.

Proof We may assume that L is the line $x_3 = 0$ and that L is not a component of $\mathcal{D}$. If $P \in C \cap L$, we have (from the assumptions)

$$\begin{aligned} \mathbf{I}(P, C \cap \mathcal{D}) &\geq \mathbf{I}(P, C \cap L), \\ F(x_1, x_2, 0) &= \prod_j (\alpha_j x_2 - \beta_j x_1)^{r_j}, \\ G(x_1, x_2, 0) &= \prod_j (\alpha_j x_2 - \beta_j x_1)^{r_j} Q(x_1, x_2). \end{aligned}$$

The latter follows from the foregoing lemma. Thus we obtain

$$G(x_1, x_2, x_3) - F(x_1, x_2, x_3) Q(x_1, x_2) = x_3 H(x_1, x_2, x_3)$$

for some $H(x_1, x_2, x_3)$, because the first member vanishes for $x_3 = 0$. Hence

$$G(x_1, x_2, x_3) = F(x_1, x_2, x_3) Q(x_1, x_2) + x_3 H(x_1, x_2, x_3).$$

Then

$$\mathcal{D} \circ C = \mathbf{V}(F + X_3 H) \circ C = \mathbf{V}(x_3 H) \circ C = \mathbf{V}(x_3) \circ C + \mathbf{V}(H) \circ C,$$

i.e., $\mathcal{D} \circ C - L \circ C = \mathcal{H} \circ C$ where $\mathcal{H} = \mathbf{V}(H)$. From Bezout's theorem it follows that $\deg H = m - 1$ because $\deg(\mathcal{D} \circ C - L \circ C) = n(m-1)$.
□

2.7. Proposition Let E be a nonsingular cubic curve in $\mathbf{P}^2$. Then $(E, \odot)$ is an abelian group.

Proof We only have to establish associativity of $\odot$, i.e., for $P, Q, R \in E$ we want $(P \odot Q) \odot R = P \odot (Q \odot R)$.
call $P \odot Q = N$, $N \odot R = T$, $Q \odot R = U$. We may illustrate this in a picture representing the consecutive additions on E (we do not claim to draw a real elliptic curve here).

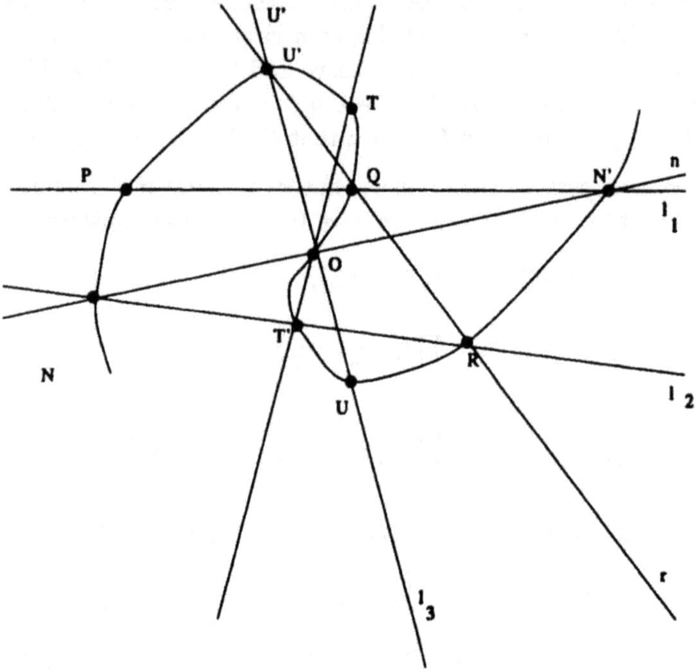

Figure 14

l_1	the line through	N, O, N'
l_2	the line through	N, T', R
l_3	the line through	U, O, U'
n	the line through	N, O, N'
t	the line through	T', O, T
r	the line through	R, Q, U'

We now have to prove that $P \odot U = T$. The line $\overline{PU}$ intersects E in a point T'' and the problem consists in proving that $T'' = T'$, or in other words, are P, U and T' colinear?

Consider E as the curve $F = 0$ and take $G = l_1 l_2 l_3$. By definition we see that the intersection divisor

$$\begin{aligned} F \circ (l_1 l_2 l_3) &= P + Q + N' + N + T' + R + U' + O + U \\ &= (N + O + N') + (U' + Q + N) + P + T' + U. \end{aligned}$$

We apply Proposition 2.6 to F, G and $L = n$ and find a conic H such that $F \circ H = U' + Q + R + P + T' + U$. Applying Proposition 2.6 again, this time to F, H and $L = t$, then we find a *line* H_1 such that $P + T' + U = F \circ H_1$, and consequently P, T' and U are colinear as they are on H_1! □

For the general form of M. Noether's theorem we refer to ([Ful] P.120). In loc. cit. the theorem is given modulo the assumption that the so-called "Noether-conditions" hold; these are fulfilled in many situations, in particular in the situation of our Proposition 2.6.

We now go on to show that one may use $\odot$ to establish that E is not birationally equivalent to a line. First some preparatory results, with interest in their own right, are necessary.

2.8. Proposition Let E be as before, a nonsingular cubic curve, and $\mathcal{D} = \mathbf{V}(G)$ a curve of degree m in $\mathbf{P}^2$ (G need not be irreducible unless stated), such that E is not a component of $\mathcal{D}$.
Put $\mathcal{D} \circ E = \sum_{i=1}^{3m} P_i$ and let O and O' be as in Lemma 2.4. Then we have

$$\sum_{i=1}^{3m} P_i = mO',$$

where we write mQ for $Q \odot Q \odot \cdots \odot Q$, m-times.

Proof If $m = 1$ then it follows from the construction that $P_1 \odot P_2 \odot P_3 = O'$. We go on by induction on m, assuming the result has been established for $\mathcal{D}$ of degree $m-1$ or lower. Now look at $\mathcal{D}$ of degree m, $\mathcal{D} \circ E = \sum_{i=1}^{3m} P_i$. The line $\overline{P_1 P_2}$ intersects E in a third point $P_{1,2}$; the line $\overline{P_3 P_4}$ intersects E in a third point $P_{3,4}$, the line $r = \overline{P_{1,2} P_{3,4}}$ intersects E in a third point R. We have

$$\begin{aligned} E \circ (\mathcal{D} \cup r) &= \sum_{i=1}^{3m} P_i + P_{1,2} + P_{3,4} + R \\ &= \sum_{i=3}^{3m} P_i + (P_1 + P_2 + P_{1,2}) + P_{3,4} + R. \end{aligned}$$

According to Proposition 2.6 there exists an $\mathcal{H}$ of degree m such that $\mathcal{H} \circ E = \sum_{i=3}^{3m} P_i + P_{3,4} + R$; there also exists an $\mathcal{H}'$ of degree $m-1$ such that $E \circ \mathcal{H}' = \sum_{i=5}^{3m} P_i + R$. From the induction hypothesis we infer that $\sum_{i=5}^{3m} P_i + R = (m-1)O'$. We now calculate

$$\sum_{i=1}^{3m} P_i \odot P_{1,2} \odot P_{3,4} \odot R = (m-1)O' \odot 2O',$$

because P_1, P_2, $P_{1,2}$ are colinear and P_3, P_4, $P_{3,4}$ are colinear (so using the case $m = 1$ twice). However, from

$$(*) \qquad \sum_{i=1}^{3m} P_i \odot P_{1,2} \odot P_{3,4} \odot R = (m+1)O'$$

it then follows that $\sum_{i=1}^{3m} P_i = mO'$, because $P_{1,2} \odot P_{3,4} \odot R = O'$ as R is on the line ν and we may cancel O' in both members of $(*)$ because $(E, \odot)$ is a group. □

2.9. Corollary (i) If $\mathcal{D} = \mathbf{V}(G)$ and $\mathcal{D}_1 = \mathbf{V}(G_1)$ are curves of degree m in $\mathbf{P}^2$ such that $E \circ \mathcal{D} = \sum_{i=1}^{3m-1} P_i \odot P$, $E \circ \mathcal{D}_1 = \sum_{i=1}^{3m-1} P_i \odot P'$, then $P = P'$.
(ii) If $P \in E$ then there cannot exist a $\frac{\phi}{\psi} \in k(E)$ such that $\left(\frac{\phi}{\psi}\right)_\infty = P$.

Proof (i) Obvious.
(ii) Suppose there exists a rational function $\frac{\phi}{\psi}$ with pole divisor $\left(\frac{\phi}{\psi}\right)_\infty$ exactly P, say $\frac{\phi}{\psi} = \frac{G_1}{G_2}$ where G_1 and G_2 are homogeneous of same degree m. We obtain the divisor $\mathbf{div}\left(\frac{\phi}{\psi}\right)$ as $\mathbf{div}\left(\frac{\phi}{\psi}\right) = E \circ \mathbf{V}(G_1) - E \circ \mathbf{V}(G_2)$. Put

$$E \circ \mathbf{V}(G_1) = \sum_{i=1}^{3m} P_i, \quad E \circ \mathbf{V}(G_2) = \sum_{i=1}^{3m-1} Q_i \odot P,$$

where P does appear because we assumed that $\left(\frac{\phi}{\psi}\right)_\infty = P$. Since P should be the unique pole, the Q_i have to equal to the P_i up to renumbering them. From (i) above it then follows that $P = P_{3m}$. But then **div** $\left(\frac{\phi}{\psi}\right) = 0$ or $\frac{\phi}{\psi}$ is constant, a contradiction. □

2.10. Proposition An elliptic curve E is not birationally equivalent to a line.

Proof Suppose there is a birational transformation τ: $E \to L$, where L is a line, and consider $P \in E$, $\tau(P) \in L$ for some P where τ is defined. On the line L we may always find $\frac{\phi}{\psi} \in k(L)$ such that $\left(\frac{\phi}{\psi}\right)_\infty = \tau(P)$ (e.g. if $\tau(P) = (\lambda, 1)$ take $\frac{\phi}{\psi} = \frac{\xi - \mu}{\xi - \lambda}$ where $\mu \neq \lambda$ in k, ξ the image of x_1 in the coordinate ring). If τ is a birational equivalence we have an induced isomorphism τ^*: $k(L) \to k(E)$. The rational function on E obtained as $\tau^*\left(\frac{\phi}{\psi}\right)$ satisfies

$$\tau\left(\left(\tau^*\left(\frac{\phi}{\psi}\right)\right)_\infty\right) = \left(\frac{\phi}{\psi}\right)_\infty = \tau(P)$$

(τ respects local rings, see CH.VI). Consequently $\left(\tau\left(\frac{\phi}{\psi}\right)\right)_\infty = P$ but that would contradict to the foregoing Corollary 2.9(ii). □

Let us recollect the meaning of the Riemann-Roch theorem for elliptic curves (this follows from CH.VII Theorem 6.9, Corollary 6.10, but is worth restating in its own right).

2.11. Observation For a divisor D on an elliptic curve E we have.
(i) $\ell(D) = 0$ if $\deg D < 0$.
(ii) $\ell(D) = 0$ if $\deg D = 0$ and D not equivalent to the zero divisor.
(iii) $\ell(D) = 1$ if $\deg D = 0$ and $D \sim 0$.
(iv) $\ell(D) = \deg(D)$ if $\deg D > 0$.

Recall from CH.VII Observation 4.4 that we have inclusions of abelian groups:

$$\mathbf{Div}_0(E) \subset \mathbf{Div}^0(E) \subset \mathbf{Div}(E).$$

We may use these abelian groups to obtain a new description of the addition on E.

2.12. Lemma There is a surjective map π: $\mathbf{Div}^0(E) \to E$.

Proof We have fixed $O \in E$. Consider $D \in \mathbf{Div}^0(E)$. Then $D + O$ is of positive degree, hence $\ell(D+O) = 1$ (see above). Pick $f = \frac{\phi}{\psi} \in L(D+O)$, $f \neq 0$; by definition $\mathbf{div}(f) + D + O \succ 0$ and moreover $\deg(\mathbf{div}(f) + D + O) = 1$, hence there is a $P \in E$ such that $\mathbf{div}(f)+D+O = P$. For another $g \in L(D+O)$ we have $g = cf$ for some $c \in k$ and $\mathbf{div}(g) + D + O = P$ holds again. Hence there is a unique point P of E such that $D + O$ is equivalent to P. Put $\pi(D) = P$. For any $Q \in E$ we have that $Q - O \in \mathbf{Div}^0(E)$ and of course $\pi(Q - O) = Q$. If $\pi(D_1) = \pi(D_2)$ for divisors D_1 and D_2 of degree zero, then

$$D_1 + O \sim P \sim D_2 + O.$$

In $\mathbf{Div}^0(E)$ we then have $D_1 \sim D_2$. This proves that π factorizes over a bijection $\overline{\pi}$: $\mathbf{Div}^0(E)/\mathbf{Div}_0(E) \to E$, hence a bijection $\mathbf{Pic}(E) \to E$ in the terminology of CH.VII Observation 4.4. □

We may define a group structure on E by transfer of the group-structure of $\mathbf{Pic}(E)$ via the bijection $\overline{\pi}$.

2.13. Proposition The operation induced on E via the bijection π: $\mathbf{Pic}(E) \to E$ coincides with $\odot$, hence $(E, \odot) \cong \mathbf{Pic}(E)$ as groups.

Proof Consider $P, Q \in E$ and $R = P \odot Q$. In $\mathbf{Div}(E)$

$$(P - O) + (Q - O) \sim R_1 - O$$

for some $R_1 \in E$, i.e., $P + Q - O \sim R_1$. We look at the two lines

$$\begin{array}{lll} L & \text{through} & P, Q, R' \\ L' & \text{through} & R', O, R \end{array}$$

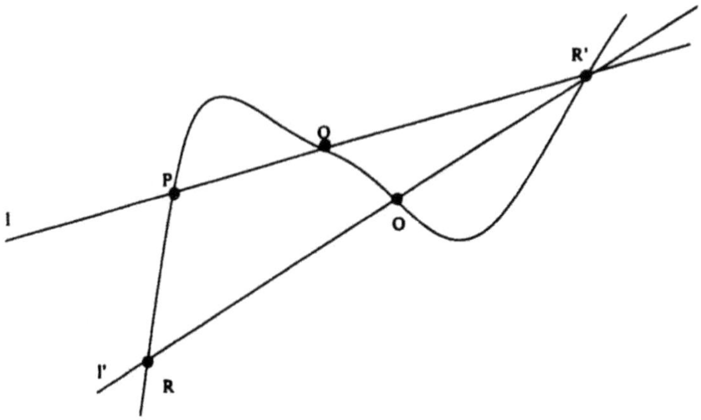

Figure 15

we have

$$\mathrm{div}\left(\frac{L}{L'}\right) = (P + Q + R') - (R' + O + R) = (P + Q) - (O + R).$$

Consequently $P + Q - O \sim R$, hence $R = R_1$. This proves the statement. □

2.14. Corollary The group structure on E is not depending on the choice of O.

□

It is customary to put O at the unique "point at ∞" of E, i.e., we think of E, defined over a subfield k of $\mathbb{C}$, as the affine curve

$$y^2 = 4x^3 - g_2x - g_3$$

together with the point O at infinity. If K is an intermediate field, $k \subset K \subset \mathbb{C}$, then we write $E(K)$ for the K-rational points of E, i.e., the solutions of $x_2^2 = 4x_1^3 - g_2x_1 - g_3$ in $\mathbf{A}_K^2$. When $K = \mathbb{C}$ we just obtain $E(\mathbb{C}) = E$. The graph of $E(\mathbb{R})$ looks differently depending whether $4x_1^3 - g_2x_1 - g_3$ has three real roots or just one real root, see Figure 16 resp. 17 hereafter.

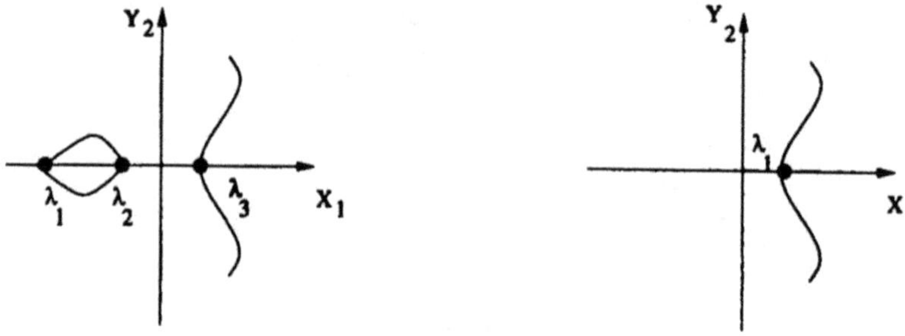

Figure 16–17

For $P \in E$ we define the *order* of P to be the order of P in the abelian group $(E, \odot)$; we say that P is a *torsion point* of E if it has finite order. We write $E(K)_{\mathrm{tors}}$ for the torsion subgroup of $E(K)$. In view of Theorem 2.1, $E(\mathbb{Q})_{\mathrm{tors}}$ is a finite group; in fact a theorem of B. Mazur states that $|E(\mathbb{Q})_{\mathrm{tors}}| \le 16$, moreover $E(\mathbb{Q})_{\mathrm{tors}}$ is necessarily one of the following fifteen groups:

$$\mathbb{Z}/m\mathbb{Z} \text{ with } 1 \le m \le 10 \text{ or } m = 12, \text{ or}$$

$$(\mathbb{Z}/2\mathbb{Z}) \times \mathbb{Z}/2m\mathbb{Z} \text{ for } 1 \le m \le 4.$$

The general theory of abelian groups yields that

$$E(\mathbb{Q}) = \mathbb{Z}^r \oplus E(\mathbb{Q})_{\mathrm{tors}}.$$

We call r the *rank* of $E(\mathbb{Q})$.

Up to now there is no algorithm for calculating the rank of an arbitrary elliptic curve over $\mathbb{Q}$. For some examples one has a good knowledge, let us just mention some cases here.

2.15. Example (i) For $x_2^2 = x_1^3 - x_1$ we have $r = 0$ and $E(\mathbb{Q}) \cong (\mathbb{Z}/2\mathbb{Z}) \times (\mathbb{Z}/2\mathbb{Z})$. The only rational points on this curve are: $O, (0,0), (1,0), (-1,0)$.

(ii) For $x_2^2 = x_1^3 + x_1$ we also have $r = 0$ but $E(\mathbb{Q}) = \{O, (0,0)\}$.

The following theorem is often useful in calculating $E(\mathbb{Q})_{\text{tors}}$ or when checking whether a certain $P \in E(\mathbb{Q})$ has finite order.

2.16. Theorem (Lutz-Nagell) Let E be defined by

$$x_2^2 = x_1^3 + Ax_1 + B,$$

with $\Delta = -4A^3 - 27B^2 \neq 0$ and $A, B \in \mathbb{Z}$. If $P = (\xi, \eta)$ in $E(\mathbb{Q})$ has finite order, then ξ, $\eta \in \mathbb{Z}$ and either $\eta = 0$ or x_2^2 divides Δ.

§3. Elliptic Functions and Weierstrass Theory

Let V be a finite dimensional vector space over $\mathbb{R}$. An additive subgroup Γ of V is called a (discrete) *lattice* if
(i) Γ is discrete and V/Γ is compact in the quotient topology of the real topology,
(ii) Γ generates V over $\mathbb{R}$, and
(iii) there is an $\mathbb{R}$-basis $\{e_1, ..., e_n\}$ for V that is a $\mathbb{Z}$-basis for Γ, i.e., $\Gamma = \mathbb{Z}e_1 \oplus \cdots \oplus \mathbb{Z}e_n$.

Note that the axioms above are certainly not independent.

We write $\mathcal{L}_V$ for the set of lattices in V, and if V is fixed and unambiguous we just write $\mathcal{L}$. In the sequel we *restrict* attention to $V \cong \mathbb{R}^2$. Observe that $\mathcal{L}$ is an $\mathbb{R}$-vector space too.
Define

$$M = \left\{(\omega_1, \omega_2) \;\middle|\; \omega_1, \omega_2 \in \mathbb{C}^*,\ \text{Im}\left(\frac{\omega_1}{\omega_2}\right) > 0\right\},$$

where Imstands for taking the imaginary part of a complex number. To $(\omega_1, \omega_2) \in M$ we associate

$$\Gamma(\omega_1, \omega_2) = \mathbb{Z}\omega_1 \oplus \mathbb{Z}\omega_2 \in \mathcal{L}$$

and so we obtain a surjective map $M \to \mathcal{L}$.

If $G = \begin{pmatrix} a & b \\ c & d \end{pmatrix} \in \mathrm{SL}_2(\mathbb{Z})$ and $(\omega_1, \omega_2) \in M$, then we look at

$$\omega_1' = a\omega_1 + b\omega_2, \quad \omega_2' = c\omega_1 + d\omega_2.$$

Since $\det G = 1$ it is clear that $\{\omega_1', \omega_2'\}$ is again a $\mathbb{Z}$-basis for $\Gamma(\omega_1, \omega_2)$.
Putting

$$z = \frac{\omega_1}{\omega_2}, \quad z' = \frac{\omega_1'}{\omega_2'},$$

then (check!) for

$$z' = \frac{az+b}{cz+d} = Gz,$$

we have

$$\mathrm{Im}(z') > 0,$$

hence $(\omega_1', \omega_2') \in M$.
By letting $G \in \mathrm{SL}_2(\mathbb{Z})$ act on the uper half plane H in $\mathbb{C}$ as above, i.e.,

$$Gz = \frac{za+b}{cz+d} \text{ if } G = \begin{pmatrix} a & b \\ c & d \end{pmatrix}$$

we obtain an action of $\mathrm{SL}_2(\mathbb{Z})$ on

$$H = \left\{ z \in \mathbb{C} \;\middle|\; \mathrm{Im}(z) > 0 \right\}.$$

3.1. Proposition Two elements of M defined the same lattice in $\mathcal{L}$ if and only if they are congruent modulo the $\mathrm{SL}_2(\mathbb{Z})$-action defined on M (i.e., $G(\omega_1, \omega_2) = (\omega_1', \omega_2')$ as above).

Proof The "if" part has been observed above.
Now consider (ω_1, ω_2) and (ω_1', ω_2') in M defining the same $\Gamma(\omega_1, \omega_2)$ in $\mathcal{L}$. From $\omega_1 = A\omega_1' + B\omega_2'$, $\omega_2 = C\omega_1' + D\omega_2'$ with $A, B, C, D \in \mathbb{Z}$ it follows that there is a matrix in $GL_2(\mathbb{Z})$ with determinant ± 1, such that it trasforms the first $\mathbb{Z}$-basis into the second $\mathbb{Z}$-basis. If the determinant would be equal to -1, then $\mathrm{Im}\left(\frac{\omega_1'}{\omega_2'}\right)$ would have opposite sign of $\mathrm{Im}\left(\frac{\omega_1}{\omega_2}\right)$, contrary to the assumptions. □

From the foregoing proposition it follows that

$$\mathcal{L} = M \bmod \mathrm{SL}_2(\mathbb{Z})$$

is the orbitspace of the action of $\mathrm{SL}_2(\mathbb{Z})$ on M. We let $\mathbb{C}^*$ act on $\mathcal{L}$ (and on M in the obvious way) by transforming $\Gamma(\omega_1, \omega_2)$ to $\Gamma(\lambda\omega_1, \lambda\omega_2)$, $\lambda \in \mathbb{C}^*$. The orbitspace for this $\mathbb{C}^*$-action, $M/\mathbb{C}^*$, may be identified to H via the association

$$(\omega_1, \omega_2) \mapsto z = \frac{\omega_1}{\omega_2} \in H.$$

The $\mathrm{SL}_2(\mathbb{Z})$-action on M is then transfered into an action of

$$G = \mathrm{PSL}_2(\mathbb{Z}) = \mathrm{SL}_2(\mathbb{Z})/\{\pm \mathrm{Id}\}$$

on H. So we have estabilished the following

3.2. Proposition The mapping $(\omega_1, \omega_2) \mapsto \dfrac{\omega_1}{\omega_2}$ induces a bijection

$$\mathcal{L}/\mathbb{C}^* \to H/G;$$

in other words, a G-orbit in H may be identified with a lattice up to complex homothety.

□

A function $F\colon \mathcal{L} \to \mathbb{C}$ is said to have *weight* $2k$ if for $\lambda \in \mathbb{C}^*$, $\Gamma \in \mathcal{L}$, $F(\lambda\Gamma) = \lambda^{-2k} F(\Gamma)$. We abbreviate notation as follows:

$$F(\Gamma(\omega_1, \omega_2)) = F(\omega_1, \omega_2).$$

Then the foregoing definition just reads as follows:

$$(**) \qquad F(\lambda\omega_1, \lambda\omega_2) = \lambda^{-2k} F(\omega_1, \omega_2).$$

Viewing F as a function on M via $M \to \mathcal{L} \xrightarrow{F} \mathbb{C}$, we see that F is invariant for the action of $\mathrm{SL}_2(\mathbb{Z})$ exactly because it is defined via $\mathcal{L}$. Taking $\lambda = \omega_2^{-1}$ in $(**)$ yields that $F(z, 1) = \omega_2^{2k} F(\omega_1, \omega_2)$ only depends on $z = \dfrac{\omega_1}{\omega_2}$, therefore there exists a function f on H such that

$$F(\omega_1, \omega_2) = \omega_2^{-2k} f\left(\frac{\omega_1}{\omega_2}\right).$$

Expressing the $\mathrm{SL}_2(\mathbb{Z})$-invariance of F yields, for $g = \begin{pmatrix} a & b \\ c & d \end{pmatrix} \in \mathrm{SL}_2(\mathbb{Z})$,

$$F(a\omega_1 + a\omega_2, c\omega_1 + d\omega_2) = F(\omega_1, \omega_2),$$

or

$$\omega_2^{-2k} f\left(\frac{\omega_1}{\omega_2}\right) = (c\omega_1 + d\omega_2)^{-2k} f\left(\frac{a\omega_1 + b\omega_2}{c\omega_1 + d\omega_2}\right),$$

or

$$f(z) = (cz + d)^{-2k} f(gz).$$

The latter expresses that f is a modular function ($G = \mathrm{PSL}_2(\mathbb{Z})$ is callled the modular group) but we do not go deep in the theory of modular functions here. There is a lot of interesting theory available for further reading if one is interested in the interplay between geometry, number theory and analytical methods, cf. [A. Ogg], [M. Knopp], [J-P. Serre].

By a *lattice* in $\mathbb{C}$ we shall mean a lattice in $\mathbb{C} \cong \mathbb{R}^2$. For $k > 1$ in $\mathbb{Z}$ we define

$$G_k(\Gamma) = \sum_{\gamma \in \Gamma}{}' \gamma^{-2k},$$

where $\sum'$ means that $\gamma = 0$ is excluded from the summation. The series $G_k(\Gamma)$ is called the ***Eisenstein series of index k***.

3.3. Lemma For any $\Gamma \in \mathcal{L}$, the series $\sum'_{\gamma \in \Gamma} |\gamma|^{-\sigma}$ is convergent for $\sigma > 2$.

Proof Let $C(R)$ be the inner part of the circle with radius R. Then

$$\sum_{|\gamma| \geq R} |\gamma|^{-\sigma} \leq \int\int_{\mathbb{C} - C(R)} \frac{dxdy}{(x^2 + y^2)^{\frac{\sigma}{2}}}.$$

Rewriting in polar coordinates yields

$$\sum_{|\gamma| \geq R} |\gamma|^{-\sigma} \leq \int_R^\infty \int_0^{2\pi} \frac{r dv d\varphi}{r^\sigma}.$$

In case $\sigma > 2$, the right hand member is less than $\alpha |r^{2-\sigma}|_R^\infty$, hence less than $\alpha R^{2-\sigma}$.

For all ε there is an $N(\varepsilon)$ such that $R \geq N(\varepsilon)$ yields $\alpha R^{2-\sigma} \leq \varepsilon$, this proves that the series $\sum_\gamma' |\gamma|^{-\sigma}$ converges for $\sigma > 2$. □

We may view G_k as the function defined on M by

$$G_k(\omega_1, \omega_2) = \sum_{(m,n)\neq(0,0)} (m\omega_1 + n\omega_2)^{-2k}.$$

The function on the upper half plane H corresponding to this is

$$G_k(z) = \sum_{(m,n)\neq(0,0)} (mz + n)^{-2k}.$$

3.4. Proposition The Eisenstein series $G_k(z)$, for $k > 1$ in $\mathbb{Z}$, is a modular form of weight $2k$. Moreover $G_k(\infty) = 2\zeta(2k)$, where ζ is the classical zeta-function.

Proof For $\lambda \in \mathbb{C}^*$ we have $G_k(\lambda\Gamma) = \lambda^{-2k}G_k(\Gamma)$ as one easily verifies. Hence $G_k(z)$ satisfies the modularity condition mentioned before Lemma 3.3. In the definition of a modular form holomorphicity everywhere is contained; let us just state here that this holomorphicity follows from the lemma. For holomorphicity of G_k at ∞ we have to verify that $G_k(z)$ has a limit when $\mathrm{Im} z \to \infty$. In view of the convergence properties such a limit may be calculated by summing the limit of each term; terms $(mz + n)^{-2k}$ with $m \neq 0$ will tend to zero but the other terms yield $\sum_{-\infty}^{\infty} n^{-2k}$ (except for $n = 0$). Hence we obtain

$$G_k(\infty) = z\sum_{n=1}^{\infty} n^{-2k} = 2\zeta(2k)$$

where the latter is just the definition of the zeta function. □

It is conventional to put $g_2 = 60G_2$, $g_3 = 140G_3$ and $\Delta = g_2^3 - 27g_3^2$, $j = 1728\frac{g_2^3}{\Delta}$.

If Γ is a lattice in $\mathbb{C} \cong \mathbb{R}^2$, it defines a compact group $\mathbb{C}/\Gamma$. A meromorphic function f on $\mathbb{C}$ is said to be *elliptic* or *doubly periodic* with respect to the lattice Γ, if for all $\gamma \in \Gamma$ we have $f(z+\gamma) = f(z)$. An elliptic function can be considered as a meromorphic function on the analytical space $\mathbb{C}/\Gamma$. The latter space may be viewed as a flat torus T:

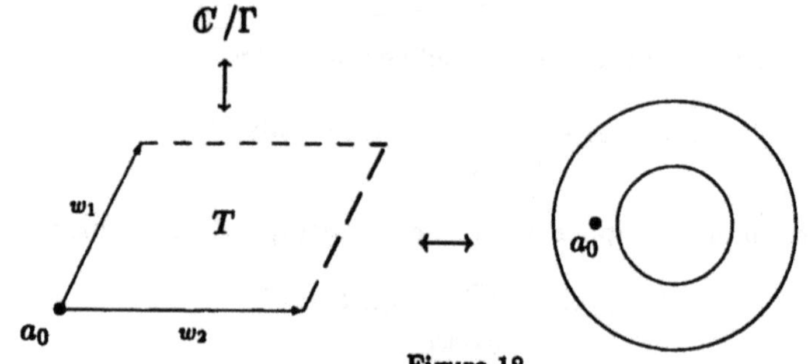

Figure 18

where in T the parallel edges are being identified in order to obtain a compact Riemann surface of genus one.
The set of elliptic functions with respect to Γ forms a field with respect to pointwise addition and multiplication. Moreover, if f is elliptic then clearly its derivative f' is elliptic too.

Consider an elliptic function f. For $a \in \mathbb{C}$ we may use Laurent expansion of f at a and define $\mathbf{ord}_a(f)$ to be the smallest index with nonzero Laurent coefficient in the expansion of f at a. Obviously $\mathbf{ord}_a(f)$ is negative if and only if a is a pole of f.
The periodicity of f yields

$$\mathbf{ord}_{a+\gamma}(f) = \mathbf{ord}_a(f)$$

for all $\gamma \in \Gamma$; so we may view $\mathbf{ord}_a(f)$ as a number associated to a in $E = \mathbb{C}/\Gamma$.
Since $\mathbb{C}/\Gamma$ is compact, any meromorphic function f on $E = \mathbb{C}/\Gamma$ has only finitely many poles and zeros; in fact equally many of each when counting multiplicities. The formal sum

$$\mathbf{div}(f) = \sum_{a \in \mathbb{C}/\Gamma} \mathbf{ord}_a(f) \cdot a$$

has only finitely many nonzero coefficients; we may view $\mathbf{div}(f)$ as an element in the free abelian (additive) group $\mathbf{Div}(\mathbb{C}/\Gamma)$ generated by the elements of $E = \mathbb{C}/\Gamma$.

3.5. Proposition Let f be a non-constant elliptic function with respect to Γ and write $\mathbf{div}(f) = \sum n_i(a_i)$, then

(i) f is not everywhere holomorphic.
(ii) $\sum_i \text{Res}_{a_i}(f) = 0$, where $\text{Res}_{a_i}(f)$ is the residue of f at a_i.
(iii) $\sum_i n_i = 0$.
(iv) $\sum n_i a_i = 0$ in $\mathbb{C}/\Gamma$.

Proof (i) If f is holomorphic then it has to be bounded in any compact set, in particular in a period parallelogram. The periodicity of f then entails that f is bounded in the whole of $\mathbb{C}$ and by Liouville's theorem this can only happen for a constant f.
(ii) Choose $a \in \mathbb{C}$ such that f has neither a pole nor a zero on the edges of the period parallelogram P_a with vertices a, $a + \omega_1$, $a + \omega_2$, $a + \omega_1 + \omega_2$. Write dP_a for the boundary of P_a oriented in the anti-clockwise way.
Consider

$$\int_{dP_a} f(z)dz.$$

Periodicity of f entails that opposite edges of P_a cancel their contribution in the integral. Hence, the integral equals $2\pi i \sum \text{Res}_{a_i}(f) = 0$, where the a_i are the poles and zeros in P_a (but these are all of them in view of the choice of P_a).
(iii) If f is elliptic then f' and $\frac{f'}{f}$ are elliptic too. Apply (ii) to f' and (iii) follows.
(iv) Look at the function $\frac{zf'}{f}$ and integrate it over dP_a as above. This time contributions from parallel edges do not cancel but

$$\left(\int_a^{a+\omega_1} - \int_{a+\omega_2}^{a+\omega_1+\omega_2}\right) \frac{zf'(z)}{f(z)}dz = -\omega_2 \int_a^{a+\omega_1} \frac{f'(z)}{f(z)}dz.$$

Since f has period ω_1 the image $f([a, a+\omega_1])$ is a closed curve C_1 and we obtain

$$\int_a^{a+\omega_1} \frac{f'(z)}{f(z)}dz = \oint_{C_1} \frac{d\theta}{\theta} = 2\pi i \text{Index}(0, C_1) = 2\pi i m_1$$

for some $m_1 \in \mathbb{N}$.
A similar argument applies to the other pair of adges and then adding both similar terms we have calculated the integration in two ways, yielding:

$$\sum n_i a_i = m_2\omega_1 - m_1\omega_2 \in \Gamma.$$

Note that in $\sum n_i a_i$ we use the sum in $\mathbb{C}$ which is obviously completely different from the formal sum $\sum n_i(a_i)$ in $\mathbf{div}(f)$. □

3.6. Corollary If f and g are elliptic with repsect to Γ such that $\mathbf{div}(f) = \mathbf{div}(g)$, then $\frac{f}{g}$ has no pole, consequently it must be a constant! This states that the divisor of an elliptic function determines that function up to a constant. From Proposition 3.5(ii) it follows that it is impossible for f to have only one simple pole. The simplest situation would be for f to have a double pole with zero residue or two simple poles with opposite residues. From Proposition 3.5(iii) it follows that an elliptic function f has as many poles as zeros. If f is not a constant then $f(z) - f(z_0)$ and $f(z)$ have the same poles so they also have the same number of zeros. The latter then states that f takes all values the same number of times. The latter number is called the *valence* of f.

□

Let us introduce some examples of elliptic functions.
The Weierstrass $\wp$-function is defined by

$$\wp(z) = \frac{1}{z^2} + \sum_{\gamma\in\Gamma}{}' \frac{1}{(z-\gamma)^2} - \frac{1}{\gamma^2},$$

where $\sum'$ means that $O \in \Gamma$ is not allowed in the summation. For a specific term in $\wp(z)$ we obtain

$$\left|\frac{1}{(z-\gamma)^2} - \frac{1}{\gamma^2}\right| = \left|\frac{2z\gamma - z^2}{\gamma^2(z-\gamma)}\right| = \frac{2|z|\left|1-\frac{z}{2\gamma}\right|}{|\gamma|^3\left|1-\frac{z}{\gamma}\right|^2} \leq \frac{c}{|\gamma|^3}$$

for some constant c, if z varies in a bounded closed set D not intersecting the lattice Γ. In such a set D, $\sum' \frac{1}{(z-\gamma)^2} - \frac{1}{\gamma^2}$ is uniformly convergent. It follows that $\wp(z)$ is meromorphic but holomorphic everywhere except in the lattice points where $\wp(z)$ has poles of order 2.

From the uniform convergence observed before we retain that $\wp'(z)$

may be calculated by term by term differentiation. Hence we ontain

$$\wp'(z) = \frac{-2}{z^3} + \sum_{\gamma\in\Gamma}{}' \frac{-2}{(z-\gamma)^3} = \sum_{\gamma\in\Gamma} \frac{-2}{(z-\gamma)^3}.$$

In view of Lemma 3.3, or by a direct verification, we know that $\wp'(z)$ is meromorphic and of course doubly periodic as well.
For $i = 1, 2$ we have $\wp'(z+\omega_i) - \wp'(z) = 0$ and thus $\wp(z+\omega_i) - \wp(z) = c-i$ for some constants $c_1, c_2 \in \wp(z) = c-i$ for some constants c_1, c_2.
Specialize $z = \frac{-\omega_i}{2}$, then the foregoing reduces to

$$\wp\left(\frac{\omega_i}{2}\right) - \wp\left(-\frac{\omega_i}{2}\right) = c_i.$$

However, $\wp$ is an even function in the sense that $\wp(z) = \wp(-z)$ for any z, thus we must have $c_i = 0$ for $i = 1, 2$.
We have proved that $\wp(z)$ is meromorphic and doubly periodic for Γ hence $\wp(z)$ is an elliptic function. Also $\wp'(z)$ is an example of an elliptic function, but $\wp'(z)$ is an odd function in the sense that $\wp'(-z) = -\wp'(z)$.

Let us now look at the Taylor expansions of $\wp(z)$ and $\wp'(z)$. We calculate (in a neighborhood of zero):

$$\begin{aligned} \wp(z) - \frac{1}{z^2} &= \sum_{\gamma\in\Gamma}{}' \left(\frac{1}{(z-\gamma)^2} - \frac{1}{\gamma^2}\right) \\ &= \sum{}' \frac{1}{\gamma^2}\left[\frac{1}{\left(1-\frac{z}{\gamma}\right)^2} - 1\right] \\ &= \sum{}' \frac{1}{\gamma^2}\left[\left(1 + \frac{z}{\gamma} + \frac{z^2}{\gamma^2} + \cdots\right) - 1\right] \\ &= \sum{}' \frac{1}{\gamma^2}\left[\frac{2z}{\gamma} + \frac{3z^2}{\gamma^2} + \frac{4z^3}{\gamma^3} + \frac{5z^4}{\gamma^4} + \cdots\right]. \end{aligned}$$

Since $\wp(z)$ is an even function we obtain

$$\begin{aligned} \wp(z) &= \frac{1}{z^2} + 3G_2z^2 + 5G_3z^4 + \cdots \\ &= z^{-2} + \sum_{k\geq 1}(2k+1)G_{k+1}z^{2k}, \end{aligned}$$

where this expansion holds in a neighborhood of zero, in fact for $0 < |z| < \min\{|\gamma| \mid \gamma \in \Gamma - \{0\}\}$. By differentiation (which may be carried out term by term)

$$\wp(z) = -2z^{-3} + \sum_{k \geq 1} (2k+1)2kG_{k+1}z^{2k+1},$$

in the same neighborhood of zero as indicated above.
Since $\wp'(z)$ is odd it vanishes at $a \in \mathbb{C}$ for which $a = -a \bmod \Gamma$ and $a \notin \Gamma$. A set of representative for these points is exactly given by $\frac{\omega_1}{2}, \frac{\omega_2}{2}, \frac{\omega_1 + \omega_2}{2}$

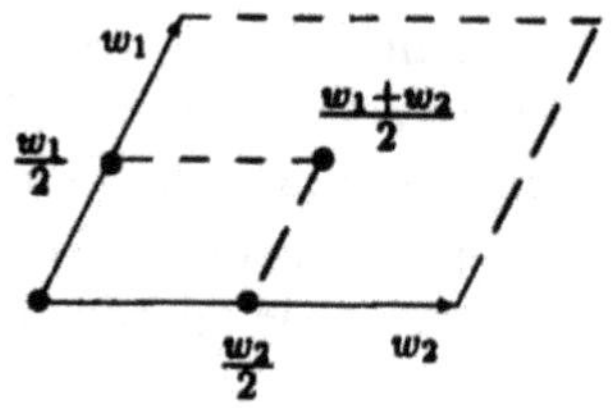

Figure 19

Since $\wp'$ has just one triple pole on $E = \mathbb{C}/\Gamma$ it can only have three zeros. Therefore we have obtained

$$\operatorname{div}(\wp'(z) = -3(O) + 1 \cdot \left(\frac{\omega_1}{2}\right) + 1 \cdot \left(\frac{\omega_2}{2}\right) + 1 \cdot \left(\frac{\omega_1 + \omega_2}{2}\right).$$

In a similar way we may observe that the elliptic function $\wp(z) - \wp(a)$ with $a \in \mathbb{C} - \Gamma$, has only a unique double pole on E, hence it has only two zeros and thus they must be a and $-a$ (because $\wp$ is even!) when $a \not\equiv -a \bmod \Gamma$. However, in case $a \equiv -a \bmod \Gamma$ then $\wp'(a)$ yields that the zero a for $\wp(z) - \wp(a)$ is at least double, so also in this case we arrive at

$$\operatorname{div}(\wp(z) - \wp(a)) = -2(O) + 1 \cdot (a) + 1 \cdot (-a).$$

Note that the two symmetric zeros of $\wp(z)$ are not explicitly known, even if we could derive their existence in such an elementary way. The foregoing observation now allow us to give a complete description of the field of Γ-elliptic functions. In fact we will show how

$(\wp, \wp')$ provide "coordinate functions" actually defining on E the structure of an elliptic curve!

3.7. Theorem The field of Γ-elliptic functions is

$$\mathbb{C}(\wp, \wp') \cong \frac{\mathbb{C}(x)[y]}{(y^2 - yx^3 + g_2 x + g_3)},$$

with $\Delta = g_2^3 - 27g_3^2 \neq 0$. The subfield of even elliptic functions is exactly $\mathbb{C}(\wp)$.

Proof Any elliptic function f may be written as $f_0 + f_1$ where f_0 is even and f_1 is odd; indeed take

$$f_0(z) = \frac{1}{2}(f(z) + f(-z)) \text{ and } f_1 = f - f_0.$$

Since $\wp'$ is odd, $f = f_0 + \wp'\left(\frac{f_1}{\wp'}\right)$, with $\frac{f_1}{\wp'}$ being an even elliptic function. This already implies that $\mathbb{C}(\wp, \wp')$ is a quadratic extension of the field of all even elliptic functions.
Suppose f is even. Put

$$\nu_i = \begin{cases} \mathbf{ord}_{a_i}(f) \text{ if } -a_i \not\equiv a_i \bmod \Gamma, \\ \\ \frac{1}{2}\mathbf{ord}_{a_i}(f) \text{ if } a_i \equiv -a_i \bmod \Gamma. \end{cases}$$

Note that ν_i is always an integer!
Consider $g = \prod(\wp - \wp(a_i))^{\nu_i}$, the product ranging over a set of representatives of the classes $\{a_i, -a_i\}$ with $a_i \notin \Gamma$. The divisors of f and g now have the same coefficients for all points (a) such that $a \in \Gamma$. From Proposition 3.5(iii) it follows that also the coefficient of (O) is the same in both divisors. Consequently $f = \text{cst} \cdot g$ and thus it is a rational function in $\wp$. It follows that $\mathbb{C}(\wp)$is the whole subfield of even elliptic functions.
It remains to find the quadratic equation for $\wp'$ over $\mathbb{C}(\wp)$. We have observed earlier

$$\begin{aligned} \wp(z) &= z^{-2} + 3G_2 z^2 + 5G_3 z^4 + O(z^6), \\ \wp'(z) &= -2z^{-3} + 6G_2 z + 20G_3 z^3 + O(z^5). \end{aligned}$$

Hence

$$\begin{aligned} \wp'(z)^2 &= 4z^{-6} - 24G_2z^{-2} - 80G_3 + O(z^2), \\ \wp(z)^3 &= z^{-6} + 9G_2z^{-2} + 15G_3 + O(z^2). \end{aligned}$$

Thus,

$$\wp'(z)^2 - 4\wp(z)^3 = -g_2\wp(z) - g_3 + h(z),$$

where $h(z)$ is Γ-elliptic, holomorphic and of type $O(z^2)$, consequently $h = 0$. We know $\mathbf{div}(\wp')$ (see observations preceding this theorem), hence we know $\mathbf{div}(\wp'^2)$ and thus we know all zeros of the cubic polynomial $y\wp^3 - g_2\wp - g_3$; in fact they are nothing but

$$e_1 = \wp\left(\frac{\omega_1}{2}\right), \quad e_2 = \wp\left(\frac{\omega_2}{2}\right), \quad e_3 = \wp\left(\frac{\omega_1 + \omega_2}{2}\right)$$

and each of these is a double zero. By comparision of the leading term we obtain

$$(\wp')^2 = 4(\wp - e_1)(\wp - e_2)(\wp - e_3).$$

It is obvious that $\mathbf{div}(\wp - e_i) = -2(O) + 2\left(\frac{\omega_i}{2}\right)$. As the points $\frac{\omega_1}{2}$, $\frac{\omega_2}{2}$, $\frac{\omega_1 + \omega_2}{2}$ are different modulo Γ, we see that all the foregoing divisors $\mathbf{div}(\wp - e_i)$ for $i = 1, 2, 3$, are different. Consequently, the functions $\wp - e_i$ are different for $i = 1, 2, 3$ and hence the e_i are different for $i = 1, 2, 3$. It follows that the discriminant

$$\Delta = \prod_{i \neq j}(e_i - e_j) \neq 0$$

and it is well known that Δ can be expressed as $g_2^3 - 27g_3^2$. Note that since the coefficient of $\wp^2$ in the cubic equation is zero, we also retain that $e_1 + e_2 + e_3 = 0$. □

let $X_0 \subset \mathbf{A}^2_{\mathbf{C}}$ be the affine curve given by the equation

$$y^2 = 4x^3 - g_2x - g_3.$$

To find singular points we should have

$$\begin{cases} 0 = \dfrac{\partial}{\partial x}\left(y^2 - 4x^3 + g_2x + g_3\right) = -12x^2 + g_2, \\[2ex] 0 = \dfrac{\partial}{\partial y}\left(y^2 - 4x^3 + g_2x + g_3\right) = 2y. \end{cases}$$

So the singular points would be $\left(\sqrt{\frac{g_2}{12}}, 0\right)$ and $\left(-\sqrt{\frac{g_2}{12}}, 0\right)$ but these points cannot be on the curve exactly because $g_2^3 - 27g_3^2 \neq 0$.
Define

$$\begin{array}{rcl} \wp_0 : \mathbb{C} - \Gamma & \longrightarrow & X_0 \\ a & \mapsto & (\wp(a), \wp'(a)) \end{array}$$

This defines a bijection between $E_0 = E - \{0\}$ and X_0; that is clear because $\wp$ assumes every value exactly twice with $\wp(z) = \wp(-z)$ but $\wp'$ separates z and $-z$. The holomorphic $\wp_0$ has holomorphic inverse $\wp_0^{-1}$: $X_0 \to E_0$ because the derivative of $\wp_0$, $z \mapsto (\wp'(z), \wp''(z))$ never vanishes (indeed $\wp'(z) = 0$ only holds for $z = \frac{\omega_1}{2}$, $z = \frac{\omega_2}{2}$, $z = \frac{\omega_1 + \omega_2}{2}$, and $\wp''$ does not vanish there because these are simple zeros of $\wp'$).

Embedding $\mathbf{A}^2_{\mathbb{C}}$ into $\mathbf{P}^2_{\mathbb{C}}$ via $(a, b) \mapsto (a, b, 1)$, we define the projective curve $\mathcal{X}$ by the homogeneous equation

$$Y^2T = 4X^3 - g_2XT^2 - g_3T^3.$$

We extend $\wp_0$ to

$$\begin{array}{rcl} \wp : \mathbb{C} & \longrightarrow & \mathcal{X} \\ z & \mapsto & (z^3\wp(z), z^3\wp'(z), z^3) \end{array}$$

actually defining an analytic isomorphism $E \cong \mathcal{X}$.

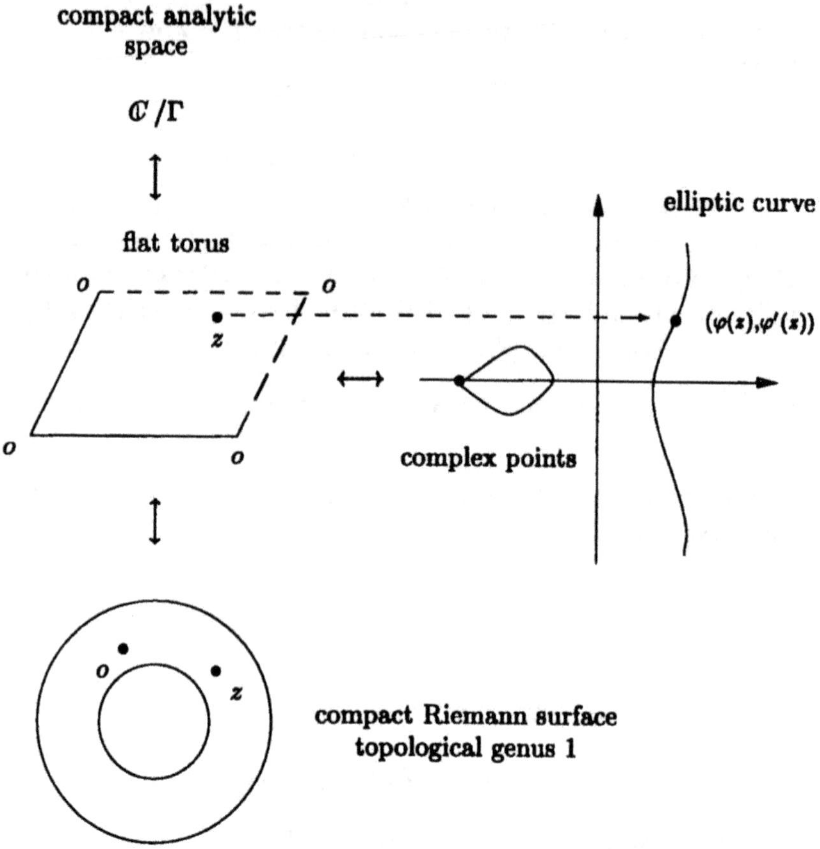

Figure 20

The additive group structure on $\mathbb{C}/\Gamma$ (induced from $\mathbb{C}$) translates into an abelian group structure on the elliptic curve given by

$$(\wp(z),\wp'(z)) \odot (\wp(z_1),\wp'(z_1)) := (\wp(z+z_1),\wp'(z+z_1))$$

where $z+z_1$ is the sum in $\mathbb{C}/\Gamma$.

In this presentation the O in $\mathbb{C}/\Gamma$ corresponds to the point at ∞ of the elliptic curve. By a calculation of the coordinates of $P \odot Q$ in terms of the coordinates of P and Q, and comparing this to the formula for $\wp$ and $\wp'$ of a sum one can "calculate the hard way" that the operation on the elliptic curve is indeed the one we have introduced before. We omit this calculation here. The analytical isomorphism $\mathbb{C}/\Gamma \cong \mathcal{X}$ is then also a group isomorphism for the "canonical" additions we have introduced.

APPENDIX I
Finiteness Conditions and Field Extensions

All rings considered in this appendix are commutative associative rings with 1.

§1. Modules, Finiteness Conditions

Let R be a ring. An *R-module* is an abelian group M (the group operation in M is written $+$; the identity of the group is 0) together with a scalar multiplication, i.e., a function from $R \times M$ to M (denote the image of (r, m) by $r \cdot m$ or rm) satisfying:

(a) $(r + s) \cdot m = rm + sm$ for $r, s \in R$, $m \in M$;

(b) $r \cdot (m + n) = rm + rn$ for $r \in R$, $m, n \in M$;

(c) $(rs) \cdot m = r \cdot (sm)$ for $r, s \in R$, $m \in M$;

(d) $1_R \cdot m = m$ for $m \in M$, where 1_R is the multiplicative identity in R.

From the definition it is easy to see that $0_R \cdot m = 0_M$ for all $m \in M$.

Example (i) A $\mathbb{Z}$-module is just an abelian group, where $(\pm r)m = \pm(m + \cdots + m)$ (r times) for $r \in R$, $r \geq 0$.

(ii) If R is a field, an R-module is the same thing as a vector space over R.

(iii) The multiplication in R makes any ideal of R into an R-module. In particular, R is in itself an R-module.

(iv) If φ: $R \to S$ is a ring homomorphism, we define $r.s$ for $r \in R$, $s \in S$, by the equation $r \cdot s = \varphi(r)s$. This makes S into an R-module. In particular, if a ring R is a subring of a ring S, then S is an R-module.

A subgroup N of an R-module M is called a *submodule* if $rm \in N$ for all $r \in R$, $m \in N$; N is then an R-module. If N is a submodule of M, then the factor group M/N can be made into an R-module by defining $r \cdot \overline{m} = \overline{rm}$ for $r \in R$, $\overline{m} \in M/N$, and is called the *factor module* of M by modulo I.

If W is a set of elements of an R-module M, the *submodule generated* by W is defined to be

$$\left\{\sum r_i w_i \;\middle|\; r_i \in R, w_i \in W\right\};$$

it is the smallest submodule of M which contains W. If $W = \{\xi_1, ..., \xi_s\}$ is finite, the submodule generated by W is denoted by $\sum_{i=1}^{s} R\xi_i$. M is said to be *finitely generated* if $M = \sum_{i=1}^{s} R\xi_i$ for some $\xi_1, ..., \xi_s \in M$. Note that this concept agrees with the notions of finitely generated abelian groups and ideals, and with the notion of a finite dimensional vector space if R is a field.
If every submodule of an R-module is finitely generated, then M is called a *Noetherian R*- module. If R itself is an R-module, then R is called a *Noetherian ring*. Fields and PID's (principal ideal domains) are Noetherian rings. By the Hilbert basis theorem (CH.I Theorem 1.8), the polynomial ring $k[x_1, ..., x_n]$ over a field k in n variables is Noetherian. If R is Noetherian, then every factor ring of R is Noetherian.

Let S be a ring and R a subring of S. Let U be a subset of

S. Recall that the subring generated by U in S is the smallest subring in S containing U. The subring generated by R and U in S is usually denoted by $R[U]$, it is defined as the set

$$\left\{\sum a_\alpha v_{i_1}^{\alpha_1}\cdots v_{i_s}^{\alpha_s} \;\middle|\; v_{ij}\in U,\ a_\alpha\in R\right\}.$$

If R and U satisfy the following two conditions:

- R is a subring of S with $1_R = 1_S$, and
- $S = R[U]$, where U is a subset of S,

then S is said to be a *ring extension* of R.

There are several types of finiteness conditions for S over R, depending on whether we consider S as an R-module, a ring, or (possibly) a field.

1.1. Definition S is said to be *module-finite* over R, if S is finitely generated as an R-module.

If R and S are fields, i.e., S is a field extension of R, then S is an R-vector space and we denote the dimension of S over R by $[S:R]$.

If $U = \{v_1, ..., v_n\} \subseteq S$, then we usually write $R[v_1, ..., v_n]$ for $R[U]$. In this case $R[v_1, ..., v_n]$ may also be obtained as follows. Considering the ring homomorphism

$$\begin{array}{rccc}\varphi: & R[x_1, ..., x_n] & \longrightarrow & S \\ & x_i & \mapsto & v_i\end{array}$$

then $\mathrm{Im}\varphi = R[v_1, ..., v_n]$.

1.2. Definition With notation as above, if there are finitely many $v_1, ..., v_n \in S$ such that $S = R[v_1, ..., v_n]$, then S is said to be *ring-finite* over R (or S is a finitely generated ring extension of R).

Suppose $R = K$, $S = L$ are fields. If $L = K(v_1, ..., v_n)$, where the latter is the quotient field of the subring $K[v_1, ..., v_n] \subseteq L$, for a finitely many $v_i \in L$, then L is said to be a *finitely generated field extension* of K.

1.3. Proposition (i) If S is module-finite over R, then S is ring-finite over R.
(ii) If L is ring-finite over K (K, L fields), then L is a finitely generated field extension of K.
(iii) Let $R \subset S \subset T$ be ring extensions.
(a) If $S = \sum_{i=1}^{n} Rv_i$, $T = \sum_{j=1}^{m} Sw_j$, then $T = \sum_{i,j} Rv_iw_j$.
(b) If $S = R[v_1, ..., v_n]$, $T = S[w_1, ..., w_m]$, then $T = R[v_1, ..., v_n, w_1, ..., w_m]$.
(c) If R, S, T are fields, and $S = R(v_1, ..., v_n)$, $T = S(w_1, ..., w_m)$, then $T = R(v_1, ..., v_n, w_1, ..., w_m)$.
So the finiteness conditions are transitive.

□

Example (i) $S = R[x]$ (the ring of polynomials in one variable x) is ring-finite, but not module-finite.

(ii) $L = K(x)$ (the field of rational functions in one variable) is a finitely generated field extension of K, but L is not ring-finite over K.

§2. Integral Elements

2.1. Definition Let $R \subset S$ be a ring extension. An element $v \in S$ is said to be *integral* over R if there is a monic polynomial $f = x^n + a_1x^{n-1} + \cdots + a_n \in R[x]$ such that $f(v) = 0$. If S and R are fields, we usually say that v is *algebraic* over R if v is integral over R.

2.2. Proposition Let $R \subset S$ be a ring extension. Suppose S is a domain. Let $v \in S$. Then the following are equivalent:
(i) v is integral over R.
(ii) $R[v]$ is module-finite over R.
(iii) There is a subring R' of S containing $R[v]$ which is module-finite over R.

Proof (i) ⇒ (ii) If $v^n+a_1v^{n-1}+\cdots+a_n=0$, then $v^n \in \sum_{i=0}^{n-1} Rv^i$. It follows by induction that $v^m \in \sum_{i=0}^{n-1} Rv^i$ for all m, so $R[v] = \sum_{i=0}^{n-1} Rv^i$.
(ii) ⇒ (iii) Let $R' = R[v]$.
(iii) ⇒ (i) If $R' = \sum_{i=1}^{n} Rw_i$and $R[v] \subset R'$, then

$$vw_i = \sum_{j=1}^{n} a_{ij}w_j, \quad a_{ij} \in R,\ i = 1, ..., n.$$

Then we have the linear system of equations

$$\sum_{j=1}^{n}(\delta_{ij}v - a_{ij})w_j = 0, \quad i = 1, ..., n.$$

If we consider this system of equations in the quotient field of S, we see that $(w_1, ..., w_n)$ is a nontrivial solution, so $\det(\delta_{ij}v - a_{ij}) = 0$. Since v appears only in the diagonal of the matrix, this determinant has the form

$$v^n + a_1v^{n-1} + \cdots + a_n, \quad a_i \in R.$$

So v is integral over R. □

2.3. Corollary The set of elements of S which are integral over R is a subring of S containing R.

Proof If a, b are integral over R, then b is integral over $R[a]$, so $R[a][b] = R[a, b]$ is module-finite over R by Proposition 2.2(iii). And $R[a \pm b]$, $R[ab] \subset R[a, b]$, so they are integral over R by the Proposition. □

We say that S is *integral* over R or an *integral (ring) extension* of R if every element of S is integral over R. R is said to be *integrally closed* if it has no proper integral (ring) extension.
It is clear that if R and S are fields and S is integral over R, then S is an *algebraic extension* of R in the language of field theory. Similarly, saying that a field is integrally closed is equivalent to saying that it is algebraically closed in the language of field theory.

2.4. Corollary Let S be a domain.
(i) Suppose S is module-finite over R, then S is integral over R.
(ii) Suppose S is ring-finite over R, then S is module-finite over R iff S is integral over R.

□

2.5. Corollary Let $K \subset L$ be a field extension.
(i) The set of elements of L which are algebraic over K is a subfield of L containing K.
(ii) Suppose L is module-finite over K, and R is a subring of L such that $K \subset R \subset L$. Then R is a field.

□

§3. Ring-finite Field Extensions

Let $K \subset L$ be a field extension. Suppose $L = K(v)$ for some $v \in L$. Then L is equal to the quotient field of $K[v]$. Moreover, considering the ring homomorphism

$$\varphi: \quad K[x] \longrightarrow K[v] \subset L$$

mapping x to v. Then we have $K[v] \cong K[x]/(f)$ for some $f \in K[x]$ (note that $K[x]$ is a PID). Hence we should distinguish:

Case 1. $f = 0$. Then $K[v] \cong K[x]$, and so $L \cong K(x)$. In this case L is not ring-finite (or module-finite) over K.

Case 2. $f \neq 0$. We may assume f is monic. Now $\langle f \rangle$ is a prime ideal since $K[v]$ is a domain, so f is irreducible and $\langle f \rangle$ is maximal. Therefore, $K[v]$ is a field, so $K[v] = K(v) = L$, i.e., L is *ring-finite* over K. Furthermore, Since $\langle f \rangle = \mathrm{Ker}\varphi$, it follows that $f(v) = 0$, or in other words, v is algebraic over K, and L is *module-finite* over K.

To finish the proof of the Nullstellensatz, we must prove the claim (∗) of CH.I §5. This states that if a field L is a ring-finite

extension of an algebraically closed field k, then $L = k$. But this is equivalent to prove that L is module-finite over k (hence algebraic over k).
The above discussion indicates that a ring-finite field extension can be module-finite. The next proposition shows that this is always true, and completes the proof of the Nullstellensatz.

3.1. Theorem (Zariski) If a field L is ring-finite over a subfield K, then L is module-finite (and hence algebraic) over K.

Proof Suppose $L = K[v_1, .., v_n]$. The case $n = 1$ is taken care of by the above discussion, so we assume the result for all extensions generated by $n-1$ elements. Let $K_1 = K(v_1)$. By induction, $L = K_1[v_2, ..., v_n]$ is module-finite over K_1. We may assume v_1 is not algebraic over K (otherwise the proof is finished already).
Each v_i satisfies an equation

$$v_i^{n_i} + a_{i1} v^{n_i - 1} + \cdots = 0, \quad a_{ij} \in K_1.$$

If we take $a \in K[v_1]$ which is a multiple of all the denominators of the a_{ij}, we get equations

$$(av_i)^{n_i} + a a_{i1} (av_i)^{n_i - 1} + \cdots = 0, \quad i = 1, ..., n.$$

It follows from the corollary to Proposition B2 that for any $z \in L = K[v_1, ..., v_n]$, there is an N such that $a^N z$ is integral over $K[v_1]$. In particular, this must hold for $z \in K(v_1)$. But since $K(v_1)$ is isomorphic to the field of rational functions in one variable over K, this is impossible (exercise 3). □

Exercises for §3

1. Let R be a subring of S, S a subring of T. If S is integral over R, and T is integral over S, show that T is integral over R.
2. Let L be a field, k an algebraically closed subfield of L.
   (a) Show that any element of L which is algebraic over k is already in k.

(b) An algebraically closed field has no module finite field extensions except itself.

3. Let k be a field, $L = k(x)$ the field of rational functions in one variable over k.

   (a) Show that any element of L which is integral over $k[x]$ is already in $k[x]$. (Hint: If $z^n + a_1 z^{n-1} + \cdots = 0$, write $z = \frac{f}{g}$, f, g relatively prime. Then $f^n + a_1 f^{n-1} g + \cdots = 0$, so g divides f.)

   (b) Show that there is no nonzero element $f \in k[x]$ such that for every $z \in L$, $f^n z$ is integral over $k[x]$ for some $n > 0$. (Hint: Let $z = \frac{1}{g}$, where g is an irreducible polynomial not dividing f.)

4. Let R be a domain with quotient field K, and let L be a finite algebraic extension of K.

   (a) For any $v \in L$, show that there is a nonzero $a \in R$ such that av integral over R.

   (b) Show that there is a basis $v_1, \ldots, v_n$ for L over K (as a vector space) such that each v_i is integral over R.

§4. Existence of a Primitive Element

In this section we give a sketch of the proof of the "theorem of the primitive element" which is used in CH.VI §§5–6.

Let K be a field of char$K = 0$ as usual. First, we recall some easily verified facts about K and the field extensions of K stemming from elementary field theory.

4.1. Proposition Let $f \in K[X]$ be an irreducible monic polynomial of degree $n > 0$. Then

(i) $L = K[X]/\langle f \rangle$ is a field, and if x is the residue of X in L, then $f(x) = 0$.

(ii) Suppose L' is a field extension of K, $y \in L'$ such that $f(y) =$

0. Then the homomorphism

$$\begin{array}{ccc} K[X] & \longrightarrow & L' \\ \sum c_i X^i & \mapsto & \sum c_i y^i \end{array}$$

induces an isomorphism of L with $K(y)$.
(iii) With L', y as in (ii), suppose $g \in K[X]$ and $g(y) = 0$. Then f divides g.
(iv) $f = (X - x) \cdot f_1$, $f_1 \in L[X]$.

□

Using Proposition 4.1, an easy induction on the degree of f yields the following

4.2. Theorem Let $f \in K[x]$ be an arbitrary polynomial. Then there is a field L containing K such that $f = \prod_{i=1}^{n}(X - x_i) \in L[X]$. L is called a *splitting field* of f.

□

Furthermore, It follows from Proposition 4.2(iii) and Theorem 4.2 that we have the following

4.3. Proposition Let f be an irreducible monic polynomial in $K[X]$ of degree $n > 0$. Let L be a splitting field of f in the sense of Theorem 4.2, so $f = \prod_{i=1}^{n}(X - x_i)$, $x_i \in L$. Then all x_i are distinct.

□

We are now ready to prove the main theorem of this section. (Note that we have assumed char$K = 0$.)

4.4. Theorem (Theorem of the Primitive Element) Let L be a finite algebraic extension over K, i.e., $L = K(x_1, ..., x_n)$ for a finite number $x_1, ..., x_n \in L$ and L is algebraic over K. Then there is a $z \in L$ such that $L = K(z)$.

Proof We first assume $L = K(x, y)$ where $x, y \in L$. Let f and g be monic irreducible polynomials in $K[T]$ such that $f(x) = 0$, $g(y) = 0$. By Theorem 4.2 and Proposition 4.3, there is a field L' in which $f = \prod_{i=1}^{n}(T - x_i)$, $g = \prod_{j=1}^{m}(T - y_j)$, $x = x_1$, $y = y_1$, $L' \supset L$. Choose $\lambda \in K$ so that

$$\lambda x + y \neq \lambda x_i + y_j, \quad \text{for all } i \neq 1, j \neq 1.$$

Let $z = \lambda x + y$, $K' = K(z)$. Then $h(T) = g(z - \lambda T) \in K'[T]$. But $h(x) = 0$, $h(x_i) \neq 0$ if $i > 0$. Therefore, $(h, f) = (T - x) \in K'[T]$. Then $x \in K'$, so $y \in K'$, and consequently $L = K'$.
Now suppose $L = K(x_1, ..., x_n)$. Then we can use induction on n to find $\lambda_1, ..., \lambda_n \in K$ such that $L = K(\sum \lambda_i x_i)$. □

Finally, we elucidate the structure of a finitely generated field extension $k \subset K$ with $\text{tr.deg}_k K = 1$, where $\text{tr.deg}_k K$ is the transcendence degree of K over k in the sense of CH.V Definition 6.5.

4.5. Proposition Suppose k is algebraically closed of char$k$ = 0. Let $k \subset K$ be as above, $x \in K$, $x \notin k$. Then
(i) K is algebraic over $k(x)$.
(ii) There is an element $y \in K$ such that $K = k(x, y)$.
(iii) If R is a domain with quotient field K, $k \subset R$, and p is a prime ideal in R, $0 \neq p \neq R$, then the natural homomorphism from k to R/p is an isomorphism.

Proof (i) Since $\text{tr.deg}_k K = 1$, we may find some $t \in K$ so that K is algebraic over$k(t)$. Thus, x is algebraic over $k(t)$, so there is a polynomial $f \in k[T, X]$ such that $f(t, x) = 0$ (clear denominators if necessary). Since x is not algebraic over k (§3 Exercise 2), t must appear in f, hence t is algebraic over $k(x)$. Then $k(x, t)$ is algebraic over $k(x)$ by Corollary 2.5, so K is algebraic over $k(x)$ (§3 Exercise 1).
(ii) Since char$k(x) = 0$, this is an immediate consequence of Theorem 4.4.
(iii) Suppose there is an $x \in R$ whose residue $[x]$ in R/p is not in k, then $[x]$ is not algebraic over k because k is alge-

braically closed by assumption. Let $y \in p$, $y \neq 0$. Choose $f = \sum a_i(X)Y^i \in k[X,Y]$ so that $f(x,y) = 0$. If we choose f of lowest possible degree , then $a_0(X) \neq 0$. Thus $a_0(x) \in p$, so $a_0([x]) = 0$. But $[x]$ is not algebraic over k, hence there is no such x. □

Exercises for §4

1. Let $L = K(x_1, ..., x_n)$ be a finite algebraic field extension of K as in Theorem 4.4. Suppose $k \subset K$, k algebraically closed, and $V \subset \mathbf{A}_k^n$, $V \neq \mathbf{A}_k^n$ an algebraic set. Show that $L = K(\sum \lambda_i x_i)$ for some $(\lambda_1, ..., \lambda_n) \in \mathbf{A}_k^n - V$.

Appendix II Localizaiton, Discrete Valuation Rings and Dedekind Domains

All rings considered in this appendix are commutative associative with 1. By a *domain* we mean a ring without nonzero divisors of zero.

§1. Localization at a Multiplicative Subset

Let R be a ring and S a *multiplicative* subset of R in the sense that $1 \in S$ and the products of elements of S are again in S. It is easy to check that

a. On $R \times S = \{(r, s) \mid r \in R, s \in S\}$ the relation

$$(r_1, s_1) \sim (r_2, s_2) \Leftrightarrow s(s_2 r_1 - s_1 - r_2) = 0 \text{ for some } s \in S$$

is an equivalence relation.

b. Write $S^{-1}R$ for the quotient set of $R \times S$ with respect to $\sim$ defined above, and write $\frac{r}{s}$ for the class of (r, s) in $S^{-1}R$. $S^{-1}R$ form a ring with identity by defining the addition and multiplication as

$$\frac{r_1}{s_1} + \frac{r_2}{s_2} = \frac{s_2 r_1 + s_1 r_2}{s_1 s_2},$$

$$\frac{r_1}{s_1} \cdot \frac{r_2}{s_2} = \frac{r_1 r_2}{s_1 s_2}.$$

$S^{-1}R$ is called the *localization* of R at S.

c. Let p be a prime ideal of R. $S_p = R - p$ is a multiplicative subset of R, and $S_p^{-1}R$ is a local ring (as defined in CH.II §6) with the maximal ideal

$$\mathcal{M}_p = \left\{ \frac{r}{s} \,\middle|\, r \in p,\ s \in S_p \right\}.$$

d. Let S be a multiplicative subset of R. $S^{-1}R = \{0\}$ if and only if $0 \in S$, and the natural map ϕ_R: $R \to S^{-1}R$ given by $\phi_R(r) = \frac{r}{1}$ is a ring homomorphism and its kernel is $\{r \in R \mid sr = 0, \text{ for some } s \in S\}$.

e. $S^{-1}R$ has the following universal mapping property: if f: $R \to A$ is a ring homorphism such that $f(s)$ is invertible in A for all $s \in S$, then there exists a unique ring homomorphism f_S: $S^{-1}R \to A$ such that $f = f_S \circ \phi_R$, where ϕ_R is as defined in above d. (Of course one can also use this property as a definition of $S^{-1}R$.)

So, one easily checkis that if $R = k[V]$ is the coordinate ring of an affine variety V and p is a prime ideal of $k[V]$, then the localization of R at S_p is nothing but the subring $S_p^{-1}k[V]$ of $k(V)$ we considered in CH.II §6.

§2. Discrete Valuation Rings

The notion of a local ring we use in this section is given in CH.II §6.

2.1. Proposition Let R be a domain which is not a field. Then the following are equivalent:
(i) R is Noetherian and local, and the maximal ideal is principal.
(ii) There is an irreducible element $t \in R$ such that every nonzero element $z \in R$ may be written uniquely in the form $z = ut^n$, u a unit in R, n a non-negative integer.

Proof (i) $\Rightarrow$ (ii): Let $\mathfrak{m}$ be the maximal ideal, t a generator for $\mathfrak{m}$. Suppose $ut^n = vt^m$, u, v units, $n \geq m$. Then $ut^{n-m} = v$ is a unit, so $n = m$ and $u = v$. Thus the expression of any $z = ut^n$ is unique. To show that any z has such an expression, we may assume that z is not a unit, so $z = z_1 t$ for some $z_1 \in R$ (note that R is a local ring, see CH.II §6 exercise 3). If z_1 is a unit we are finished, so assume $z_1 = z_2 t$. Continuing in this way, we find an infinite collection $z_1, z_2, \ldots$ with $z_i = z_{i+1}t$. Since R is Noetherian, the chain of ideals $\langle z_1 \rangle \subset \langle z_2 \rangle \subset \cdots$ must have a maximal member (CH.I §1 exercise 8), so $\langle z_n \rangle = \langle z_{n+1} \rangle$ for some n. Then $z_{n+1} = vz_n$ for some $v \in R$, so $z_n = vtz_n$ and $vt = 1$. But t is not a unit.
(ii) $\Rightarrow$ (i): $\mathfrak{m} = \langle t \rangle$ is clearly the set of non-units. It is not hard to see that the only ideals in R are the principal ideals $\langle t^n \rangle$, $n \geq 0$, so R is a PID. □

2.2 Definition An ring R satisfying the conditions of Proposition 2.1 is called a ***discrete valuation ring***, abbreviated DVR. An element t as in (ii) is called a ***uniformizing parameter*** for R.

The following consequences are easily derived from the above:

- If t is a uniformizing parameter for R, then any other uniformizing parameter is of the form ut, u a unit in R.
- Let $K = Q(R)$ be the quotient field of R. If we fix one

uniformizing parameter t, then

a. any nonzero element $z \in K$ has the form $z = ut^n$, u a unit of R, $n \in \mathbb{Z}$, and moreover this expression is *unique*;

b. if we define

$$o(z) = \begin{cases} n, & \text{if } z \neq 0 \text{ and } z = ut^n \\ \infty, & \text{if } z = 0, \end{cases}$$

then we have defined a function (which is usually called the order function):

$$\begin{aligned} o: K &\longrightarrow \mathbb{Z} \cup \{\infty\} \\ z &\mapsto o(z) \end{aligned}$$

with

$$R = \{z \in K \mid o(z) \geq 0\}$$

$$\mathfrak{m} = \{z \in K \mid o(z) > 0\}$$

and the set of units in R, denoted $U(R)$, consists of elements z of R with $o(z) = 0$;

c. the order function $o(z)$ obtained above satisfies

V1. $o(z) = \infty$ if and only if $z = 0$,

V2. $o(z_1 z_2) = o(z_1) + o(z_2)$,

V3. $o(z_1 + z_2) \geq \min\{o(z_1), o(z_2)\}$, for all $z, z_1, z_2 \in K$.

Conversely, Let v: $K \to \mathbb{Z} \cup \{\infty\}$ be a function satisfying the above V1–V3. Then v corresponds to a DVR, or more precisely, we have the following proposition.

2.3. Proposition (i) $R = \{z \in K \mid v(z) \geq 0\}$ is a DVR with the maximall ideal $\mathfrak{m} = \{z \in K \mid v(z) > 0\}$.
(ii) The quotient field of R, denoted $Q(R)$, is equal to K, and there is a chain of R-modules

$$\cdots \subset t^2R \subset tR \subset R \subset t^{-1}R \subset t^{-2}R \subset \cdots$$

such that each $z \in K$ is contained in some $t^n R$ where $n \in \mathbb{Z}$, where t is a uniformizing parameter of R. In this case, the valuation function v is nothing but the order function determined by the valuation ring R.

Proof (i) Since v satisfies V1–V3, clearly R is a ring. Moreover, it is easy to see that $v(1) = 0$, and it follows that if $z \in K$, $z \notin R$, then $z^{-1} \in \mathbf{m}$. Note that since $v(K^*)$ is a subgroup of $\mathbb{Z}$, where K^* denotes the multiplicative group of K, there is some $t \in K$ such that $v(K^*) = \mathbb{Z}v(t)$. We may assume $v(t) > 0$. Thus, for any $z \in \mathbf{m}$, $v(z) = nv(t)$ with $n > 0$. It follows that $v(zt^{-n}) = 0$, i.e., $zt^{-n} \in R$. Hence $z = t^n r \in \langle t \rangle$ where $r \in R$. Clearly t is irreducible. Hence R is a DVR by Proposition 2.1 (ii).
(ii) From the proof of (i) this is clear. □

Example (i) Let $V = \mathbf{A}_k^1$, $k[V] = k[x]$, $k(V) = k(x)$.

a. For each $a \in k = V$, the local ring $\mathcal{O}_a, V$ of V at a is a DVR with $t = x - a$ as a uniformizing parameter. Indeed, these DVR's are the only DVR's with quotient field $k(x)$ which contain k, except $k[x^{-1}]_{(x^{-1})}$ appearing hereafter, where $k[x^{-1}]_{(x^{-1})}$ is the localization of $k[x^{-1}]$ at $\{1, x^{-1}, x^{-2}, \ldots\}$.
b. The set

$$\mathcal{O}_\infty = \left\{ \frac{f}{g} \in k(x) \;\middle|\; \deg(g) \geq \deg(f) \right\}$$

forms a DVR with $t = \dfrac{1}{x}$ as a uniformizing parameter.

(ii) Let $p \in \mathbb{Z}$ be a prime number. Then

$$R_p = \left\{ r \in \mathbb{Q} \;\middle|\; r = \frac{a}{b},\ a, b \in \mathbb{Z},\ p \text{ does not divide } b \right\}$$

is a DVR with p as a uniformizing parameter. Indeed, these DVR's are the only DVR's with quotient field $\mathbb{Q}$.

Exercises for §2

1. Let R be a DVR with quotient field K, $o(z)$ the order function on K.

a. If $o(z_1) < o(z_2)$, show that $o(z_1 + z_2) = o(z_1)$.

b. If $z_1, \ldots, z_n \in K$, and for some i, $o(z_i) < o(z_j)$ (all $j \neq i$), then $z_1 + \cdots + z_n \neq 0$.

2. Let R be a DVR with maximal ideal $\mathbf{m}$, and quoitent field K, and suppose a field k is a subring of R, and that the composition $k \to R \to R/\mathbf{m}$ is an isomorphism of k with $R/\mathbf{m}$ (as in Example (i) above). Verify the following assertions:

a. For any $z \in R$, there is a unique $\lambda \in k$ such that $z - \lambda \in \mathbf{m}$.

b. Let t be a uniformizing parameter for R, $z \in R$. Then for any $n \geq 0$ there are unique $\lambda_0, \lambda_1, \ldots, \lambda_n \in k$ and $z_n \in R$ such that $z = \lambda_0 + \lambda_1 t + \cdots + \lambda_n t^n + z_n t^{n+1}$. (Hint: For uniqueness use exercise 1; for existence use a. and induction.)

3. Let R be a DVR as in Exercise 2 above. For every $n \geq 0$, viewing $\mathbf{m}^n/\mathbf{m}^{n+1}$ as a k-vector space, show that the following properties hold.

a. $\dim_k(\mathbf{m}^n/\mathbf{m}^{n+1}) = 1$ for all $n \geq 0$.

b. $\dim_k(R/\mathbf{m}^n) = n$ for all $n > 0$.

c. Let $z \in R$. Then $o(z) = n$ if $\langle z \rangle = \mathbf{m}^n$, and hence $o(z) = \dim_k(R/\langle z \rangle)$.

4. Let k be a field. The ring of *formal power series* over k, denoted $k[[x]]$, is defined to be

$$\left\{ \sum_{i=0}^{\infty} a_i x^i \;\middle|\; a_i \in k \right\}.$$

Define $\sum a_i x^i + \sum b_i x^i = \sum (a_i + b_i) x^i$, and $\sum a_i x^i \sum b_i x^i = \sum c_i x^i$, where $c_i = \sum_{j+k=i} a_j b_k$. Show that $k[[x]]$ is a ring containg $k[x]$ as a subring. Show that $k[[x]]$ is a DVR with uniformizing parameter x. Its quotient field is denoted $k((x))$.

5. let the notation be as in exercise 3 above.

a. Let R be a DVR satisfying the conditions of exercise 2. Any $z \in R$ thendetermines a power series $\sum \lambda_i x^i$, if $\lambda_0, \lambda_1, \ldots$ are determined as in exercise 2(b). Show that the map $z \mapsto \sum \lambda_i x^i$ is an injective ring homomorphism of R into $k[[x]]$. we often write $z = \sum \lambda_i t^i$, and call this the *power series expansion* of z in terms of t. Show that the homomorphism extends to a homomorphism of K into $k((x))$, and that the order function on $k((x))$ restricts to that on K.

b. Let $a = 0$ in Example (i) above, $t = x$. Find the power series expansion of $z = (1-x)^{-1}$ and of $(1-x)(1+x^2)^{-1}$ interms of t.

§3. Dedekind Domains

The notion of a Noetherian resp. a local ring we use in this section is given in CH.I §1 resp. in CH.II §6. A discrete valuation ring is defined in the last section.

3.1. Definition Let R be a domain. R is called a *Dedekind domain* if it satisfies the following three conditions:
(i) R is Noetherian;
(ii) R is integrally closed (in its field of quotients) in the sense of Appendix I §2;
(iii) Every nonzero prime ideal of R is a maximal ideal (or equivalently, every nonzero prime is a minimal nonzero prime).

Example Every PID (principal ideal domain) R is a Dedekind domain. Indeed, if R is a PID then the conditions (i) and (ii) of definition 3.1 are satisfied. To see that R is integrally closed in its quotient field, let $w = \frac{a}{b}$ be a nonzero element in the quotient field of R, where $a, b \in R$ with $b \neq 0$. Suppose that w satisfies a monic polynomial $f(x) = x^n + c_1x^{n-1} + \cdots c_{n-1}x + c_n$ in $R[x]$, i.e., $f(w) = 0$. If we put $M = R + Rw + \cdots + Rw^{n-1}$, then $b^{n-1}M \subset R$. Since R is PID, there exists some nonzero $u \in R$ such that $b^{n-1}M = Ru$. Evidently $M^2 = M$, which implies $M^2 = ((b^{n-1})^{-1}Ru)^2 = (b^{n-1})^{-1}Ru$, and consequently $M = (b^{n-1})^{-1}Ru = R$. Hence $w \in R$ and it follows that R is integrally closed.

Let R be a domain and F its quotient field. An *R-fractional ideal* (or just simply a fractional ideal) M is a *nonzero* R-submodule of F (see Appendix I §1 for the definition of a module and its submodules) such that $aM \subset R$ for some nonzero $a \in R$. It is easy to see that every finitely generated nonzero R-submodule of F is a fractional ideal for R.
If M is a fractional ideal, we define

$$M^{-1} = \{a \in F \mid aM \subset R\}.$$

It is clear that M^{-1} is an R-submodule of F and $M^{-1} \neq \{0\}$. Moreover, if $b \neq 0$ is in $M \cap R$, then $M^{-1}b \subset R$. Hence M^{-1} is a fractional ideal. If c is a nonzero element of R and M is the principal ideal Rc, Then it is clear that $M^{-1} = Rc^{-1}$.
If M is a fractional ideal, then $M^{-1}M \subset R$ so $M^{-1}M$ is an ideal of R. We say that M is *invertible* if

$$M^{-1}M = R.$$

Clearly, every nonzero principal ideal of R is invertible.

3.2. Lemma (i) Invertible fractional ideals are finitely generated as R-modules.
(ii) If M is an invertible fractional ideal satisfying $M^2 = M$, then $M = R$.

Proof Straightforward. □

3.3. Proposition Let R be a domain and F its quotient field. If every R-fractional ideal of F is invertible, then
(i) R is integrally closed (in F);
(ii) any nonzero prime ideal of R is maximal.

Proof (i) Since every fractional ideal is invertible, by lemma 3.2(ii) this may be proved as in the PID case.

(ii) Let p be a nonzero prime ideal of R. If p is not maximal, we have an ideal J of R such that $J \neq R$, $J \supset p$ but $J \neq p$. The assumption that J is invertible implies $p = JK$ where K is an ideal of R. Since $J \neq p$, we $p \not\supset J$. Also $p \not\supset K$ since $K = RK \supset JK = p$, so $p \supset K$ implies $p = K$ and $K = JK$. Then $R = KK^{-1} = JKK^{-1} = J$, contrary to $R \neq J$. Thus we have $p = JK$ with $J \not\subset p$ and $K \not\subset p$. This contradicts the assumption that p is prime. □

It follows immediately from Lemma 3.2 and Proposition 3.3 that the following theorem holds.

3.4. Theorem A domain R is Dedekind if and only if every R-fractional ideal of F is invertible.

□

For the proofs of Theorem 3.5–3.8 below we refer to [Jac] CH.10.

3.5. Theorem A domain R is Dedekind if and only if every proper ideal of R can be written in one and only one way (up to reordering) as a product of prime ideals.

□

Let R be a Dedekind domain and F its quotient field. If we write $\mathcal{F}(R)$ for the set of all R-fractional ideals of F, then $\mathcal{F}(R)$ is a group with the (multiplicative) structure defined as follows: for $M, N \in \mathcal{F}(R)$,

$$MN = \{\sum a_i b_i \mid a_i \in M, b_i \in N\},$$

M^{-1} is defined as before. The identity element is R. Let $\mathcal{P}(R)$ denote the set of principal R-fractional ideals of F. Then $\mathcal{P}(R)$ is a subgroup of $\mathcal{F}(R)$. The *ideal class group* (or Picard group) of R, denoted $\mathcal{I}(R)$, is defined as $\mathcal{F}(R)/\mathcal{P}(R)$.

3.6. Theorem Let R be a Dedekind domain. The following conditions are equivalent:

(i) $\mathcal{I}(R) = 1$;
(ii) R is a UFD (unique factorial domain);
(iii) R is a PID.
These three conditions hold if R has a finite number of prime ideals.

□

3.7. Theorem Let R be a domain.
(i) If R is Dedekind, then the localization $S^{-1}R$ of R at every multiplicative subset $S \subset R$ (see §1) is Dedekind.
(ii) If R is Dedekind and p is a nonzero prime ideal in R, then the localization $S_p^{-1}R$ of R at $S_p = R - p$ is a discrete valuation ring in the sense of §2.
(iii) If R is Noetherian and $S_p^{-1}R$ is a discrete valuation ring for every maximal ideal p of R, then R is Dedekind.

□

3.8. Theorem Let R be a Dedekind domain and A an integrally closed domain and an integral ring extension of R, finitely generated as an R-module (i.e., module-finite). Then A is a Dedekind domain.

□

The hypotheses in Theorem 3.8 hold in the following situation: R is a Dedekind domain with quotient field F. L a *separable* field extension of F (e.g., L is separable in case $\text{Char}L = 0$). A the integral cloure of R in L (i.e., all the elements of L which are integral over R).

References

[BW] T. Becker and V. Weispfenning, *Gröbner Bases*, A computational approach to commutative algebra, Graduate Texts in Math., 141, Springer-Verlag, 1993.

[CLO'] D. Cox, J. Little and D. O'shea, *Ideals Varieties and Algorithms*, Springer-Verlag, Berlin, 1991.

[Ful] W. Fulton, *Algebraic Curves*, W. A. Benjamin, New York, 1969.

[Har] R. Hartshorne, *Algebraic Geometry*, Springer-Verlag, Berlin, 1977.

[Jac] N. Jacobson, *Basic Algebra*, Vol.II, San Francisco, 1974–1980.

[Jen] W.E. Jenner, *Rudiments of Algebraic Geometry*, Oxford University Press, 1963.

[Ken] K. Kendig, *Elementary Algebraic Geometry*, Springer-Verlag, 1977.

[Lan] S. Lang, *Algebra*, Addison-Wesley, Reading, Massachusetts, 1965.

[LVO] Li Huishi and F. Van Oystaeyen, *Zariskian Filtrations*, Monograph, Kluwer Academic Publishers, 1996.

[Mat] H. Matsumura, *Commutative Algebra*, (Second edition), The Benjamin/Cummings Publishing Company, Inc., 1980.

[Ser] J.P. Serre, *Cours à Arithmetique*, P.U.F.

[Sha] I.R. Shafarevich, *Basic Algebraic Geometry*, Springer-Verlag, 1974.

[SZ] P. Samuel and O. Zariski, *Commutative Algebra*, Vol.I-II, Van Nostrand, Princeton, 1958–1960.

Index

Milton Keynes UK
Ingram Content Group UK Ltd.
UKHW021822071024
449327UK00021B/1391

9 780367 398965